Fundamentals of Production Logistics

Peter Nyhuis · Hans-Peter Wiendahl

Fundamentals of Production Logistics

Theory, Tools and Applications

Translated by: Rett Rossi

With 178 Figures and 6 Tables

Springer

Univ.-Prof. Dr.-Ing. habil. Peter Nyhuis
Univ.-Prof. Dr.-Ing. Dr. h.c. mult. Hans-Peter Wiendahl

IFA Institut für Fabrikanlagen und Logistik
Leibniz Universität Hannover
An der Universität 2
30823 Garbsen
Germany

nyhuis@ifa.uni-hannover.de
wiendahl@ifa.uni-hannover.de

www.ifa.uni-hannover.de

Rett Rossi
rettrossi@rettrossi.de
www.rettrossi.de

ISBN 978-3-540-34210-6 e-ISBN 978-3-540-34211-3

DOI 10.1007/978-3-540-34211-3

Library of Congress Control Number: 2008926597

Cover design: Frido Steinen-Broo, eStudio Calamar, Spain

Printed on acid free paper

9 8 7 6 5 4 3 2 1

springer.com

Preface to the 1st English Edition

In recent years, the Logistic Operations Curves Theory has raised increasing attention in the international scientific community of operations management. This has encouraged us to present an English translation of the second German edition. The Logistic Operations Curves Theory continues to be developed and will be further expanded.

Our sincere thanks is addressed to Daniel Berkholz, a research engineer at the Institute of Production Systems and Logistics (IFA), Leibniz University of Hannover, as well as the translator Rett Rossi for their enthusiasm and never ending efforts to help us find a sound, scientific path through the jungle of new variables and terms not used in traditional literature.

We wish our readers a fruitful discourse with our ideas and look forward to receiving any feedback you may have.

Garbsen, December 2007

Peter Nyhuis
Hans-Peter Wiendahl

Preface to the 2nd German Edition

The first edition of this book was met with very positive resonance. Numerous questions and suggestions provided an incentive for us to continue working on the *Logistic Operating Curves (LOC)* and to let the new results flow into this second edition.

The range of validity for the Logistic Operating Curves Theory has been extended by Schneider, who developed the *Manufacturing System Operating Curve (MSOC)*. The MSOC makes it possible to establish Logistic Operating Curves for manufacturing areas with randomly networked workstations.

The *Schedule Reliability Operating Curves (SROC)* developed by Yu are a completely new approach. Yu has succeeded in deriving an approximation equation for describing the schedule reliability of a workstation.

The *Storage Operating Curves (SOC)* were expanded by Lutz to include the so-called *Service Level Operating Curves (SLOC)*. Based on them and in analogy to the Bottleneck Oriented Logistic Analysis for production areas the *Logistic Storage Analysis* was developed. Quantifying the logistic potential in a storage area and identifying the measures necessary for exploiting it are essential elements of this method of analysis.

By linking the Bottleneck Oriented Logistic Analysis with the Logistic Storage Analysis the logistic interactions within a supply chain can also be understood. The potential can be determined with regards to the service and stock levels including over a number of value adding stages. Moreover, these can be aggregated and expressed as a total potential. Thus, there is now a fundamental and consistent analysis method available for quantifying the inherent relations between the logistic objectives in production systems, in the different storage stages, and in the entire supply chain.

We would like to express our heartfelt appreciation to Stefan Lutz, Michael Schneider and Kwok-Wai Yu for their support in developing the new sections.

To all of our readers, including both those who work in research as well as practitioners, we hope to provide continued inspiration and practical support in overcoming their logistic problems. We welcome the chance to receive your constructive criticism, suggestions and any experiences you may have in applying the Logistic Operating Curves Theory.

Hannover/Munich, Summer 2002

Peter Nyhuis
Hans-Peter Wiendahl

Preface to the 1st German Edition

For many production enterprises, the possibility of distinguishing themselves from their competitors is frequently possible due to a shorter delivery time and higher delivery reliability. This requires firmly controlling the internal throughput times and schedule adherence. At the same time cost relevant goals such as stable and high utilization as well as low stock levels in the raw material, semi-finished and finished goods stores cannot be forgotten. Solving this well known dilemma of operations planning is the object of countless efforts from researchers and practitioners alike. In the 1960s, great hope was set in the methods of Operations Research, in particular in queuing theory. However, due to the complex boundary conditions of job shop and series production, queuing theory was unable to establish itself. Even simulations did not provide the hoped for breakthrough due to the large amount of effort required especially for a company's already running operations.

The Funnel Model and the Throughput Diagram that is derived from it, developed by Prof. Hans Kettner and his assistants at the Institute of Production Systems and Logistics, Leibniz University of Hannover in the 1970s was thus met with great interest. In particular, it attracted attention because the logistic objectives throughput time, WIP, utilization and schedule reliability could for the first time be presented visually and conclusively. The Load Oriented Order Release method that arose from there and the further developed Load Oriented Manufacturing Controls were widely accepted in job shop production.

The Logistic Operating Curves (LOC), developed later within the context of simulation analyses, quantitatively described the impact of the WIP on the utilization and throughput time also for the first time. Due to the huge efforts required for the underlying simulations, the LOC were impractical and therefore limited to theoretical applications.

It was Nyhuis' habilitation in the early 1990s that initially made it possible to simply calculate the Logistic Operating Curves based on the combination of an ideal manufacturing process model suggested by von Wedemeyer with experimental and empirical supported analyses. Consequently in the years following, an extensive field of application opened up both in research and on the production floor.

For the first time, the models of the Logistic Operating Curves for production and storage processes are comprehensively described in this book. In addition, the necessary formulas are derived step by step, and a comparatively simple computational scheme using data that is standard in manufacturing and storage controls is developed from there. Thorough tests with field data and extensive simulations prove how the individual equation parameters influence the order and capacity structures. Thus making it possible to estimate the accuracy of the information even when the original data is inaccurate or contains errors – as is frequently the case on the shop floor. Through a comparison with queuing theory and simulations, both the advantages and the limitations of Logistic Operating Curves are clearly identified.

The usefulness of the Logistic Operating Curves is evident in numerous theoretical and application based projects conducted by the Institute of Production Systems and Logistics. Currently the Logistic Operating Curves are mainly applied: in dimensioning WIP buffers and WIP areas when planning factories; conducting a Logistic Positioning for manufacturing areas and stock levels with respect to the throughput time, utilization and WIP; in production control in order to continually improve the logistic objectives; in parametrizing lot size determination, throughput scheduling, and order release in PPC systems as well as in Bottleneck Oriented Logistic Analyses for developing the hidden logistic potential of the throughput time and WIP. Further foreseeable application possibilities include guiding construction and development areas, extending the Logistic Operating Curves to include schedule reliability, cost-wise evaluating production processes with different WIP situations and evaluating supply chains beyond the manufacturer's borders.

This book is based on a large amount of theoretical and empirical work at the Institute of Production Systems and Logistics, some of which goes back twenty or more years. Most notable here are the dissertations from Bechte, Dombrowski, Dräger Erdlenbruch, Fastabend, Gläßner, Lorenz, Ludwig, Möller, Penz, Petermann, Scholtissek, Springer and Ullmann. Each of which have focused on various aspects of modeling, planning, and controlling production based on Throughput Diagrams and Logistic Operating Curves. Every one of the authors has thus contributed to the Logistic Operating Curves Theory.

We would like to wish all of our readers, including both those who work in research as well as in the industry, inspiration and practical use in overcoming their logistic problems. Furthermore, we would be grateful for constructive criticism, suggestions and experiences in applying the Logistic Operating Curves Theory.

Hannover, Summer 1999

Peter Nyhuis
Hans-Peter Wiendahl

Translator's Notes

Translating the Fundamentals of Production Logistics has provided a unique and extremely pleasurable challenge – one that I hope we have successfully met. Seeing that the theory, tools and applications presented here have only been partially exposed to an English language audience, the terminology for discussing them has up to now not fully existed. This of course is often the case when introducing new concepts. However, Prof. Dr. Nyhuis and Prof. Dr. Wiendahl also had a goal in mind when choosing these terms: They wanted to target an as broad as possible international audience who are interested in logistics and curious about what approaches to a logistic theory are being made elsewhere. What this meant for the translation was trying to find clear, self-explanatory terms and implementing them consistently. We hope that we have managed to achieve this. Nonetheless, just as the authors hope to hear from readers with regards to their experiences and thoughts I too hope that you will also let the authors know if mistakes or areas lacking clarity are found. After all, this edition of the Fundamentals of Production Logistics is meant to initiate discussion and is therefore hopefully just a beginning.

As this is a translation of the German edition and not a new English edition, the majority of bibliographical references are also German. Those that are in English have been marked with an asterisk e. g. [Wien-95a*], in order to be easily identified.

Berlin, December 2007

Rett Rossi

Contents

Index of Abbreviations and Variables

General Terms (as index or suffix)

act	actual value
AP	angulation point
i	indices
is	intersection point
m	mean value
max	maximum value
med	median value
min	minimum value
mw	mean weighted value
n	number of events, orders
nd	normal distribution
s	standard deviation
t	mean value calculated by LOC Theory (as a funtion of the running variable t)
(T)	value at time T
tar	target value
v	coefficient of variation

Dimensions

unit	**meaning**
%	percent
hrs	hours
hrs · SCD	area (hours · shop calendar days)
hrs/SCD	hours per shop calendar day

	Q	number of
	SCD	shop calendar day
	units	unit (e. g. items, m^2, kg)
	units	unit

Variables

symbol	**unit**	**meaning**
a	–	lower bound
A_{SL}	units · SCD	stock level, area of
A_{SO}	units · SCD	stockout area
A_{TTP}	hrs · SCD	throughput time area
A_{WIP}	hrs · SCD	WIP level, area of
b	–	upper bound
C	–	C value
CAP	hrs/SCD	capacity
CAPD		capacity per day
CAP_m	hrs/SCD	available machine capacity
CAP_{op}	hrs/SCD	available operator capacity
DD	SCD	delivery delay store output
DD_0	SCD	delivery delay limit
DD_1	SCD	adjusted delivery delay limit
IN	hrs	input
L	SCD	lateness
L–	SCD	negative lateness (early)
L+	SCD	positive lateness (late)
L_{in}	SCD	input lateness
L_{out}	SCD	output lateness
LP	%	loading percentage
L_{rel}	SCD	relative lateness
LS	units	lot size
LS_D	units	lot size per demand
LS_{DO}	units	lot size per demand served on time
m	–	number of demands served on time
MFC	–	material flow coefficient
n	–	number of events
N	–	number of orders in system (Little's law, queuing models)
OD	–	overlapping degree
OUT	hrs	output
P	hrs/SCD	performance (Little's Law)
PE; I	SCD	reference period; Interval
Q	evtl. units	quantity

QD	units	quantity demand
QIN	units	store input quantity per input event
QOUT	units	store output quantity
QOUT	units	store output quantity per output event
R	SCD	range
RD	units/SCD	demand rate
ReD	–	delivery reliability
ReS	%	schedule reliability
RF	–	flow rate
RL	SCD	labor rate
ROUT	hrs/SCD	output rate
$ROUT_{max}$	hrs/SCD	maximum possible output rate
ROUTO	1/SCD	mean output rate in number of orders
$ROUT_{store}$	units/SCD	store output rate
SERL	%	service level
SL	units	stock level
SL_0	units	lot stock level; minimum stock level
SL_1	units	practical minimal stock level
$SL_m(t)$		mean lot stock level
SPL		split lots
SSL	units	safety stock level
t	–	running variable in the Logistic Operating Curves Theory
$t_{e,op}$	SCD	end of operation processing
$t_{e,ord}$	SCD	end of order processing
$t_{e,prop}$	SCD	end of pre-operation processing
TIO	SCD	inter-operation time
TOP	SCD	operation time
TP	min/unit	processing time
TQ		mean queuing time (queuing model)
TRP	SCD	replenishment time
TS	min	setup time
$t_{s,ord}$	SCD	start of order processing
TTP	SCD	throughput time
TTP_{order}	SCD	order throughput time
TTP_{vir}	SCD	virtual throughput time
TTR	SCD	TIO^min transport time, also minimum inter-operation time
TW	SCD	time waiting
U	%	utilisation
w		number of workstations
WC	hrs	order work content
WIP	hrs	work in process (WIP) level
WIP_A	hrs	active WIP level (during active processing)
WIP_b	hrs	WIP buffer

$WIPI_{min}$	hrs	ideal minimum WIP level
WIPO	–	WIP level in number of orders
$WIPIO_{min}$	–	ideal minimum WIP level in number of orders
WIP_P	hrs	passive WIP level
WIP_{rel}	%	relative WIP level
z	–	number of intervals during reference period
α	–	stretch factor in the Logistic Operating Curves Theory
δ	–	utilization rate (queuing model)
λ	–	mean arrival rate (queuing model)
μ	–	service rate
ρ	–	utilization rate
$\Phi(u)$	–	distribution function for the standard deviation

Abbreviations

BN	bottleneck
FCFS	first come – first served
FIFO	first in – first out
KPI	key performance indicator
LPT	longest processing time (rule)
MS	manufacturing system
OP	operation
Slack	minimum slack (rule)
SPT	shortest processing time (rule)
WS	workstation

Logistic Operating Curves

COC	Cost Operating Curve
DDOC	Delivery Delay Operating Curve
ITOC	Inter-Operation Time Operating Curve
LOC	Logistic Operating Curves
LOC Theory	Logistic Operating Curves Theory
MSOC	Manufacturing System Operating Curves
OROC	Output Rate Operating Curve
PCOC	Production Cost Operating Curve
RLOC	Relative Lateness Operating Curve
ROC	Range Operating Curve
SLOC	Service Level Operating Curve
SOC	Storage Operating Curve
SROC	Schedule Reliability Operating Curve

TROC	Transport Operating Curve
TTOC	Throughput Time Operating Curve
UOC	Utilization Operating Curve

Methods and Analyses Based on the LOC Theory

BOLA	Bottleneck Oriented Logistic Analysis
FROS	Flow Rate Oriented Scheduling
LOOR	Load Oriented Order Release
LOSA	Logistic Oriented Storage Analysis
TOLS	Throughput Oriented Lot Sizing

Chapter 1
Introduction

Change is both a typical and necessary characteristic of the evolutionary process. Although companies frequently consider it a catalyst for critical situations, there is more to it than problems, risks and dangers. A company opens up new possibilities by positioning itself actively and early, consciously grasping these factors and bearing them in mind when planning its future. Doing so it distinguishes itself positively from its competitors and thus creates new potential.

Working proactively during economically stable times is particularly important in this case. The risk that effective measures cannot be introduced quickly enough often originates in not being able to recognize relevant changes. Companies are ready to make alterations, especially in times of crisis. However, they often no longer have the energy or reserves, or are in a position where they need to make considerable cuts. The demand to be permanently innovative regarding products and processes is thus continuously and emphatically present ([Zahn-94], [Warn-93]). Companies need to develop strategies oriented on the future and possible solutions based not only on knowledge of their weaknesses and previous mistakes, but also in consideration of their business goals. Once established, these need to be pursued resolutely. Due to ever decreasing product lifecycles, increasing product diversity, unstable production plans, market globalization and numerous other factors, a company has to be as flexible and adaptable as the market itself.

1.1 Logistic Key Performance Indicators for Manufacturers

In order to attain and maintain a competitive edge it is necessary to constantly improve and revaluate products and production processes (see e. g., [Bull-92], [Port-92], [Port-93], [Warn-93], [Mert-96], [Milb-97]). Nevertheless, almost every advantage can be copied sooner or later. Therefore, a company has to become a moving target, developing new advantages at least as quickly as the old ones can be copied. Selectively improving performance is thus generally not enough to sustainably fortify a company's position. It generally provides only short-term improved results and instead of leading to a substantial change in competitive relations it at best gains time [Wild-98].

Sustainable advantages are only attainable when a strategic master plan is developed based on an analysis of corporate strengths and weaknesses and customer demands. Furthermore, it has to be built on coordinated measures and a comprehensive examination in order to not only be able to design and implement it, but to also be able to control it with regards to the desired success.

In addition to high quality standards and the price of products, the logistic factors *delivery time* and *delivery reliability* take on progressively more importance as possibilities with which a company can distinguish itself within the market (Fig. 1.1) ([Voig-90], [Kear-92], [Baum-93], [Gott-95]). Production, as the primary function for fulfilling orders, is thus increasingly called upon to improve effectiveness [Zahn-94]. The goal therefore, is to organize the entire material flow in the supply chain, from procuring raw materials and preliminary products, through the entire production process including all of the interim storage stages, up to supplying distributors or as the case may be, external customers in such a way that the firm can react to the market in the shortest time span. Since production logistics decisively influence these performance indicators, intense effort is invested both in research and in the industrial setting in order to expertly design and operate logistic systems.

The fundamental goal of production logistics can thus be formulated as the pursuance of greater delivery capability and reliability with the lowest possible logistic and production costs. Here, the logistic factor *delivery capability,* expresses the degree to which it is possible for a company, in consideration of the production situation, to commit to the customers preferred delivery date. *Delivery reliability* on

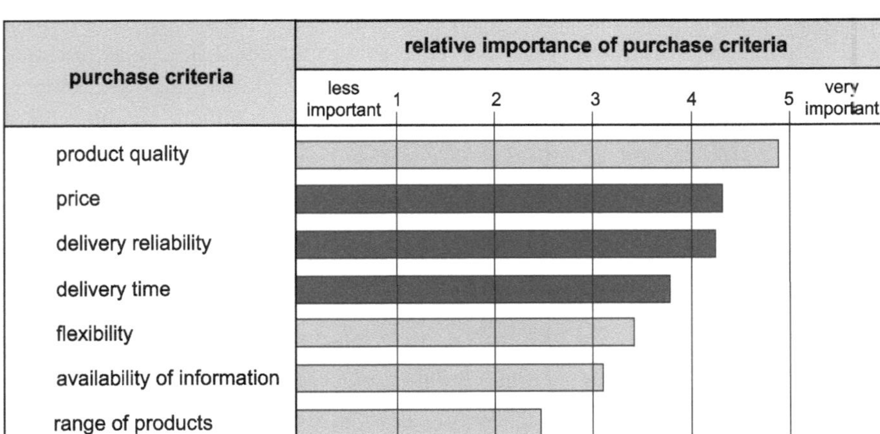

Fig. 1.1 Criteria for Purchasing Products (Siemens Inc.)

the other hand, depicts the extent to which the promised dates for the placed orders can be met. In addition to marketable production costs, delivery capability and delivery reliability are critical to a company's long-term market success (Fig. 1.2).

Designing structures for products, production and purchasing which permit market geared delivery dates under suitably applied production planning strategies makes it possible not only to develop and ensure a greater delivery capability, but also to guarantee the company's *logistic process capability*. When the targeted delivery capability based on the implemented structures is essentially possible, it is then the process controls' job to fully develop the inventory management and operative control of the created logistic potential in order to achieve a greater *logistic process reliability*. The low throughput times, realizable due to the product and production structures, should be attained during the ongoing process and maintained at a stable level. As a result, a high delivery reliability will be continuously provided.

Finally, by designing and controlling the operating logistic processes the interactions between the performance and cost objectives should be constantly monitored so as to be able to ensure the production's *economic efficiency*. In order to achieve marketable production costs it is necessary on the one hand, to strive for a maximal utilization of the available capacities and on the other hand, to reduce the storage and WIP levels as much as possible so that the costs of tied-up capital are minimized.

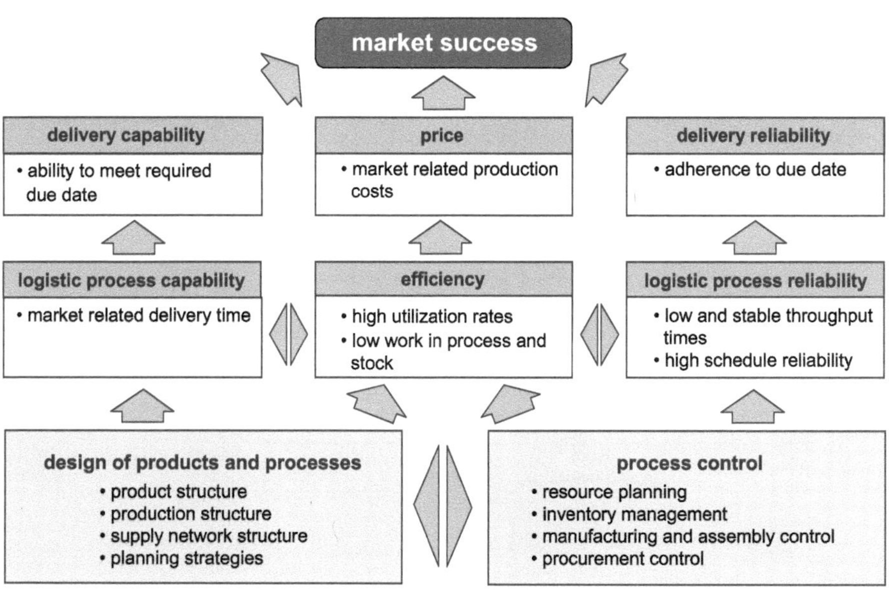

Fig. 1.2 Logistic Key Performance Indicators for Production Firms (Gläßner, IFA)

1.2 Dilemma of Operations Planning

The endeavor to strengthen targeted logistic key performance indicators (KPI) is complicated by existing conflicts between the objectives. The logistic objectives and requirements that need to be taken into consideration are neither consistent nor locally and temporally constant. Thus, a high level of work in process (WIP) is required to ensure a high level of utilization. However, a high WIP level results in longer throughput times. Nevertheless, high and based on experience, fluctuating throughput levels contradict high reliability. The tendency for these objectives to conflict is generally known as the 'dilemma of operations planning' [Gute-51]. Therefore, there is not just one target whose value has to be maximized or minimized, rather, the impact of the measures on each of the sub-goals has to be simultaneously considered. The fact that these sub-goals can be prioritized quite differently in the various stages of the production process makes the situation even more problematic.

Figure 1.3 depicts the objectives that are emphasized depending on the existing storage strategies and position of the observed processing step in relation to the *customer order decoupling point*: As long as the production is not conducted on the basis of concrete customer orders, the companies objectives i. e., maximal utilization and minimal WIP will be pursued, because these two objectives (even when contradictory) influence the economic efficiency of the production. The schedule reliability and throughput time are for the most part of secondary importance. Nonetheless, these parameters indirectly impact the storage goals. The poorer the schedule reliability is and the greater the throughput time for the related production is, the more stock there has to be in-store in order to attain a defined

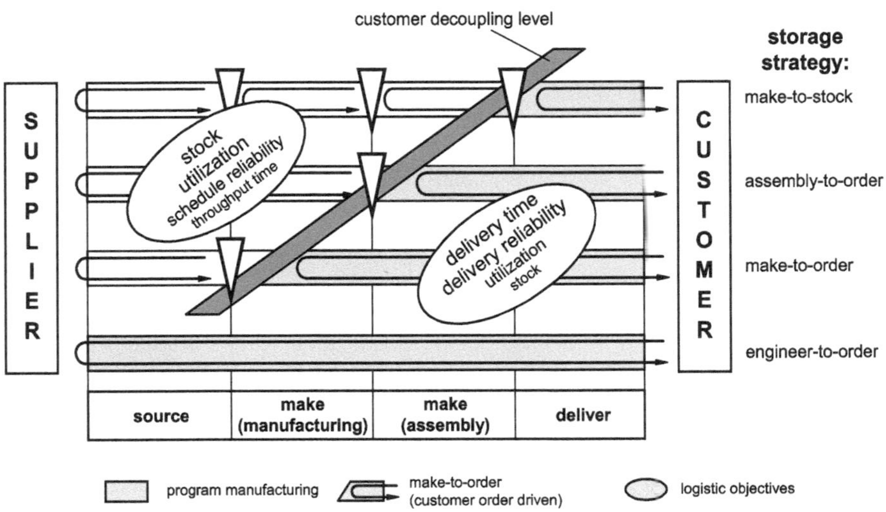

Fig. 1.3 Weighting of the Logistic Objectives with Various Storage Strategies (based on Eidenmüller)

level of service. If production is driven by the customer, the relation between the prioritized objectives is turned around. The promised delivery times and dates are then clearly given more weight, because the customer is directly impacted by a failure to do so. Whether or not it is permissible to obtain shorter delivery times through poorer utilization of the respective production lines, can only be decided on an individual basis. The required extended capacities (machines and operators) quickly increase the unit cost. The customer then has to decide if and under what circumstances they are prepared to accept higher costs in exchange for shorter delivery times. If one also considers that all of the changes in the company's environment can also impact the weighting of the objectives, then it becomes clear that there is no real 'optimum' on which a company can orient itself.

The uncertainties resulting from the above described problems often leads to the so-called *'vicious cycle of production, planning and control'* (Fig. 1.4). Empirical decision theory shows that decision-makers predominantly try to tread the supposed 'safer way' [Knol-87]. This is particularly true, if they make a bad impression by not fulfilling routine jobs (e. g., through unsatisfactory level of delivery service) and are not offered any incentives for being prepared to take risks (e. g., maintaining lower storage levels). Material planners therefore, usually orient themselves not only on average throughput times, but from a risk point of view also consider the throughput times' variance. If due to poor scheduling adherence the values for the lead time and throughput scheduling are now larger, then the orders arrive earlier than before in the production. The stock in front of the workstations then grows as do the queues. This means on average longer waiting times and thus longer order throughput times in conjunction with more widely distributed throughput times. As a result the schedule adherence is worse and the most

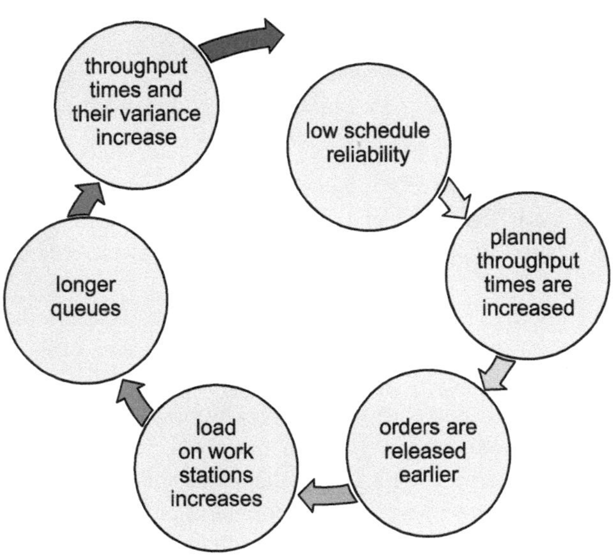

Fig. 1.4 Vicious Circle of PPC (Mather)

important contracts can only be completed by rush orders and costly special actions. The vicious cycle of PPC becomes a spiral of errors that only stabilizes when the throughput time reaches a very high level ([Math-77*], [Wien-97]).

In order to break such a vicious cycle and to make the dilemma of operations planning controllable through actively designing the processes and production it is necessary to qualitatively clarify and quantitatively describe the interdependencies between the logistic objectives as well as the possibilities for influencing them. Due to the diversity of processes in the field as well as external and internal influences these connections are not easily detected or described. It becomes increasingly clear that in order to cope with the design and control tasks, models that can be employed for describing and evaluating the processes have to be implemented. Only by applying an appropriate model is it possible to understand the complex operational sequences and their dependencies, and thus, to enable a continuous adjustment of the processes, even in the case of changing targets ([Hopp-96*], [Nyhu-96]).

1.3 Model Based Problem Solving Process

Previously, a model was primarily understood as a *representation of reality*. The term 'model' however, has been expanded in numerous fields of application. Models no longer exclusively facilitate understanding through visual means, instead they are also meant to help:

- make the primary situation and the forms which those problems may take understandable and comprehensible,
- uncover the causes and effects of the problems,
- deliver the basis of information required to derive measures,
- support specifically influencing or designing systems
- attain a fundamental understanding of a system's static and dynamic performance.

Operational systems are almost exclusively modeled using mathematical models. Even when in individual cases iterative steps and loop-like repetitions occur within a few phases, the application of such models are generally characterized by the procedures depicted in Fig. 1.5.

The first step in a *model application* is to clearly and precisely describe the problem. Furthermore, the goal of the analysis has to be explicitly defined. Since the effort and expense required for modeling and interpreting increases disproportionately to the complexity, it is particularly important in both phases to minimize the complexity of the object of examination as well as the goals. It is therefore often wise to breakdown a large problem into smaller, manageable sub-problems.

Once the problem has been formulated, a suitable model has to be chosen and if necessary adapted. If no appropriate model exists, then a mathematical one has to be developed. Once the presentation of the problem has been transferred onto the

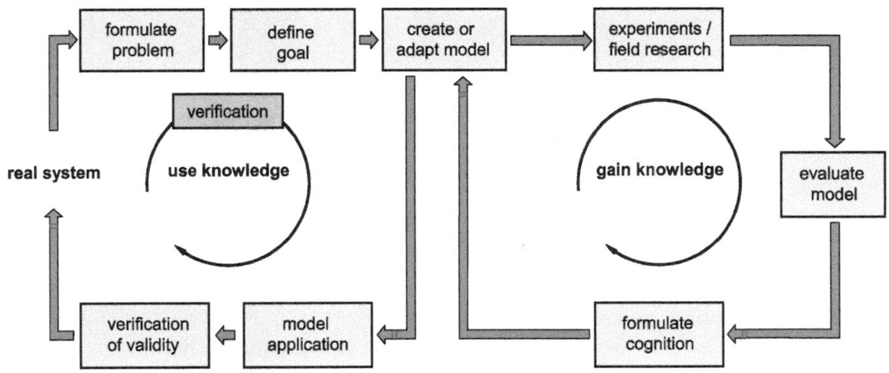

Fig. 1.5 Steps in the Model Based Problem Solving Process

model, alternative solutions can then be subsequently derived and evaluated. During this process it should be kept in mind that no knowledge beyond that which was previously incorporated into designing the model and in choosing the conditions can be gained by using models. Moreover, the results of the model application can only be as good as the underlying data.

In this respect, the next step in applying the model is also particularly important: Both the model as well as the solution that was derived should be continuously examined, especially while the solution is being applied under real conditions. This is a step which all too often is not adequately considered. The model must reflect the real system's behavior correctly and accurately enough. In addition to the technical precision and validity of the behavior (i. e., the model and the real system deliver comparable results) it is also particularly important to question the appropriateness of the cost-benefit ratio. The only way to guarantee that the model and the underlying parameters provide a sufficiently good basis for making decisions under changing conditions, is to examine it during the *model use* phase.

If in special cases no appropriate model exists, then the following basic requirements should be taken into consideration in the *model design* [Oert-77]:

- Direct Relationship with Reality: The model should illustrate the situation of interest for the actual targeted system as realistically as possible.
- Greater General Validity: The model should be applicable to a variety of real systems either directly or with minimal adaptation.
- Clear and Comprehensible Information: Based on the underlying objectives of the model application, the facts of interest should be simply but concisely presented. In particular, graphics or mathematical notation often depict information clearer than lists or tables.
- Limited to the Essentials: For practical reasons, it is important that the model is restricted to portraying and providing information only about the essential aspects of the real system.

Generally, in order to comply with these requirements the complex events found on the production floor, have to be strongly simplified by at least temporarily factoring out many of the secondary aspects. Only through reducing (abandoning insignificant features) and idealizing (simplifying vital features) is it possible to develop straightforward models which can be mathematically formulated and then often transferred to other similar applications. Whether or not the simplifications are feasible can be verified during both the *model validation* phase and while the model is being applied.

If connections between the relevant events can be derived during the course of the modeling or the model application, which can be determined absolutely and repeatabley under similar conditions and using similar methods, then the corresponding formulation can also be described as a law. The range of application for such a law increases to the extent that the formulation can be freed from special individual cases.

Models can be generally considered an abstract but very concentrated means of presenting descriptions of real or imagined systems. A common and at the same time typical property of all mathematical models is that they cannot in principle provide an absolutely true depiction of the original process, nor in the rule should they. Rather, they should be tailored to a specific application purpose and reproduce important characteristics relevant to it with sufficient accuracy. Therefore, a model can also only be chosen or judged from the perspective of its function. Figure 1.6 [Prof-77] shows the correlation between the effort required to develop and apply a model and the achievable *model quality*.

Accordingly, economical considerations also have to be integrated into the requirements for the exactness of the model. Generally speaking the more routine

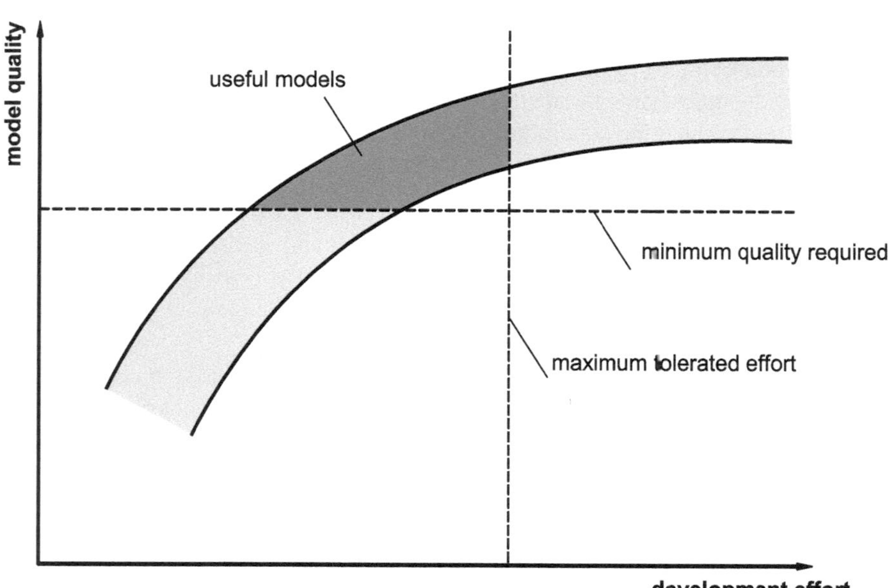

Fig. 1.6 Correlation between the Model Quality and the Amount of Effort Required for Development (Profos)

the applications is, the smaller the permissible effort to use it is. The goal of the analysis as well as the desired accuracy of the results and the necessary model particularization should be determined according to the motto "as general as possible and as accurate as necessary" [VDI-93]. The exact, qualitative and quantitative definition of the problem and goals thus becomes more significant. The wide corridor shown in the above graphic, represents a loose correlation between the depicted parameters. It also indicates that the choice or design of the model can strongly influence the cost-benefit ratio.

This book introduces a model, which can be employed to describe the previously outlined dilemma of operations planning qualitatively and quantitatively, as well as with regards to specific operational conditions. The *Logistic Operating Curves (LOC)* which the model is based on meet the abovementioned basic model requirements and make it possible to deduce the so-called *Basic Laws of Production Logistics.* Before the Logistic Operating Curves are introduced, it is first necessary to clearly define the objectives which underlie the model derivation.

1.4 Objectives of Production Logistics

In the following section, we assume that it is neither useful nor possible to completely describe the logistic processes within a complex production with one single model. It is therefore necessary to break the problem down. In order to do so, we will refer to the primary reference processes, which were defined by Kuhn [Kuhn-95]: *production and testing, transportation,* and *storage and supply*. Generally speaking these can be used to describe every production process from a logistic viewpoint.

The operational objectives depicted in Fig. 1.7 are derived from the logistic key performance indicators illustrated in Fig. 1.2. As can be seen, the objectives are contradictory and thus describe the specific problems which pertain to the individual reference processes.

Production is concerned with short throughput times and a high schedule reliability in order to on the one hand, fulfill customer demands and on the other hand, increase planning reliability. Furthermore, with shorter throughput times the risk of changes being made to orders in progress decreases. However, from a business perspective, it is preferred that the available production equipment is highly utilized and that there is the lowest possible WIP. In this way, the costs which are influenced by the production logistic can be minimized. It is thus obvious that whereas some of these sub-goals support one another, others contradict each other.

The objective for the stores has to be to keep the storage level and the related storage duration as low as possible given the store in/output. Nevertheless, at the same time a high level of delivery service has to be met for the areas requiring supplies by ensuring a minimal delivery delay. Thus, there are also competing objectives in the stores area. The objective output rate or utilization, which is critical for both production and testing as well as transportation, are generally not

		production reference processes		
		production and testing	transport	storage and supply
logistic objectives	schedule adherence	high schedule reliability	high schedule reliability	low delivery delay
	throughput time	short throughput time	short transport throughput time	short storage time
	output rate	high utilization	high utilization	
	inventory	low work in process	low work in process	low stock
	costs	low costs per unit	low costs per transport operation	low storage costs

Fig. 1.7 Logistic Objectives for the Production Reference Processes

determined for the storage processes unless the performance of inventory personal and facilities are being investigated.

The illustrated conflict between the objectives of the three reference processes is well-known, however up until now and particularly in the industry it has been only quantified with difficulty. Thus, conducting a targeted *Logistic Positioning* has been almost impossible. Previously, knowledge gained through experience has generally been relied upon in order to provide the market geared, target values for the necessary delivery times, existing work content/capacity structures etc. used in designing and operating the production processes. However, based alone on the complexity of the processes in the production and the interdependencies between the logistic objectives it is highly improbable that the best possible compromise can be found using this method. If while increasingly pursuing one of the sub-goals, another one is infringed upon (which is almost inevitable given the contradictions) then many companies fall back to the initial state. Thus it is quite common for companies to have actions to reduce inventory. Difficulties in fulfilling delivery obligations though often lead to the stock piling up again to the point which had originally initiated the inventory reduction sales (see [Jüne-88], [Eide-95]).

In order to break through this cycle, it is desirable to also be able to quantitatively represent the interdependencies between the logistic objectives and the possibilities for influencing them. Doing so allows various strategies to be pursued according to the operational and market situations. The Logistic Operating Curves, presented in the following, are useful for this purpose.

1.5 Logistic Operating Curves – an Explanatory Model for Production Logistics

A Logistic Operating Curve visually represents the correlation between a specific parameter of interest (the *objective* or dependent variable) and an independent variable. Therefore, for every value of the independent variable – which can be changed by external conditions – at least one value can be determined for the objective.

In preparation for this description, it is first necessary to determine the independent variables and objectives based on these considerations. In principle it can be said that all of the previously mentioned logistic objectives can also be seen as an objective in a LOC. This however, is not the case for independent variables: The costs caused by the production logistics, result from a process but are in no way an independent variable that can be directly changed by external factors. This is also true for the schedule adherence, which occurs at the end of a process and can therefore be influenced only indirectly. The WIP, throughput time, and output rate however, remain as potential independent variables. All three of these can in principle be seen as both an independent variable and as a result.

In the following chapters we have chosen to use the WIP as the independent variable. This decision is supported by the fact that the WIP can be actively influenced through targeted control measures in all three reference processes i. e., it can be ensured that the input is temporarily greater, less than or equal to the output of the observed process. The remaining logistic objectives thus represent the dependent variables. Since the parameters WIP, throughput time and output rate can however be mathematically transformed into each other (see Chaps. 2 and 3), it is possible to declare one of the other parameters as an independent variable according to the particular problem.

In Fig. 1.8 the various LOC for the three reference processes are illustrated as schematic diagrams. The curve for production and testing shows that the throughput time of a production system mainly increases proportional to the growing WIP. The throughput time however, cannot fall short of a specific minimum which arises from the technology dependent operation time during the order processing and where applicable, the transportation time between two operations. Short throughput times are thus generally linked to a smaller variance of the throughput time. The greater planning certainty that results from this leads to a greater schedule reliability. Nevertheless, with the escalating WIP and throughput times, and as experience has shown the resulting increased throughput time variance, the planning reliability decreases. With a high WIP level, the Output Rate Operating Curve for a workstation is largely independent of the WIP. Should the WIP fall below a particular level though, there will be output losses due to a temporary lack of work materials.

Finally, the Cost Operating Curves (COC) make it possible to draw clear conclusions about the optimal costs of the operating sector. When the WIP level is low there is a loss of utilization which in turn increases the costs due to the higher

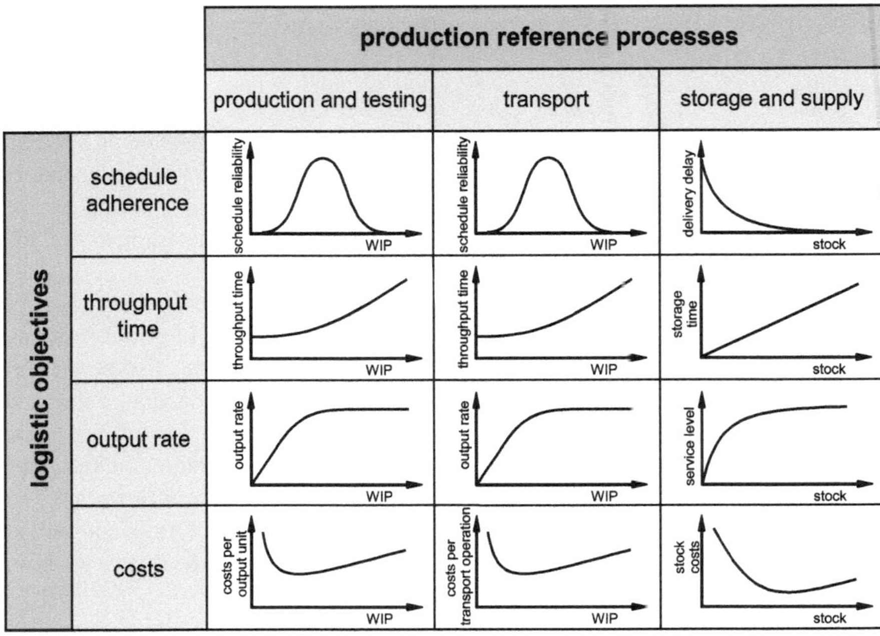

Fig. 1.8 Logistic Operating Curves for Production Reference Processes

expenses related to the idle time of the available capacity. Higher WIP levels however, also give rise to greater stock related costs. The Output Rate and Throughput Time Operating Curves are often plotted together in one diagram and then referred to as Logistic Operating Curves ([Bech-84], [Nyhu-91], [Wien-93], [Wien-97]).

Since the Logistic Operating Curves for the transportation reference process generally behave very similar to the production and testing curves we will not explain them separately. The individual Transportation Operating Curves (TROC) are essentially influenced by the mode and number of employed transport means, the implementation strategy and the resulting percentage of double move or empty trips as well as the integration of the transport system in the organization of the process.

The LOC for the storage and supply reference processes represents the time an item or group of items spends in storage and is delayed in delivery as a function of the mean stock. These Logistic Operating Curves are also referred to as Storage Operating Curves (SOC) ([Gläs-95], [Nyhu-95]) and can be influenced by all the parameters that impact the behavior of the store input or output. In the Cost Operating Curve for this reference process, the value of the article as well as the costs of shortages are included.

The Logistic Operating Curves sketched here, very clearly illustrate the prevailing conflict between the goals of the logistic objectives for all three reference processes. The LOC make it possible to decide which characteristic needs to be weighted the most depending on the system specific boundary conditions and the operational and/or market situation. At the same time, they indicate what consequences are to be expected with regards to the other characteristics. They are thus well suited for supporting the previously mentioned Logistic Positioning within the area of conflict between the relevant competing logistic objectives.

Generally speaking, it is possible to construct the Logistic Operating Curves for different basic conditions and then compare them with each other in order to assess the effects of interventions on the logistic aspects of the manufacturing process. Alternative plans can thus undergo a logistic oriented evaluation which can then be used as a basis for managerial decisions. The LOC can therefore provide an important contribution so that a company can concertedly align themselves with the logistic key performance indicators.

1.6 Goals and Structure of the Book

The overriding goal of this book is to provide a model that can be applied both in the field as well as in research, which enables quick and reliable statements to be made about where the most important points for tapping into potential logistic improvements are in a firm. Furthermore, this model can be used to determine which specific measures should be introduced in order to exploit this potential. The Logistic Operating Curves play a key role here.

The endeavor to develop Logistic Operating Curves such as these and to employ the applications outlined here has existed for a long time. Their familiar beginnings are generally based upon both queuing and simulation theory. Thus, after explaining and defining the key logistic objectives (Chap. 2) these modeling techniques will be introduced and discussed in relation to our general goals (Chap. 3). As a result, it will be shown that both of these approaches exhibit considerable difficulties and weaknesses with regards to the problems presented here. They can therefore only be limitedly implemented in evaluating and designing real job shop production processes. Whereas the conditions required for applying queuing theory are generally not given in small series and single item production processes, simulations are very laborious. Moreover, the objective quality of the simulation results, which appears to be a given, absolutely needs to be questioned. Experimental models such as simulation also exhibit "no physical meaning" i. e., in no way can general statements be made about the internal operations and connections between the objects of investigation. Instead, they are restricted to describing the connections between the input and output values included in the experiment.

These problems are prevented by a mathematical approach, which more accurately calculates the LOC based on less data with the help of an approximation equation. The theory underlying the approximation equation, which is extensively explained in Chap. 4, supports the fundamental functions of a model. It is thus possible to mathematically predict measured behaviors of production systems for the logistic objectives (system analysis) as well as to support the system conception and lay-out (system synthesis).

In order to make it possible to confidently apply the Logistic Operating Curves Theory (LOC Theory) as well as when necessary, to enable further developing it, we will comprehensively discuss the fundamentals of this theory. The model evaluation is conducted based on simulations as well as through the analysis of real processing data. The analysis of the real data is particularly significant as the described procedure can also be practically applied in order to regularly examine and where necessary adjust the LOC parameters.

A special feature of the LOC Theory is that it also contributes to gaining information. In Chap. 5, universally applicable correlations are developed and summarized in the form of the *Basic Laws of Production Logistics*. Independent of individual cases, they describe the relevant relationships between the logistic objectives and the potential for influencing them. Thus, already based on logical conclusions, they can be applied for example to assess qualitatively (and to some extent also quantitatively) the impact of individual measures or behaviors with regards to the logistic objectives.

In Chap. 6, the possible application and limitations of the Logistic Operating Curves Theory will be explained with the help of simple, comprehendable examples. Understanding the model, takes on a critical role here. Only when the basic concept of the model and the significance of the individual variables and parameters are correctly understood, can the concept be transferred failure-free to a more complex application.

The conclusions derived from the Logistic Operating Curves Theory are in principle resource oriented. However, further methods of analysis are required in order to be able to describe the logistic efficiency of a complex production. The Bottleneck Oriented Logistic Analysis (BOLA) is a tried and tested approach. This controls technique combined with the LOC Theory is used to analyze the through-

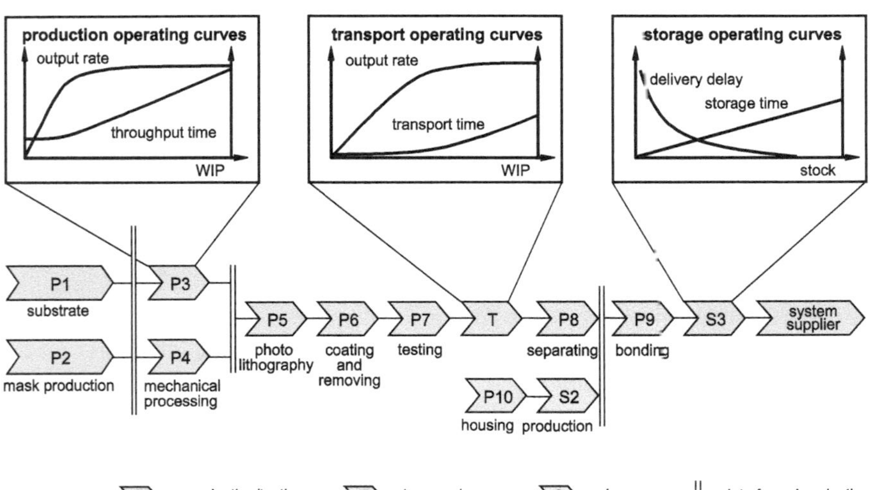

Fig. 1.9 Logistic Operating Curves for Production Reference Processes (Example of a Process Chain)

put time, WIP and material flow. Moreover, it supports a logistic oriented modeling, evaluation and design of interconnected production processes. Based on two practical applications of this analysis method we will demonstrate how the production process can be transparently represented based on operational feedback data. The examples also demonstrate how this analysis can be employed to indicate logistic bottlenecks and weak points in the material flow and how measures for eliminating them can be developed and evaluated (Chap. 7). At the same time, the BOLA allows concrete statements to be made, for example, about what percent the average order throughput time can be reduced by.

Finally in Chap. 8, we will describe how the model that is developed for the reference process production and testing can be transferred over to storage processes – despite a totally different framework. Furthermore, we will present the possible applications that result from this for the procurement and inventory management areas.

Meanwhile, the LOC Theory has also been transferred to the transportation reference process ([Egli-01], [Wien-96a]). From a logistic perspective, a complex production network can theoretically also be evaluated and designed with a universal model application. The principle of this is illustrated based on the example of a processing chain for the manufacturing of micro-electronic parts above in Fig. 1.9.

Chapter 2
Basic Principles of Modeling Logistic Operating Curves

In order to understand the theoretical background of Logistic Operating Curves it is necessary to explain the *fundamentals of modeling* them as well as to define the basic terms and performance figures. In this chapter we will first concentrate on the terms required for deriving and interpreting the Throughput Time Operating Curves and Output Rate Operating Curves (TTOC and OROC) for the production process. The Funnel Model, the Throughput Diagram and Little's Law will form the foundation for this discussion.

2.1 Funnel Model as a Universal Model for Describing Production Processes

The *Funnel Model* and the resulting *Throughput Diagram* are a model recognized among experts and commonly used for describing production processes. Before introducing this model, we will first describe the basic definitions and the corresponding performance figures. We will limit our remarks here to those performance figures that are particularly important for Logistic Operating Curves. Further information can be gleaned from the bibliography (see e. g., [Wien-95a*], [Wien-97], and the bibliographies provided in them).

2.1.1 Work Content and Operation Times

The *work content* WC is a key parameter for the Funnel Model as well as all modeling approaches based on it. The work content is a measure of the planned time for an operation on a workstation. According to Eq. 2.1, it results from the setup time for each lot, the processing time per piece and the lot size:

$$WC = \frac{(LS \cdot t_p + t_s)}{60} \tag{2.1}$$

where WC work content [hrs],
LS lot size [units],
t_p processing time per piece [min/piece],
t_s setup time per lot [min].

In the industrial setting, the work content of orders on a workstation can have very different values. Figure 2.1 shows the distribution of the work content from 377 orders processed on one workstation during the course of approximately one year. What is both noticeable and typical of the *work content distribution* is the left skew and large variance of the individual values. Thus, approximately 25% of all operations indicate a work content of up to two hours. At the same time though, there are also a few operations with a work content of more than thirty hours. Without statistical parameters it is almost impossible to formulate a feasible, continuative description of such a distribution which would also enable comparisons. Here, both the mean value and the standard deviation have proven to be particularly useful for the general characterization of the work content's distribution.

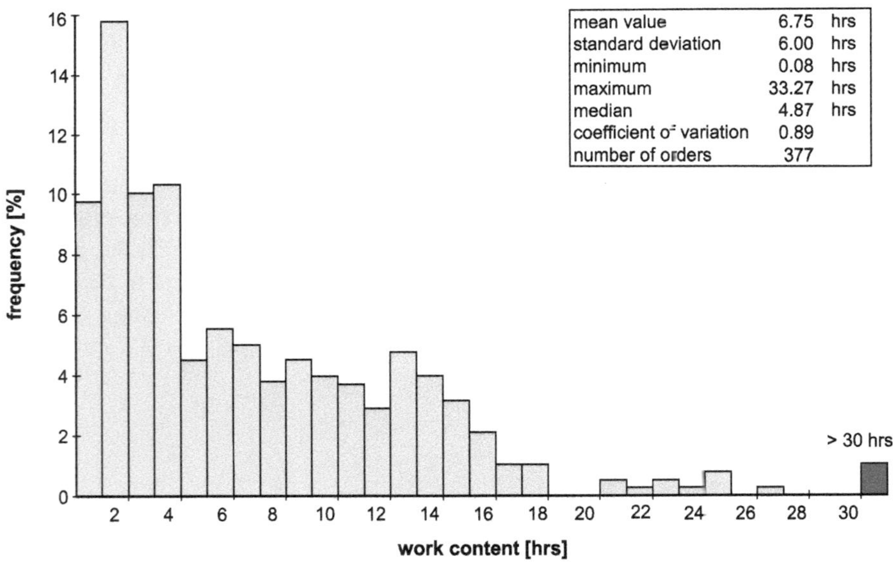

Fig. 2.1 Distribution of the Work Content per Operation (Practical Example)

The mean value of the work content is 6.75 hours and is calculated as

$$WC_m = \frac{\sum_{i=1}^{n} WC_i}{n} \tag{2.2}$$

where WC_m mean work content [hrs],
WC_i individual work content per operation [hrs],
n number of operations reported back [-].

The *standard deviation* describes the range of the distribution:

$$WC_s = \sqrt{\frac{\sum_{i=1}^{n} (WC_m - WC_i)^2}{n}} \tag{2.3}$$

where WC_s standard deviation of the work content [hrs],
WC_m mean work content [hrs],
WC_i individual work content per operation [hrs],
n number of operations reported back [-].

When comparing the variability of different distributions it is helpful to determine the *coefficient of variation*. This dimensionless number describes the variance of the distribution in relation to the mean:

$$WC_v = \frac{WC_s}{WC_m} \tag{2.4}$$

where WC_v coefficient of variation for the work content [-],
WC_s standard deviation of the work content [hrs],
WC_m mean work content [hrs].

For the example in Fig. 2.1, the coefficient of variation assumes the value of 0.89. This value is typical for many manufacturers. In Fig. 2.2 the distributions of the variation coefficients for all of the workstations from two studies are presented. The first enterprise is involved in manufacturing electronics (in this case printed circuit boards) and has 69 workstations. In the second company, a mechanical engineering firm, a mid-volume production area with a total of 33 workstations was analyzed. In both companies the mean coefficient of variation for the work content is approximately 1. Furthermore, it is also determinable that the structure of the work content to be completed on all of the workstations is more or less heterogeneous; the variation coefficient is less than 0.5 in only a few exceptional cases. The values stated here have also been confirmed in comprehensive studies from other authors (see [Ludw-95] and the bibliography cited there).

The work content is usually stated in *planned hours*. For a series of problems however, it is necessary to express how long the workstation is occupied in *shop*

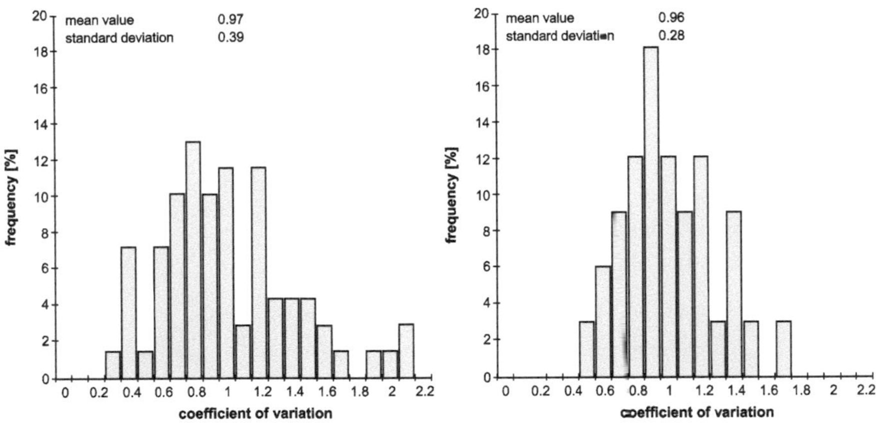

a) Electronic Manufacturer (69 workstations) b) Mechanical Engineering Firm (33 workstations)

Fig. 2.2 Coefficient of Variation Distributions for the Work Content (Results of Two Industrial Case Studies)

calendar days. This parameter known as the *operation time*, is calculated by dividing the work content by the workstation's maximum possible output rate:

$$TOP = \frac{WC}{ROUT_{max}} \tag{2.5}$$

where TOP operation time [SCD],
WC work content [hrs],
$ROUT_{max}$ maximum possible output rate [hrs/SCD]

The *maximal possible output rate* basically results from the capacity. In principal, the upper limit of a workstation's output rate is determined by the limiting capacity factor (i. e., facilities or operators). It should also be taken into consideration that the maximal possible operating time of a resource is decreased by disruptions e. g., machine failure and that also the capacity of the operators is influenced by a number of different factors. Therefore, because of failures and disrupted procedures the scheduled operating time of an operator is generally less than that which they actually have to work. This maximum available operation time however, can be increased through overtime and holiday shifts. Finally, the labor utilization rate, which describes the impact of the work flow intensity on the actual operation time, also needs to be considered (for an extensive discussion about concepts of capacity see [Kalu-94]).

In previous studies, the system capacity is often used in a simplified manner instead of the maximal possible output rate in the abovementioned Eq. 2.5 (see e. g., [Wien-95a*], [Wien-87], [Ludw-95]). The rational behind this is that the operation time is generally small in comparison to the throughput time and thus the inaccuracy can be accepted in favor of the reduced effort required to acquire data and make calculations. This argumentation can generally be followed as long as the analysis of the throughput time is prioritized. Nevertheless, since in this book the operation time will also be used for modeling the production processes and for comparing the modeling rules, greater precision is required.

The equations used for calculating the mean (2.2), standard deviation (2.3) and variation coefficient (2.4) of the work content are, similarly, also applicable to the corresponding distribution parameter for the operation time. As long as only one workstation is observed, and its maximal possible output rate is constant or at least is independent from the work content of the individual operations, it can be said that

$$WC_v = TOP_v \tag{2.6}$$

where WC_v coefficient of variation for the work content [-],
TOP_v coefficient of variation for the operation time [-].

2.1.2 Throughput Time

Another fundamental aspect of the Funnel Model – its definition of the throughput element – is presented in a simplified form in Fig. 2.3. The upper part describes the throughput of a production order consisting of two manufacturing orders (further referred to as just "order") and an assembly order. When the manufacturing is performed in lots an order will be transported to the subsequent workstation after the completion of an operation and possible waiting time at the corresponding station. Once there, the lot usually meets a queue and has to wait until the orders to be completed ahead of it are processed. Provided the capacities required to handle the order are available, the workstation can be reset and the processing of the lot can be carried out. This cycle continues until all the operations for the order have been completed.

The steps of an operation are presented in the lower part of the figure. According to it, the throughput time for an operation is defined as the time span an order requires from the completion of the previous operation or from the order's point of input (at the start of an operation) until the end of the observed operation's process.

$$TTP = t_{e,op} - t_{e,prop} \tag{2.7}$$

where TTP throughput time [SCD],
$t_{e,op}$ end of operation [SCD],
$t_{e,prop}$ end of preceding operation [SCD].

The *post-processing waiting time* as well as the *transport time* and *pre-processing waiting time* is assigned according to this definition to the observed operation. Since in the broader sense these time slices do not need to be differentiated, they are consolidated into the *inter-operation time* TIO. Next to the lot's operation time (consisting of both the setup time and the processing time) the inter-operation time is the second component of the throughput time.

The definition of the individual *throughput time components* can also be transferred to the level of the parts to be manufactured (see e.g., [Kett-76], [Wien-95a*], [Wien-87]) and is thus also applicable to continuous flow manufac-

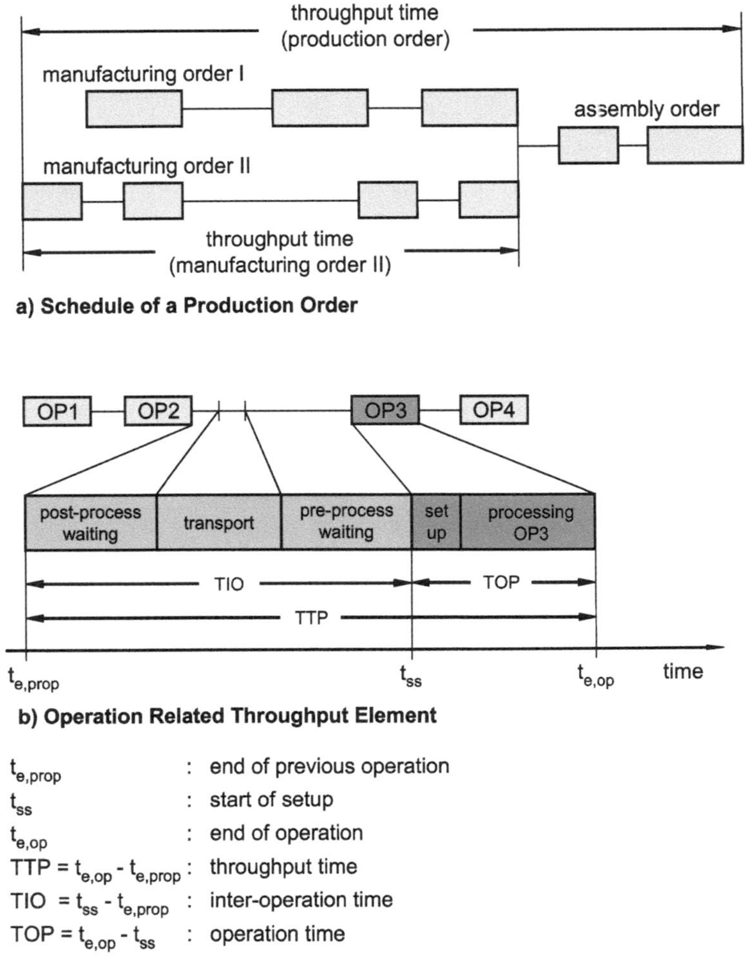

Fig. 2.3 Throughput Time Components and Throughput Element

turing. For further considerations however, it should initially be assumed that the individual lots are transported and processed as closed lots. All other comments that are based on the thus defined throughput time components, also apply only under these conditions – unless it is otherwise explicitly stated.

The *operation time* can be calculated in the same way as the throughput time provided that the time at which the setup began is known. This time is nevertheless, often not recorded. Furthermore, it is also necessary for the subsequent modeling steps to establish the operation times for those operations not yet processed and for which the required completions notes thus cannot yet exist. As an alternative to determining them via feedback data, the operation time can be calculated through the work content and the maximal possible output rate of the observed system according to Eq. 2.5.

After determining the operation time it is then possible to derive the *inter-operation time* between two operations by calculating the difference:

$$TIO = TTP - TOP \tag{2.8}$$

where TIO inter-operation time [SCD],
TTP throughput time [SCD],
TOP operation time [SCD].

The relationship between the three time parameters, presented in Eq. 2.8, applies not only to the individual values of each operation, but also to their respective means:

$$TIO_m = TTP_m - TOP_m \tag{2.9}$$

where TIO_m mean inter-operation time [SCD],
TTP_m mean throughput time [SCD],
TOP_m mean operation time [SCD].

The arithmetic mean of the throughput time is calculated as

$$TTP_m = \frac{\sum_{i=1}^{n} TTP_i}{n} \tag{2.10}$$

where TTP_m mean throughput time [SCD],
TTP_i work content per operation [SCD],
n number of the operations which have been reported back [-].

Corresponding equations for the mean operation time and inter-operation time can also be formulated.

For the throughput time, the *flow rate*, which is defined as the proportion of throughput time to operation time (Eq. 2.11), can be applied as a relative measure:

$$RF_m = \frac{TTP_m}{TOP_m} \tag{2.11}$$

where RF_m mean flow rate [-],
TTP_m mean throughput time [SCD],
TOP_m mean operation time [SCD].

2.1.3 Lateness

If, in addition to the completion notes the corresponding target data is also collected, then both the target throughput as well as the *lateness* can be calculated and depicted (Fig. 2.4). The difference between the actual processing end and the target processing end corresponds to the *output lateness*. Whereas, a positive difference indicates that the order has been reported back later than planned, a negative

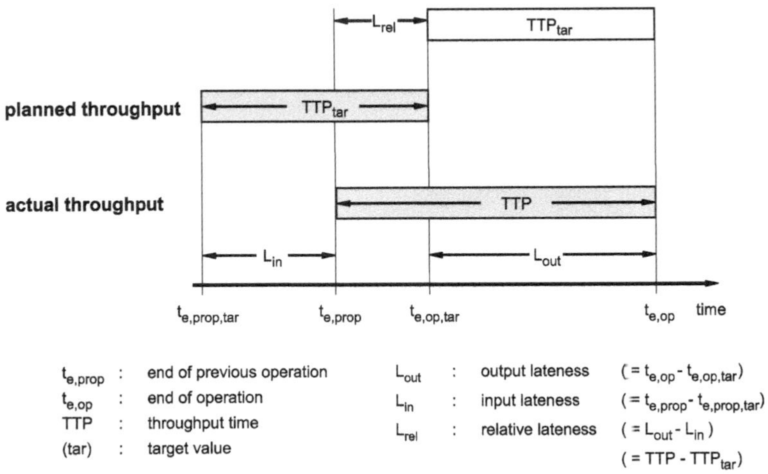

Fig. 2.4 Different Types of Lateness for an Operation

difference indicates that it was completed too early. The input situation can be similarly expressed: Instead of the output dates, the measured and planned input dates would then be compared with each other. The *relative lateness* is the result of the difference between the actual throughput time and the target throughput time. It can be used to identify whether the output schedule situation worsened or improved in comparison to the input schedule. A positive value here means that the operation was delayed in comparison to the plan, while a negative value indicates that the throughput time was shorter than planned.

With the help of the above definitions, the operational throughput time and lateness as well as their components can be calculated and statistically evaluated. Moreover, the definitions also form the basis for both the Funnel Model and the Throughput Diagram derived from it.

2.2 Logistic Objectives in a Throughput Diagram

In the *Funnel Model,* it is assumed that similar to illustrating flow processes in chemical engineering, the throughput behavior of any production related capacity unit can be comprehensively described through the *input, WIP* and *output.* Each capacity unit independent of whether it has to do with an individual workstation, a cost center or the entire production can be represented as a funnel (Fig. 2.5a). The lots arriving at the workstation together with the lots already there form the queued orders. After being processed these drain out of the funnel. The funnel opening therefore symbolizes the output rate (also referred to in practice as simply the 'throughput', 'output' or 'yield') which can vary within the limits of the maximum capacity.

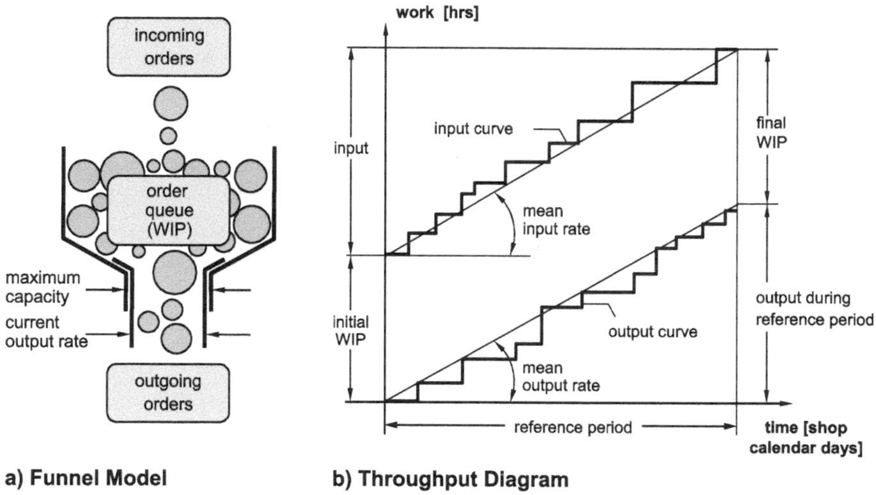

Fig. 2.5 Funnel Model and Throughput Diagram of a Workstation (according to Bechte)

The funnel events can be transferred to the so-called *Throughput Diagram* (Fig. 2.5b). In order to do so, the completed orders with their work content (in planned hours) are cumulatively plotted over their completion dates (*output curve*). Similarly, the *input curve* is developed by plotting the incoming orders with their work content over the input date. The beginning of the input curve is determined by the WIP found on the workstation at the beginning of the reference period (*initial WIP*). The *final WIP* can be read from the diagram at the end of the reference period. Whereas, the average slope of the input curve is called the *mean input rate*, the average slope of the output curve corresponds to the *mean output rate*. It can generally be said that when a system with a relaxed state is observed over a long period the input and output curves have to run parallel to each other.

The Throughput Diagram accurately describes the dynamic system behavior qualitatively and chronologically. It illustrates the impact of the logistic objectives on each other and makes it possible to express them mathematically.

2.2.1 Output Rate and Utilization

Figure 2.6 illustrates a Throughput Diagram for a workstation. In addition to the order input and output, the *mean output rate* as well as the trend of the WIP during the reference period is plotted. The mean output rate $ROUT_m$ results from the ratio between the output OUT and the duration of the reference period PE.

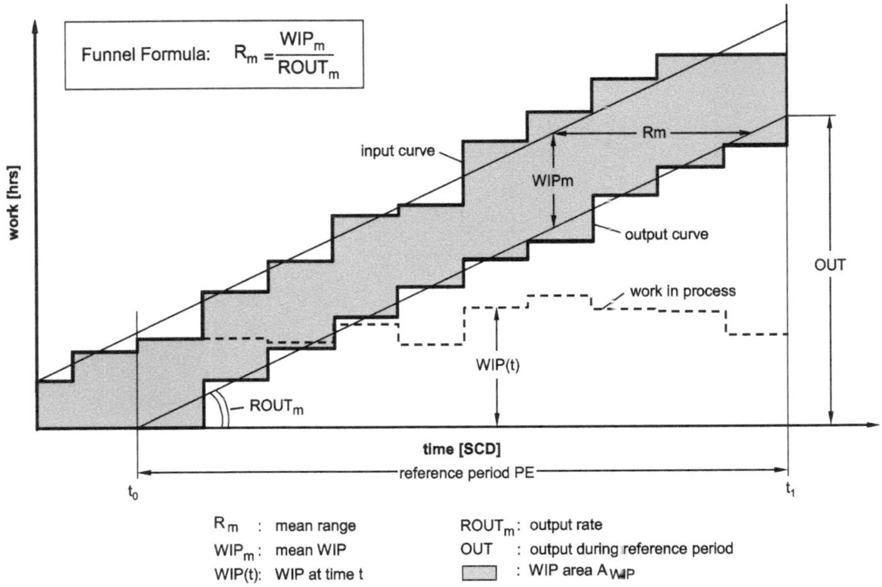

Fig. 2.6 WIP, Range and Output Rate in the Throughput Diagram (according to Bechte)

The output corresponds to the sum of the work content (OUT) which was reported back, therefore,

$$ROUT_m = \frac{\sum_{i=1}^{n} TO_i}{PE} \qquad (2.12)$$

where $ROUT_m$ mean output rate [hrs/SCD],
WC_i work content (in planned hours) per operation [hrs],
n number of operations which have reported back [-],
PE reference period [SCD].

The above comments are based on the assumption that, corresponding to common practice, there is only one completion note produced per operation. Nevertheless, it should be kept in mind that short reference periods in particular, can lead to inaccurate depictions (see also [Ludw-95], [Penz-96]). Thus, the mean output rate calculated using Eq. 2.12, could be influenced by the possibility that another order is being processed that will not be completed until the next reference period. At the same time, it is also possible that at the beginning of the reference period an order will be reported back, which was processed mainly during the previous period. Given long enough reference periods, which should be the basis of planning and constructing production systems, these effects compensate for one another. Using sufficiently long enough reference periods therefore allows the imprecision of the problem to be ignored.

The *utilization* of a system is defined as the ratio of mean output to possible output (for more information on the term 'maximal possible output rate' see also Sects. 2.1.1 and 4.1.2):

$$U_m = \frac{ROUT_m}{ROUT_{max}} \cdot 100 \tag{2.13}$$

where U_m mean utilization [%],
$ROUT_m$ mean output rate [hrs/SCD],
$ROUT_{max}$ maximum possible output rate [hrs/SCD].

2.2.2 Work in Process (WIP)

The *work in process* (WIP), represented by the vertical distance between the output and input curves, corresponds to the work content of the orders queued for processing as well as those being processed. The mean WIP can be calculated by dividing the *WIP area* AWIP by the *reference period* PE. The WIP area itself can be calculated by determining the difference between the area below the input curve and the area below the output curve (during the time interval $[t_0,t_1]$). The calculation of the mean WIP is generally described as

$$WIP_m = \frac{\int_{t_0}^{t_1} IN(T)dT - \int_{t_0}^{t_1} OUT(T)dT}{t_1 - t_0} \tag{2.14}$$

where WIP_m mean WIP [hrs],
IN(T) input (cumulative work content of the incoming operations as a function of time) [hrs],
OUT(T) output (cumulative work content of the outgoing operations as a function of the time) [hrs],
t_0 beginning of the reference period,
t_1 ending of the reference period.

Numerous industrial studies have demonstrated that it is sufficient to base the completion notes of an operation on discrete time segments. The completion notes (also referred to here as feedback or responses) therefore, normally occur at the end of a shift or a work day. In this case, the calculation of the mean WIP is noticeably simplified. It can then be determined by dividing the sum of the individual WIP values for each time segment (e. g., per day) by the number of time segments contained within the reference period:

$$WIP_m = \frac{\sum_{t=t_0}^{t_1} WIP(T)}{z} \tag{2.15}$$

where WIP_m mean WIP [hrs],
$WIP(T)$ WIP during the time segment T [hrs],
z number of intervals during reference period [-].

The ratio of the output area, or mean WIP to mean output rate corresponds to the work in process' *mean range* R_m. This relationship is also referred to as the *Funnel Formula*:

$$R_m = \frac{A_{WIP}}{OUT} = \frac{WIP_m}{ROUT_m} \tag{2.16}$$

where R_m mean range of the WIP [SCD],
A_{WIP} area of the WIP [hrs · SCD],
OUT output during reference time [hrs],
WIP_m mean WIP [hrs],
$ROUT_m$ mean output rate [hrs/SCD].

If the shape of the input and output curves are idealized as a straight line, the vertical distance corresponds to the mean WIP, while the horizontal distance corresponds to the mean range of the WIP (see Fig. 2.6).

2.2.3 *Weighted Throughput Time and Range*

In order to represent the throughput time, the throughput element for each completed operation is plotted in addition to the input and output curves in the Throughput Diagram (Fig. 2.7). This element represents the throughput of one order through the workstation. Based on these throughput elements, conclusions about a station's processing behavior can be drawn. When the orders are strictly processed according to the First In – First Out principle, the throughput elements lay exactly between the input and output curves. In turn, when the start of the throughput element deviates more or less strongly from the input curve it indicates a change in the order sequencing: Some orders are processed on the system directly after the input, while others are constantly deferred and consequently indicate a much longer throughput time.

In Fig. 2.7, it can be seen that throughput elements exhibit a second dimension. In the horizontal direction – corresponding to the explanation about the operation based throughput element (see Fig. 2.3) – the elements are formed by the order's input and output dates and in the vertical direction by the order's work content. This *weighting* produces two dimensional elements for each of the implemented operations. The sum of these results in the so-called throughput time area ATTP. If in analogy to the WIP area, the throughput area is set in relation to the output (that is the sum of the work content which was reported back) when calculating the range, then a further parameter for the throughput time results as a comple-

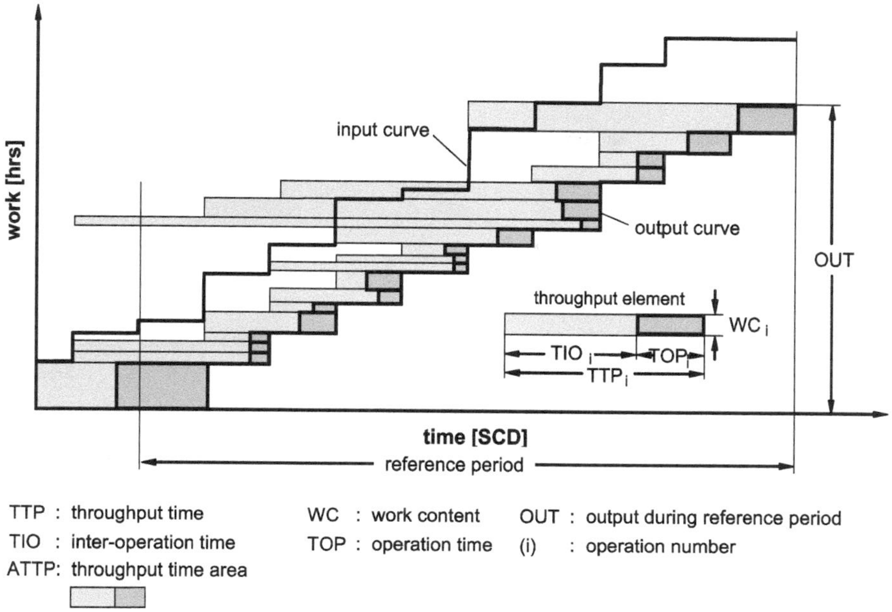

Fig. 2.7 Throughput Elements in the Throughput Diagram (according to Bechte)

ment to the arithmetic mean. This parameter is called the weighted mean throughput time:

$$TTP_{mw} = \frac{\sum_{i=1}^{n}(TTP_i \cdot WC_i)}{\sum_{i=1}^{n} WC_i} \tag{2.17}$$

where TTP_{mw} weighted mean throughput time [SCD],
TTP_i throughput time of operation i [SCD],
WC_i work content of operation i [hrs].

Between the arithmetic mean throughput time, the weighted mean and the range a number of important mathematical relationships can be derived that allow conclusions about the processing behavior and the development of the WIP to be made.

If the ratio between the WIP and the output rate remains the same (the input/output curves are parallel) and the sequencing rules have a negligible influence on the throughout time area ATTP (Fig. 2.7), then the ATTP corresponds to the AWIP (Fig. 2.6). That is, when calculating the throughput time area, portions of throughput time areas that have accumulated before the reference period are taken into consideration. At the same time however, those portions of orders that will be

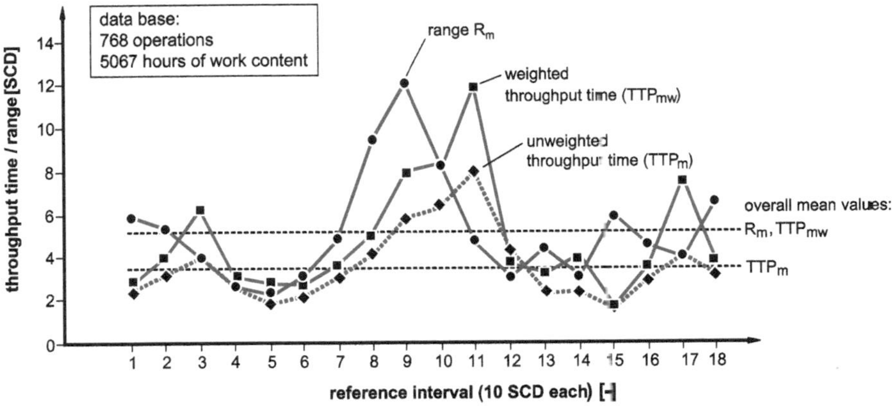

Fig. 2.8 Chronological Behavior of Throughput Time Measures (Pick and Place Machine)

finished in the following period are not considered. Thus, given the abovementioned conditions the two area portions compensate for one another. Therefore,

$$R_m \cong TTP_{mw} \tag{2.18}$$

where R_m mean range [SCD],
TTP_{mw} weighted mean throughput time [SCD].

Generally speaking, with longer reference periods these prerequisites are with good approximation a given. Especially during shorter reference periods though, the performance figures could demonstrate a different behavior over time. Whereas, the range reacts directly to changes in the input caused by increasing the WIP (when the output stays the same), both the weighted and unweighted throughput time are essentially determined by the WIP level at the time of the order input – when the processing is executed according to the FIFO principle. The specific value of the range thus reacts more quickly to changes in the WIP than the value of the throughput time does. Furthermore, when the sequencing rules are independent of the work content, the unweighted throughput time is always less than the weighted throughput time. We will examine the reasons for these circumstances and their mathematical expression more closely in Sect. 4.3.2.

The explanation for the correlations between the throughput time parameters can also be seen in studies conducted in industrial settings. Figure 2.8 illustrates the behavior of the three parameters during a timeframe of 18 intervals (each of which were two weeks long) as measured on a pick and place machine. It confirms that the value of the mean range and the weighted throughput time are the same, while the simple mean is less. Furthermore, it can be seen that the mean values of the range always change earlier than the values of the throughput time, whereas the output rate during the entire investigation period was almost constant. This fact has to be taken into consideration when interpreting the analysis results, but can also be used quite purposefully for the diagnosis and early detection of undesirable trends.

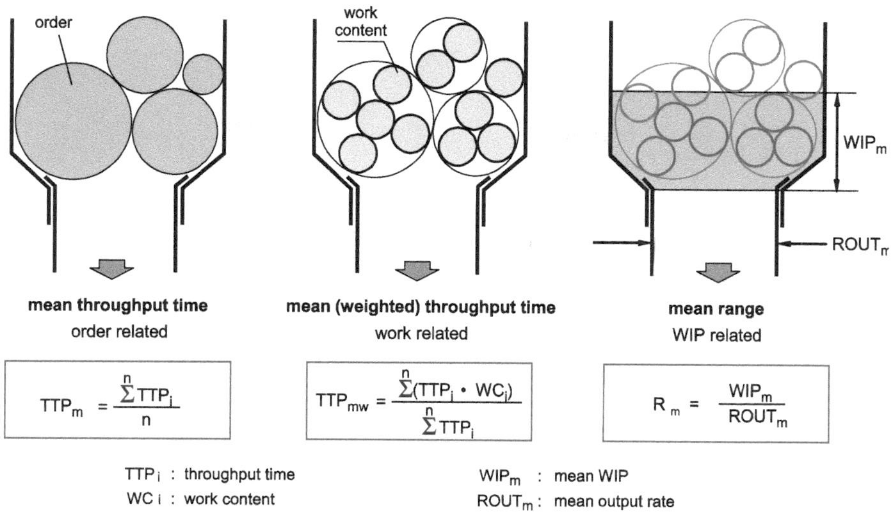

Fig. 2.9 Comparison of Performance Figures for a Workstation's Throughput Time

Figure 2.9 presents a comparative summary of these three throughput time parameters. The graph emphasizes that the calculation of the mean unweighted throughput time is based upon the number of orders. This value indicates how long on average an order stays on a workstation. With the mean weighted throughput time and the range however, the WIP (either through the throughput time area of the completed orders or directly from the mean WIP) is considered when calculating the respective throughput time values. With the weighted throughput time, the reference to the order is broadened because the work content is also taken into consideration. The range though, does not directly refer to the individual orders. However, over a long reference period the values for the mean weighted throughput time and the range have to largely coincide with one another on a relaxed system because both the WIP and the output are derived from the order's work content.

2.3 Little's Law

A second model that can be used to calculate the throughput time is *Little's Law* (see e. g., [Conw-67*]). This model, which is frequently implemented in queuing theory, is quite similar to the Funnel Model (see Sect. 3.1). In the following section we will elaborate the differences between these two modeling approaches and establish the basis for a synthesis of these two models and the Logistic Operating Curves Theory introduced in Chap. 4.

Little's Law states that the mean throughput time of a workstation can be calculated by dividing the mean number of orders in the system N_m by the mean arrival rate λ

$$TTP_m = \frac{N_m}{\lambda} \tag{2.19}$$

where TTP_m mean throughput time [SCD],
N_m mean number of orders in the system [-],
λ mean arrival rate (orders per shop calendar days) [1/SCD].

The mean performance P_m of a workstation is derived according to Eq. 2.20 via the arrival rate λ, the mean work content c_m and the number n of parallel workstations within the observed system:

$$P_m = \frac{\lambda \cdot c_m}{n} \tag{2.20}$$

where P_m mean performance [hrs/SCD],
λ mean rate of arrival (orders per shop calendar days) [1/SCD],
c_m mean work content [hrs],
n number of parallel workstations within a system [-].

By substituting Eq. 2.20 in Eq. 2.19 the mean throughput time is calculated as

$$TTP_m = \frac{N_m \cdot c_m}{P_m \cdot n} \tag{2.21}$$

where TTP_m mean throughput time [SCD],
N_m mean number of orders in the observed system [-],
c_m mean work content [hrs],
P_m mean performance of a workstation [hrs/SCD],
n number of parallel workstations within the observed system [-].

The range of application for the preceding formula is very broad. The nomenclature is therefore often not defined in the literature; in special cases the variables are to be assigned according to the problem. Only then is it possible to determine what dimensions should be used. Thus, not only shop calendar days can be used as the time unit to express the mean arrival rate, but also for example hours or minutes. However, here we will apply the abovementioned dimensions in order to facilitate the comparison between Little's Law and the Funnel Formula. Furthermore in Eq. 2.21 the following substitutions will be made:

$TTP_m \rightarrow TTP_{vir}$: mean virtual throughput time [SCD],
$N_m \rightarrow WIPO_m$: mean WIP level in number of orders [-],
$c_m \rightarrow WC_m$: mean work content [SCD],
$P_m \rightarrow ROUT_m$: mean output rate of the workstation [hrs/SCD].

The variable TTP_m will not be used here, because according to Eq. 2.10, TTP_m is the arithmetical mean of the individual throughput times. Equation 2.21 is not based on individual events, but rather on several mean values so that the value that is calculated through it instead represents an apparent or *virtual* mean value. Therefore, the variable *mean virtual throughput time* TTP_{vir} will be introduced.

It needs to be pointed out with particular emphasis here that in previous studies ([Bech-84] and other cited studies based on it), the term 'virtual throughput time' is used for the mean range calculated using the Funnel Formula. This nomenclature is not followed here as the range R_m of the expected value corresponds to the *weighted mean throughput time* TTP_{wm}. Consequently, it has to be called *virtual weighted throughput time*.

Based on the above substitutions Eq. 2.21 is then written

$$TTP_{vir} = \frac{WIPO_m \cdot WC_m}{ROUT_m} \tag{2.22}$$

where TTP_{vir} mean virtual throughput time [SCD],
$WIPO_m$ mean WIP level in number of orders,
WC_m mean work content [hrs],
$ROUT_m$ mean output rate [hrs/SCD].

The supposition that $WIPO_m \cdot WC_m = WIP_m$ (in planned hours) would seem obvious: In this case Eq. 2.22, which is derived from Little's Law, would be identical to the Funnel Formula (2.16) and the values for the mean range and the mean virtual throughput time would be the same, at least within the conditions of the model application (see below). However, as will be shown in Sect. 4.6 this is not generally the case.

Nevertheless, it is possible to describe the mean output rate (using the dimension [hrs/SCD]) through the product of the mean work content WC_m and the mean number of orders completed per shop calendar day $ROUTO_m$:

$$ROUT_m = ROUTO_m \cdot WC_m \tag{2.23}$$

where $ROUT_m$ mean output rate [hrs/SCD],
$ROUTO_m$ mean output rate in number of orders [1/SCD],
WC_m mean work content [hrs].

Equation 2.22 can thus be simplified as

$$TTP_{vir} = \frac{WIPO_m}{ROUTO_m} \tag{2.24}$$

where TTP_{vir} mean virtual throughput time [SCD],
$WIPO_m$ mean WIP level in number of orders [-],
$ROUTO_m$ mean output rate in number of orders [1/SCD].

This equation is directly derived from Eq. 2.19, with the condition that the arrival rate λ corresponds to the mean output rate in number of orders ($ROUTO_m$).

In [Conw-67*], the following are named as prerequisites for applying Little's Law:

- The completion notes for the individual operations are collected with enough accuracy.
- There is sufficient enough planning data to determine the planned operation times.
- There are long periods of investigation.
- There is no overlapped manufacturing allowed.

The analogy of Little's Law to the Funnel Formula is obvious and can be visualized in a Throughput Diagram (Fig. 2.10). In order to calculate the value of the throughput time, the WIP on a workstation is put in relation to the output rate in both the Funnel Formula and the equation derived from Little's Law (2.24). The difference first lies in the dimensions used to indicate the base values. Whereas the Funnel Model measures the WIP in planned hours, Little s Law measures it by the number of orders. This is also true for the output rate, which is accordingly given by hours per shop calendar day and number of orders per shop calendar day respectively. The resulting values for the mean range and mean virtual throughput time are in both cases given in the same dimension, however, different correlations are expressed through Eqs. 2.16 and 2.24: The Funnel Formula states the average length of time required to completely process the WIP on a workstation when the mean output rate is a constant and no new orders are begun in the meantime. Little's Law in contrast, shows the average length of time that an order newly arriving on the workstation has to stay until its completion.

A further fundamental difference between the Funnel Formula and Little's Law is the degree to which the respective parameters can be influenced by the completion behavior (in particular by the implemented sequencing rule) on the system

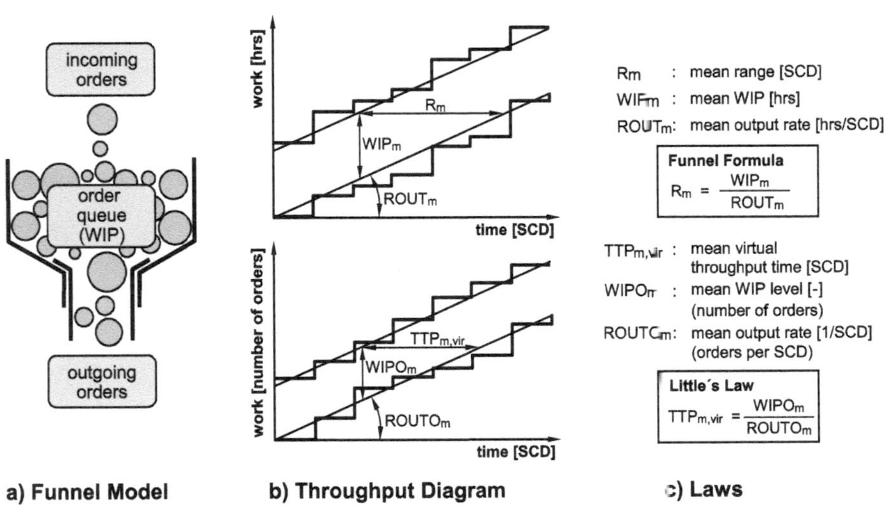

Fig. 2.10 Comparison of the Funnel Formula and Little's Law

and the heterogeneity of the work content. Thus, specific processing sequences have an impact on the mean number of orders on a workstation ($WIPO_m$) and with that also the mean virtual throughput time. By applying the SOT rule (SOT: shortest operation time) orders with less work content are prioritized for processing. As a result, fewer orders on average are found on the system, therefore the mean throughput time is also low. Conversely, in applying the LOT sequence (LOT: longest operation time) orders with large work content are prioritized and many small orders have to wait. Consequently the $WIPO_m$ and the virtual throughput operation time TTP_{vir} are correspondingly higher/longer. The greater the work content's variance is, the more prominent the effects are.

In contrast, these factors have no influence in the Funnel Formula. With the given input and a sufficiently long reference period, the WIP (in planned hours) and consequently the range, can only be influenced by the system output: Neither the sequence of the processing nor the distribution of the work content have an impact.

2.4 Logistic Operating Curves for Production Processes

Using the performance figures and diagrams introduced above, fundamental information can be gained about a production system and applied in order to analyze a variety of problems. Tracing the causes of scheduling deviations and deriving appropriate control measures is thus extensively supported. The interactions between the logistic parameters however is only partially described when at all. The following questions thus remain unanswered:

- What is the lowest throughput time that can be attained given the present processing and order structures?
- How high does the WIP level have to at least be in order to avoid a loss of output?
- What potential for improvement can be developed with which measures?

Here, *Logistic Operating Curves* (LOC) can be helpful. Logistic Operating Curves depict the correlations between the logistic objectives: output rate, throughput time, inter-operation time and range, as a function of the WIP. Due to the central importance of the LOC we will explain them in detail in the following.

The Funnel Model, the Throughput Diagram and the performance figures that result from them describe a specific stationary operating state. In Fig. 2.11a, three fundamentally different operating states are shown in simplified Throughput Diagrams. These different operating states can now be aggregated and depicted in the form of Logistic Operating Curves (Fig. 2.11b). To accomplish this, the output rate and the three throughput time parameters are plotted as a function of the related WIP. The Output Rate Operating Curve (OROC) indicates that the output rate of a workstation only changes negligibly above a specific WIP level. From this point on, there is continuously enough work available to ensure that there are

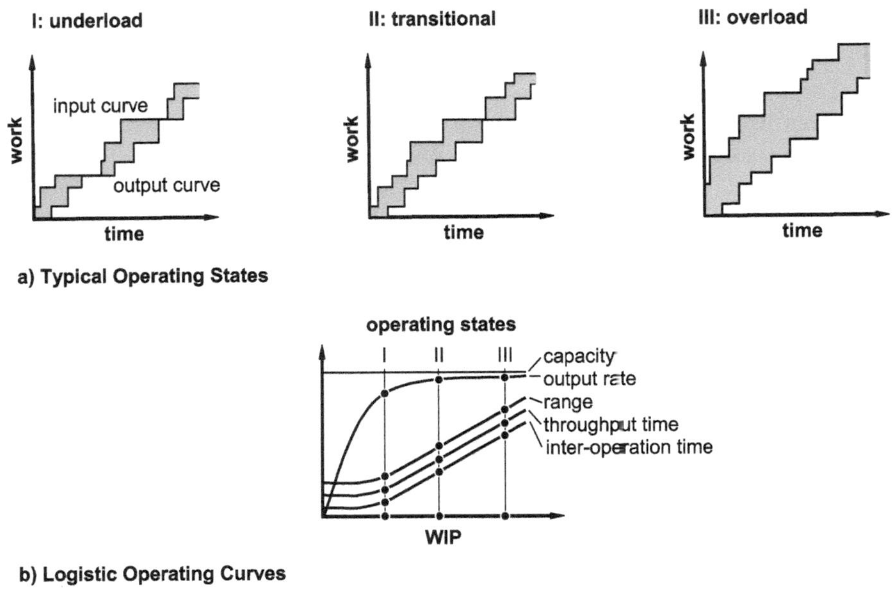

Fig. 2.11 Representation of Different Operating States in Logistic Operating Curves

no WIP dependent breaks in production. Below this WIP level however, there are increasing losses due to the intermittent lack of work in the queue. The (unweighted) throughput time though, generally increases above this critical level proportional to the WIP. When the WIP levels are reduced, the throughput time decreases. Nevertheless, the throughput time cannot fall below a minimum level derived from the order's mean operation time and where necessary, transport time. The Inter-Operation Time Operating Curve (ITOC) generally exhibits a similar behavior, approximating the transport time when the WIP levels are low. Finally, the Range Operating Curve (ROC) is derived according to the Funnel Formula, from the ratio of the WIP to output rate.

It should be pointed out that the momentary status of a system always corresponds to only one operating point on the Logistic Operating Curve. The LOC represent how the observed system with otherwise unchanged boundary conditions behaves when a different WIP level is set. They thus characterize the logistic behavior of a production process when the WIP level changes. Moreover, it is also possible to create Logistic Operating Curves for different production or order structures, to compare them with each other and to thus assess the impact of interventions on the production process from a logistic perspective.

It can easily be seen that the general form of the Logistic Operating Curves is applicable to any production system: Reduced WIP levels lead to reduced throughput times, but can also lead to material flow disruptions and thus to losses of utilization. Nonetheless, the specific shape of the LOC for the corresponding observed workstations are dependent upon the various boundary conditions such as the capacity, the orders that are to be processed (in particular their mean and variability) and how the system is integrated into the material flow.

For obvious reasons it is not possible to determine the LOC through a series of tests in an actual plant. On the one hand, this is because of the costs that accumulate when the WIP levels are extremely high or extremely low. On the other hand, due to changes in the structural conditions (e. g., a change in the product or order mixture) no comparable ratios result and thus, the individual operational states can also not be directly compared.

As mentioned in the introductory remarks, it is clear that in such cases models that can emulate the original system of interest should be implemented. In the next chapter we will describe and analyze three models that describe the correlations between the relevant logistic objectives and thus can help solve the dilemma of operations planning. Two of these, queuing theory and simulation, are traditional models; we will illustrate the possibilities and limitations of these model applications in the following. Subsequently in Chap. 4, we will introduce a new method for calculating Logistic Operating Curves using an approximation equation.

Chapter 3
Traditional Models of Production Logistics

Economic systems are generally modeled mathematically ([Prof-77], [Wöhe-90], [VDI-93]). In this chapter, we will examine some of the basic methods common to these models:

Deductive modeling methods (deductivus [lat.]: to derive the particulars from the general or universal) are distinguished from other approaches in that they are based on a qualitative conception of the influencing factor's effects. A number of relationships derived from the problem and goals are theoretically isolated. Abstracting the system specific characteristics reduces them to such to an extent that an image of the original system is developed that is limited only to the essentials. By applying basic laws the mathematical description of the interdependencies between the input and output values are derived.

The most important and at the same time most difficult step in deductive modeling is generally not the mathematical formulation, but rather the more or less intuitive identification of the relevant correlations. Due to over extensive abstraction or lack of knowledge about the process and faulty premises there is a risk that the constructed model's ability to provide meaningful information could be inadequate for practical concerns. The main advantages of deductive models are twofold: First, the model can generally be transferred as long as the prerequisites for its application are met and second, the model can in many cases be adjusted to different boundary conditions.

A typical example of a deductive model is the traditional approach for determining lot sizes according to Andler (see e. g., [Müll-62] and [Nyhu-91]). Here, the minimums are sought for two lot size dependent, but opposite types of costs, namely those for order changes and for stored stock. This model can usually be applied in every enterprise – as long as the fundamental conditions of the model can be fulfilled with a sufficient accuracy. The model is adjusted solely through the parameters (in this case the costs or the demand data); intervening in the model itself is not necessary. It is also clear that a deeper understanding of the process can be developed through the model e. g., how the batch-size influences the observed cost component can be clearly illustrated without having to use any special cases.

Empirical or experimental modeling (generally called *simulation*) develops models based on qualitative knowledge of the process. The model's structures and parameters are directly based on the relevant characteristics of the real system. Subsequently, experiments in which the input values or model structures are specifically manipulated can then be conducted on this simulated system. After the trial run, the results can be evaluated and interpreted However, the information derived directly from the experimental model is limited to describing the relationship between the input/output values and only individual results are obtained, not any universal knowledge.

Experimental models are used when no appropriate deductive model exists for the presented problem. If for example, Andler's rule for determining batch-size is not appropriate for a particular application case because the capacity load or the capital commitment in the production area requires special consideration, then one can try to accommodate these demands using simulations. In order to do so, the area of the company that is being observed first needs to be depicted as a so-called resource model. Experiments in which the system load is varied using an adjusted planning and control model (in this case for example with corresponding strategies for batch-size) can then be conducted. By analyzing the results, the best possible solution can be chosen from those tested.

This brief description clearly illustrates both the possibilities as well as the limitations of experimental models. On the one hand, it is possible to investigate problems, which due to there complexity cannot be depicted and described, with deductive models in that specific conditions can often be concretely considered. On the other hand, the model has to be developed as well as validated for each and every new application. Furthermore, the results can only be interpreted as the relationship between the input and output; the results themselves do not have any essential significance and can therefore, not be easily transferred to other situations.

In the following, we will present illustrative examples of deductive and experimental models that can be drawn upon to describe the dilemma of operations planning. In our discussion we will concentrate on the basic possibilities and limitations of applying the model and in particular on their practical relevance, the accuracy of the information they provide and the effort required to apply them. We will not attempt to provide a comprehensive overview of the two types of models, but instead will limit our remarks in each case to a specific approach. However, the general comments are also applicable to other examples of the respective model categories.

3.1 Queuing Models

Queuing models are an example of a traditional deductive model for describing the relationships between logistic parameters. Queuing or waiting line models make it possible to consider the random influences that surface in the field while planning and controlling real processes. They are mainly used to dimension the size of bottlenecks which can occur whenever an object of any sort arrives regulated or randomly at one or more workstations and is served with irregular or fixed processing times. By using mathematical approaches and given information known about the input (in particular the average arrival and completion/service rate of the objects on the operating system) the events that take place during the actual processing should become theoretically comprehensible and therefore predictable. Queuing models thus essentially allow statements about the probable relationships between the waiting time and the length of the queue as well as the service station's utilization.

Queuing theory, which was initially developed by Erlang (see [Gned-71], [Köni-76], [Gros-98*]) was initially applied to telephony. However, it quickly became clear that the knowledge gained from it could also be applied to other areas. The theory has thus been continuously developed and has spread to extremely diverse fields where the dimensioning of the system's inventory or waiting time is significant and can be influenced by the input/output ratio. Determining the number and location of cashiers in a supermarket or controlling jobs in computer systems are both examples of applied queuing theory.

The spectrum of approaches to finding solutions which has evolved due to all of the special conditions that have been considered, mirrors the breadth of application areas. In one study conducted by Lorenz, he described twenty-five different queuing models that can be divided roughly into three classes [Lore-84]:

- *Analytic Solutions* deliver exact values when calculating the average waiting time.
- *Approximation Solutions* lead to approximate values.
- *Estimation Solutions* estimate the mean using an upper and lower limit.

In order to classify the models, an identification system consisting of five values has been implemented internationally (Fig. 3.1):

A / B / X / Y / Z

The first two symbols indicate the distribution of the arrival time (A) and the service or completion pattern (B) described by the probability distribution of the inter-instant time e. g., general, type k-Erlang, exponential, deterministic. Depending on the type of distribution, a further specification about the expectation (mean value) of the arrival rate λ or service rate μ is required. Where applicable, the coefficient of variation for the arrival rate (λ_v) or service rate (μ_v) may also be necessary. The symbol X stands for the number of parallel and identical service stations within a system and Y provides the restriction on the system capacity i. e., the number of waiting slots. Finally, the symbol Z is used to describe the queuing discipline, i. e., the sequencing rules.

Generally speaking, in order to apply queuing theory within the production field, it is required that the FIFO priority rule (First In – First Out) or in the case of *multiple station systems* the FCFS principle (First Come – First Served) apply i. e., no individual elements (orders) will be prioritized. Furthermore, it is usually as-

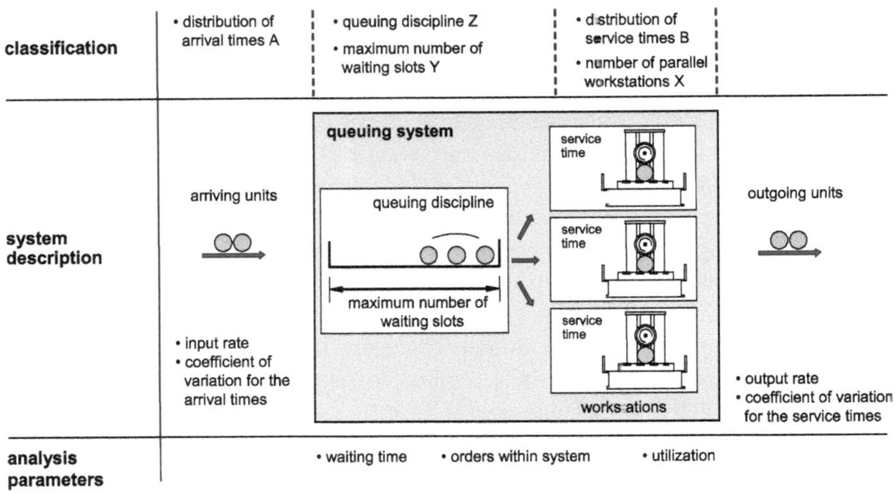

Fig. 3.1 Elements of a Queuing System for Visualizing Production Processes

sumed that there are an unlimited number of waiting slots. When these conditions are met, the classification system is then restricted to the first three parameters (A / B / X).

More in-depth information about these values, can be found in among others [Gros-98*] as well as in the bibliography provided therein.

Once a queuing model that addresses the problem and existing conditions is chosen, the waiting time, its mean and distribution, the number of orders in a systems as well as the inventory or queuing time dependent utilization can for example be calculated. This approach is thus ideal for describing the dilemma of operations planning and where necessary also for evaluating alternative approaches to dealing with it depending on the situation ([Sain-75], [Hopp-96]).

3.1.1 M/G/1 Model

In the following discussion we will use the M/G/1 model developed by Pollaczek and Khinchine as a comparative model. It can be considered a standard queuing theory model and belongs to the class of analytical approaches.

This model allows a universal output process, that is the service times are arbitrarily distributed but stochastically dependent on each other. When we apply this within production, it means that randomly distributed processing times are permitted. From this perspective, it is therefore a widely applicable model. Previous publications on queuing theory do not explicitly mention that the capacity of a system is assumed to be a constant or that it is thus independent from the corresponding system's state, however, this information is relevant to the application discussed here.

For the input, the model assumes a so-called Markov Process with a Poisson frequency distribution of the arrivals, which exponentially distributes the inter-instant times. The Markov Property states that the input process has to be stationary, ordinary and have no after-effect. By stationary, it is implied that the probability of the input events is exclusively dependent on the length of the observed time intervals and not on the intervals location on the time axis. Seasonal affects on the input or cyclical order boosts are consequently not permitted. Ordinary means here, that a simultaneous occurrence of two or more events (in this case: inputs) is practically impossible. Finally, to be independent of after-effects requires that the processing behavior with a steady state and at a specific point in time is not dependent on the state which it had assumed before this point. This property is known as "lack of memory" [Arno-95] or "memorylessness" [Gros-98*]. In particular it means that the number of events in non-overlapping time intervals are independent random values [Gned-71]. Furthermore, it is assumed that the input and output processes are stochastically independent [Stro-77]. Lastly, we assume that the system consists of exactly one workstation with its own unlimited WIP buffer, which is processed according to FIFO.

For an M/G/1 model the *mean queuing time* TQ_m is calculated

$$TQ_m = \frac{\rho}{\mu \cdot (1-\rho)} \cdot \left(\frac{1+\mu_v^2}{2} \right) \tag{3.1}$$

where TQ_m mean queuing time
ρ utilization rate
μ service rate
$1/\mu$ mean service time
μ_v coefficient of variation for the service time

We will not discuss how this formula was derived as it has been extensively addressed elsewhere (see e. g., [Gros-98*], [Hell-78]).

In order to utilize this queuing model to describe the logistical behavior of a workstation the corresponding terminology, chosen in Chap. 2, is substituted in Eq. 3.1 as follows:

$TQ_m \rightarrow TIO_m$: mean inter-operation time [SCD],
$\rho \rightarrow U_m/100$: mean utilization of the work system [-],
$1/\mu \rightarrow TOP_m$: mean operation time [SCD],
$\mu_v \rightarrow TOP_v$: coefficient of variation for the operation time [-].

Generally, the dimension of the time parameters, inter-operation time and operation time, can be freely selected and can thus be adapted to each case. Nevertheless, it is important that the dimension of the two parameters is always the same or when necessary that one of the two be transformed into the other using a conversion factor. In this way, the mean service time can also be understood as the mean work content. Then however, the inter-operation time is also calculated using hours. Nevertheless, since measuring and stating the throughput time and inter-operation time in *shop calendar days* SCD proved to be useful for the job shop production, it will also be used in the following.

Applying the substitutions mentioned above, Eq. 3.1 is written as:

$$TIO_m = \frac{TOP_m \cdot \frac{U_m}{100}}{\left(1 - \frac{U_m}{100}\right)} \cdot \left(\frac{1 + TOP_v^2}{2}\right) \tag{3.2}$$

where TIO_m mean inter-operation time [SCD],
TOP_m mean operation time [SCD],
TOP_v coefficient of variation for the operation time [-],
U_m mean utilization [%].

In order to compare different approaches to modeling, we will draw extensively upon *normalized representation* in the following discussion. In this case, the *mean flow rate*, which describes the relation of the mean throughput time to the mean operation time (see Eq. 2.11), seems well-suited as a normalization parameter. According to Eq. 2.9 the mean throughput time is derived from the addition of the mean inter-operation time and the mean operation time, so that

$$RF_m = 1 + \frac{TIO_m}{TOP_m} \tag{3.3}$$

where RF_m mean flow rate [-],
TIO_m mean inter-operation time [SCD],
TOP_m mean operation time [SCD].

By substituting Eq. 3.3 in Eq. 3.2 the flow rate can also be described as a function of the utilization,

$$RF_m = 1 + \frac{\frac{U_m}{100}}{\left(1 - \frac{U_m}{100}\right)} \cdot \left(\frac{1 + TOP_v^2}{2}\right) \tag{3.4}$$

where RF_m mean flow rate [-],
U_m mean utilization [%],
TOP_v coefficient of variation for the operation time [-].

This notation has the advantage that it demonstrates, independent of the mean operation time, the extent to which the system utilization and the inhomogeneity of the work content (expressed by TOP_v) impact the attainable level of the throughput time (expressed by RF). Figure 3.2 illustrates this function for three different coefficients of variation. For example, the diagram shows that given a strongly fluctuating work content (here, $TOP_v = 1$) the flow rate already has to be equal to 5 in order to utilize the system 80%. The throughput time is therefore five times greater than the operation time. Furthermore, in order to obtain 90% utilization, the flow rate has to once again double.

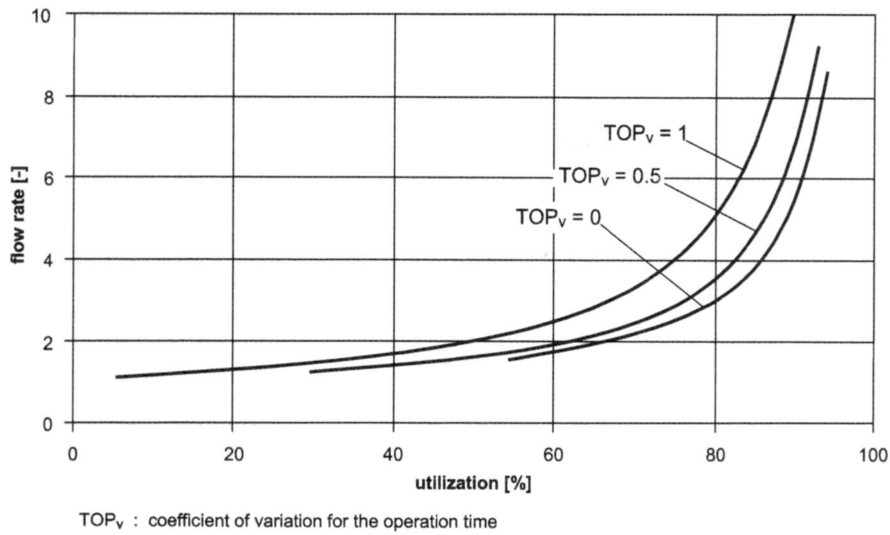

Fig. 3.2 Correlation of the Flow Rate and Workstation Utilization Based on the M/G/1 Queuing Model

3.1.2 Using Queuing Theory to Determine Logistic Operating Curves

By applying the previously described equations it is possible to describe the dependence between two logistic objectives. In order to establish further Logistic Operating Curves, it is necessary to mathematically describe not only the utilization (as a capacity value) and flow rate (as a throughput time parameter) as logistic objectives, but also the WIP. In previous studies, Little's Law (see Sect. 2.1.3) has commonly been drawn upon to supplement queuing theory for this. By applying this law (Eq. 2.22) and Eq. 2.5 (when using mean values), 2.11, 2.13, 2.25 and 3.4 the sought relationships between the mean work in process in number of orders $WIPO_m$, the flow rate RF_m and the utilization of the system U_m can be derived:

$$WIPO_m = \frac{U_m}{100} \cdot \left(1 + \frac{\frac{U_m}{100}}{1 - \frac{U_m}{100}} \cdot \left(\frac{1 + TOP_v^2}{2} \right) \right) \tag{3.5}$$

$$RF_m = \frac{100 \cdot WIPO_m}{U_m} \tag{3.6}$$

where $WIPO_m$ mean WIP (in number of orders) [-],
RF_m mean flow rate [-],
U_m mean utilization [%],
TOP_v coefficient of variation for the operation time [-].

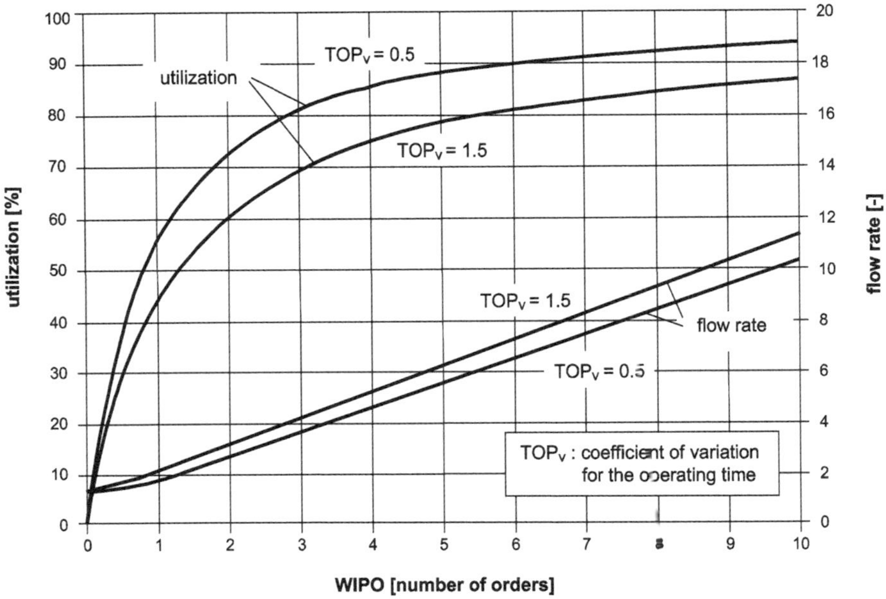

Fig. 3.3 Calculated Logistic Operating Curves Based on the M/G/1 Model and Little's Law

In Fig. 3.3, the functions resulting from these equations for the coefficient of variations 0.5 and 1.5 are depicted in the form of LOC. These key values were chosen because they represent typical upper and lower limits for the distribution parameter in the field (see also Fig. 2.2). In the diagram we can see for example, that with the underlying M/G/1 model there is a 15% loss of utilization when there is an average of four orders being processed ($WIPO_m = 4$). This applies even when the work content is comparably constant ($TOP_v = 0.5$). When the work content is more widely distributed ($TOP_v = 1.5$) the expected loss of utilization with the same WIP level is approximately 27%. Moreover, in the second case when $WIPO_m = 10$, the utilization increases to only a little less than 86%. The flow rate behaves as expected approximately proportional to the WIP with the differences arising from the utilization disparity.

Equations 3.5 and 3.6 as well as the graph derived from them are obviously only valid when the processes on a workstation meet the conditions for applying both the M/G/1Model (see above) and Little's Law (see Sect. 2.3).

3.1.3 A Critical Review of the Queuing Theory Approach

In the preceding section we have shown that Logistic Operating Curves can be easily developed using queuing models such as the M/G/1 Model and Little's Law. Furthermore, the above graph demonstrates that queuing models, as a traditional example of deductive modeling methods, distinguish themselves from others, because the knowledge derived from them is characterized by its ability to be generalized: They can be transferred to any application within the validity of the model's limitations.

Thus for example, Fromm [From-92], who with the help of queuing theory examined the influence of variability on the reduction of the throughput time in processing chains, came to the following conclusion: "Variability means uncertainty and unpredictability. It conflicts with the most important goal of process management, controlled processing. [...] variability is not only an inherently objectionable effect for the smooth running of a business process. It is also largely responsible for unnecessary waiting times".

Fromm sees three principle means of reducing waiting times:

1) Reducing the input rate. However, this is firstly, usually an exogenous parameter and can rarely be directly influenced. Secondly, it is synonymous with decreasing the utilization and thus with reducing the output.
2) Increasing the service rate through changing the system (e. g., expanding capacity with additional service stations or quicker service). Nevertheless, in order to accomplish this a capital expenditure is required, which is not manifested in a higher output, but rather also in a decreased utilization.
3) Harmonizing operation times. This alternative can be used to clearly reduce both the WIP and throughput times without requiring expenditures or having to accept losses of utilization.

Strictly speaking, since these three conclusions were found by analyzing the M/G/1 model, they are only valid for exactly these types of applications. It is easy to see though that all three of the described measures apply at least qualitatively even if the requirements for this model are not completely fulfilled.

Despite the significant contribution that the abovementioned queuing model can make to the understanding of processes, queuing theory is for the most part very critically discussed. Knolmayer [Knol-87] ascertained that there was a large gap between the models and terminology in production theory and operations research in general – queuing models in particular – and the thinking in the operational practice. Lorenz [Lore-84] came to similar conclusions. He determined that theoretical aspects clearly dominated the literature on queuing theory. The necessity alone of having to prove the specific distribution prerequisite for the input or output processes, poses a large obstacle in applying it. Indeed, newer models strive to avoid strict and in practice, difficult to verify conditions in order to become more relevant to field situations. However, the computational formulas for these are not only more complicated than the M/G/1 model discussed here, they also predominantly deal with solutions based on estimations. It is thus impossible to precisely calculate the performance figures. Furthermore, expert knowledge is required in order to be able to appropriately consider the specific characteristics of the system when choosing the method.

To some extent it is generally questioned how realistic queuing models are [Vieh-85]. Given the models' qualitative outcomes these doubts are, at least in the case of the M/G/1 model, understandable. For example, the correlations illustrated in Fig. 3.3 indicate that despite a high WIP level and the resulting high flow rate and throughput times, a full utilization of the system can barely be achieved – even when the structure of the orders' work content is very homogenous. However, in this extreme form these outcomes do not coincide with operational practice.

Nevertheless, concluding that queuing models themselves are faulty has to be challenged. Such an assumption can be extensively ruled out particularly with the basic model for queuing theory, because it deals with comparatively simple descriptions of the relations between the input and output. Even the technical correctness is ensured, not the least through the validation process conducted when developing the model. Detailed information regarding this can be found in the previously recommended reading on queuing theory (e. g., [Gros-98*]).

It is much more likely that the model's premises are violated than that the model itself is defective. In queuing theory models, the calculated system behavior is in principle determined by the shape of the distribution and the distribution parameters for the arrival/completion processes. Knowing the correct type of distribution is thus essential when using this model. Nonetheless, it requires a great deal of effort to determine it even with existing workstations. Furthermore, the procedures which should be employed to do so are greatly debated within publications on this topic (see [Lore-84] and the sources cited there). Regardless of this, the increasing dynamic between the production processes is also considered problematic because the distribution form and parameters can continuously change over time. How the necessary information can in the end be gained in not yet existing, or basically new designed systems, has to be individually decided. In any case, the results of the model's application will be influenced by this decision.

Finally, the condition that the input and output processes be stochastically independent has to be questioned, particularly with regards to production processes. In his study based on field data, Lorenz [Lore-84] was unable to positively prove this independence in any of the 90 analyzed systems. In other words, the dependency is almost always a given and is also to be expected, after all one of the key tasks of production planning and control is to coordinate the system's input and output. Thus in the operational practice, companies react to strong mid to long-term load fluctuations by adjusting the capacities (essentially by temporarily increasing or decreasing operators) or by balancing the load. Existing queuing models though, do not allow for this degree of freedom.

In summary it can be concluded that various assumptions and requirements have to be met when applying queuing theory, whereby to some degree, significant abstraction from the real conditions has to be undertaken. Although queuing models make it at least possible to successfully provide information about tendencies, this could be the reason why queuing theory has not managed to establish itself in the operational practice for designing and dimensioning production systems.

3.2 Simulation

Today, simulation is a widespread technique for researching, designing and optimizing complex systems. A simulation is an "emulation of a dynamic process in a model, in order to gain knowledge which is transferable to reality" [VDI-93]. Simulation provides the opportunity to imitate a real system with the help of a computer program and to analyze and describe its behavior by changing simulation conditions. It has often been argued that it is only possible to accurately enough assess the impact of changing the system's capacities or structures, using alternative resources, or of disturbances/changes to the processing rules by applying simulations (see for example [Kuhn-92], [Meie-93], [Ever-94], [Kluß-96]).

According to [Günz-93] and [Noch-93] simulations are well-suited particularly when:

- exact analytical computational methods cannot be implemented,
- experiments cannot be conducted on a real system in order to gain the desired knowledge
- the time-dynamic behavior of a system is to be analyzed.

A comprehensive overview of simulation components as well as the general procedures for the implementation and evaluation of simulation experiments can be found in the VDI Guideline 3633. As an aid to product and production planning examples of simulation applications are compiled in among others [Ever-87] and [Wien-93b].

3.2.1 PROSIM III Simulation System

In Fig. 3.4, the organization of the simulation system PROSIM III (Production Simulation, Version III, [Scho-96]), which was developed at the Institute of Production Systems and Logistics at the Leibniz University of Hannover, is illustrated as an example. Since its creation, PROSIM III has been successfully implemented in various fields of application e. g., designing and testing PPC functions, designing and logistically assessing production structures, training employees and developing/testing production process models (see a. o., [Nyhu-91], [Spri-92], [Gläß-95], [Ludw-95], [Pete-95], [Penz-96] and [Scho-96]). This model consists of emulating the production as well as the related material planning and order control. The production sub-model contains a concept of a universally applicable process which maps a specific operation by describing its resources, process control and order data. The amount and/or due date of the orders which need to be processed can be manipulated via the underlying demand as well as through the various production planning functions. From the resource standpoint, this simulation system can incorporate the details of everything up to and including the individual workstation. From the perspective of the orders, the smallest element is the individual operation.

The *model classes* and their essential degree of freedom are shown in the upper part of Fig. 3.4. The *load model* describes the structure of both the product and the demand (the amount and frequency distributions of the individual requirements). The *planning model* includes the planning methods implemented during the order

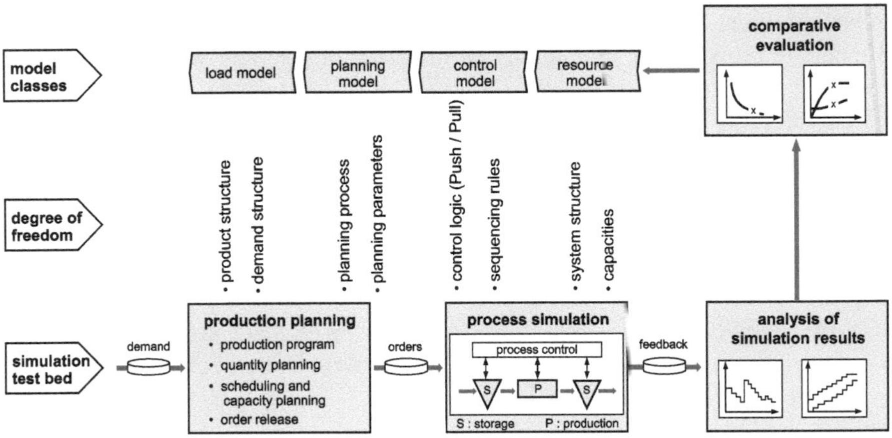

Fig. 3.4 Modeling Levels and Degrees of Freedom for the Simulation System PROSIM III (according to Scholtissek)

processing for: material planning (requirement explosion, lot size determination, order generation), schedule/capacity planning, and order release. On the order execution level, the *control model* contains various control strategies such as the priority rules that are to be implemented. Finally the *resource model* serves to describe the available resources with regards to their type (facilities, personnel) as well as their time and technical related capacities.

With the help of this type of structured simulation system, a real production system can be emulated and the behavior of the system can be analyzed by specifically changing the existing system conditions.

3.2.2 Simulation as an Aid in Determining Logistic Operating Curves

In order to be able to create Logistic Operating Curves using a simulation system, it is necessary to conduct numerous individual simulation runs during which initially only one parameter, the production's WIP, is selectively changed. All other parameters must remain unchanged within a series of simulations. Accordingly, a simulation system also has to have the potential to actively manipulate the WIP. This is for example possible when Load Oriented Order Release ([Wien-95a*], [Wien-87], [Wien-97]) is implemented as a control technique. The WIP can also be varied by utilizing a Kanban system, for example, through the number of circulating Kanban cards. A third alternative is for the orders to enter the system according to a schedule, whereby the input schedule of the individual orders is newly calculated for every simulation run using backwards scheduling via the planned throughput time (and thus indirectly through the planned WIP).

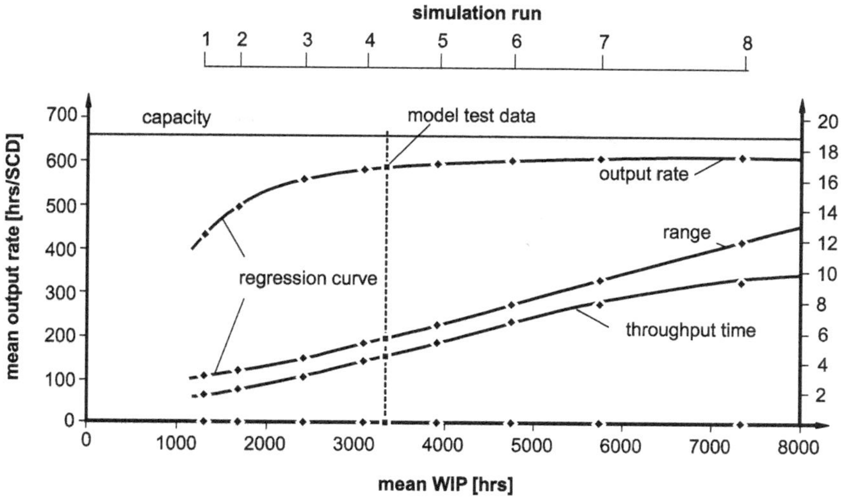

Fig. 3.5 Simulated Logistic Operating Curves for a Job Shop Production

A typical application of the Logistic Operating Curves obtained using the PROSIM III simulator and the interpretation of the results is shown in Fig. 3.5. The LOC for one of the 55 workstations in a mechanical shop are depicted. In a simulation period of 28 weeks, approximately 1,050 orders with approx. 7,500 operations were processed. Eight simulation trials were evaluated, during which the WIP was adjusted with the help of the Load Oriented Order Release algorithm (LOOR). In order to ensure a greater accuracy of the simulation's results, the simulation model and the data first had to be validated in a model test. The existing real processes were emulated, in that the simulation model was loaded with the same input data as the real system. With this type of model testing, the results that are relevant to the problem have to agree with the conditions in the field; the model should then be corrected as need be.

In the graph, it can be seen that the WIP in the model test amounted to approx. 3,300 hours. This value, as well as the key performance figures for the output rate, range and throughput time also corresponded to the field data. Further information regarding the examined company, procedures for model testing and the verification of the reproductions accuracy can be found in [Nyhu-91].

The results of the model test lay between the results of simulation runs 4 and 5. It can be determined that by reducing the WIP in runs 1 to 4, lower throughput times could be clearly realized during the model test – although only at the cost of a to some extent much lower utilization. In simulation runs 5 through 8 however, increasing the WIP only led to longer throughput times and did not lead to a significantly greater output rate in the simulated production.

The graph clarifies visually and quantitatively the dilemma of operations planning for the simulated job shop. Whereas the WIP has to be set at a minimal of 6,000 hours in order to maximize the utilization, the WIP cannot be greater than approx. 1,500 hours in order to minimize the throughput time. The model test shows that the analyzed shop has succeeded in finding a compromise between these two extremes. The results of this model test and therefore also the real production processes' are located within the range of the curve's inflection. Consequently, the company had to accept a marginal utilization loss in favor of shorter throughput times and lower WIP. Provided that the goal is to continue reducing the WIP and throughput times without sustained output rate losses the system's load, planning and control methods or resource model can now be gradually modified with the help of the simulation. After conducting a complete series of simulations (in which the WIP is changed) again and determining and comparing the Logistic Operating Curves it is then possible to quantify the effects of the interventions on the attainable throughput times and WIP. Management decisions can then be derived based upon the information recorded.

Applications of simulations such as this have previously been presented by various authors: Within the context of logistic oriented factory and organization planning, Kuhn ([Kuhn-92], [Kuhn-97]) used the Logistic Operating Curves as a general model for evaluating process chains and for explaining the impact of introduced or planned measures. Zäpfel et al. ([Zäpf-92a], [Zäpf-92b]) implemented the LOC for example to evaluate conditions for implementing production control techniques and possibilities for expanding them (e. g., Kanban, LOOR). In Rice and Gross [Rice-90*] the Logistic Operating Curves technique was implemented in the system design and optimization process for taking into consideration different order structures. Looks [Look-96] used the LOC to determine the potential for improving the throughput time and WIP, which could be developed through the use of manufacturing segments. Whereas, Larsen [Lars-92*] and Kreutzfeld [Kreu-94] investigated the influence of alternative operation sequences on a manufacturer's logistic efficiency, Meier [Meie-95] conducted a Logistic Positioning based on the simulated LOC. Finally, in work completed at the Institute of Production Systems and Logistics at the Leibniz University of Hannover (IFA) the LOC were applied: in order to calculate the number of pallets in assembly flow lines [Wink-88], to support decisions in production control [Wede-89], to evaluate methods for determining lot sizes [Nyhu-91], to logistically evaluate alternative production structures ([Brin-89], [Kupr-91]) and as a test bench for logistic concepts of production [Scho-96].

The diverse applicability of Logistic Operating Curves has lead to this technique (in addition to other types of presentation graphics) being recommended in the VDI Guideline 3633 (Part 1, Part 3 and Part 6) for the representation and interpretation of simulation results.

3.2.3 A Critical Review of Simulation

There are three particular situations in which simulation is preferred as an experimental model: First, when there is no appropriate deductive model for the presented problem, second, when the necessary experiments cannot be conducted on a real system and third, when the time-dynamic behavior is to be analyzed. One of the advantages of experimental models is that problems, which due to their complexity elude being portrayed or described using a deductive model, can also be investigated.

As a result of the increasing performance in the hardware and software field, the effort required to develop a model is in the meantime minimal in comparison with that required to derive a deductive model. Modern simulation systems are user-friendly designed. Creating even more complex models is facilitated through reusable model components, open data interfaces and modular, expandable program elements. However, this advantage is countered by a number of disadvantages which we will now discuss.

Although the procedures for developing a model are frequently quite simple, they, as well as the subsequent model validation, have to be completed for every new application.

In an ideal situation, the model test can be conducted on the basis of a comparison between the simulation results and real processing data. Even then though, it has to be considered that the model validity can generally only be verified in selective cases. It thus becomes obvious that even the applied process is the result of various – often not directly discernable – influences. Even smaller deviations in an arbitrary parameter can result in a significant change in the system behavior on the shop floor.

In addition, the model can only be validated using this method when the real system already exists. When planning a new system or strongly intervening in a system structure, the model can only be validated based on knowledge from previous experiences or through a comparison with the results of other models. The simulation results derived from it might be more comprehensive, but can in principle not be better (i. e., more accurate) than the comparison model or the estimations it is based on.

Furthermore, in many cases not all of the data that is required for the simulation application can be easily compiled, in part due to costs and/or time but also due to an inability to measure them. Thus, the system load with respect to the market dynamics and the quick product changes can only be described with difficulty, especially with regards to how it changes over time. Moreover, the available flexibility of the production's capacity can barely be depicted. It is generally managed in part with auxiliary indicators or by consciously manipulating specific variables. Employing such procedures however, includes risking that the entire results of the investigation (in the rule those which are able to be easily influenced by the simulation user) are decisively determined by these manipulations.

The following example demonstrates the problems of a simplified model. The example begins with the measuring of an order run on a system belonging to an electronic manufacturer. The data basis stems from approx. 600 completion notes, compiled within a time frame of 210 shop calendar days (SCD). The measured production process is represented in Fig. 3.6. In all, approximately 1000 hours of work content were reported back. The work in process amounted to an average of 22 hours with a mean range of 4.6 SCD. What is noticeable, is that the system

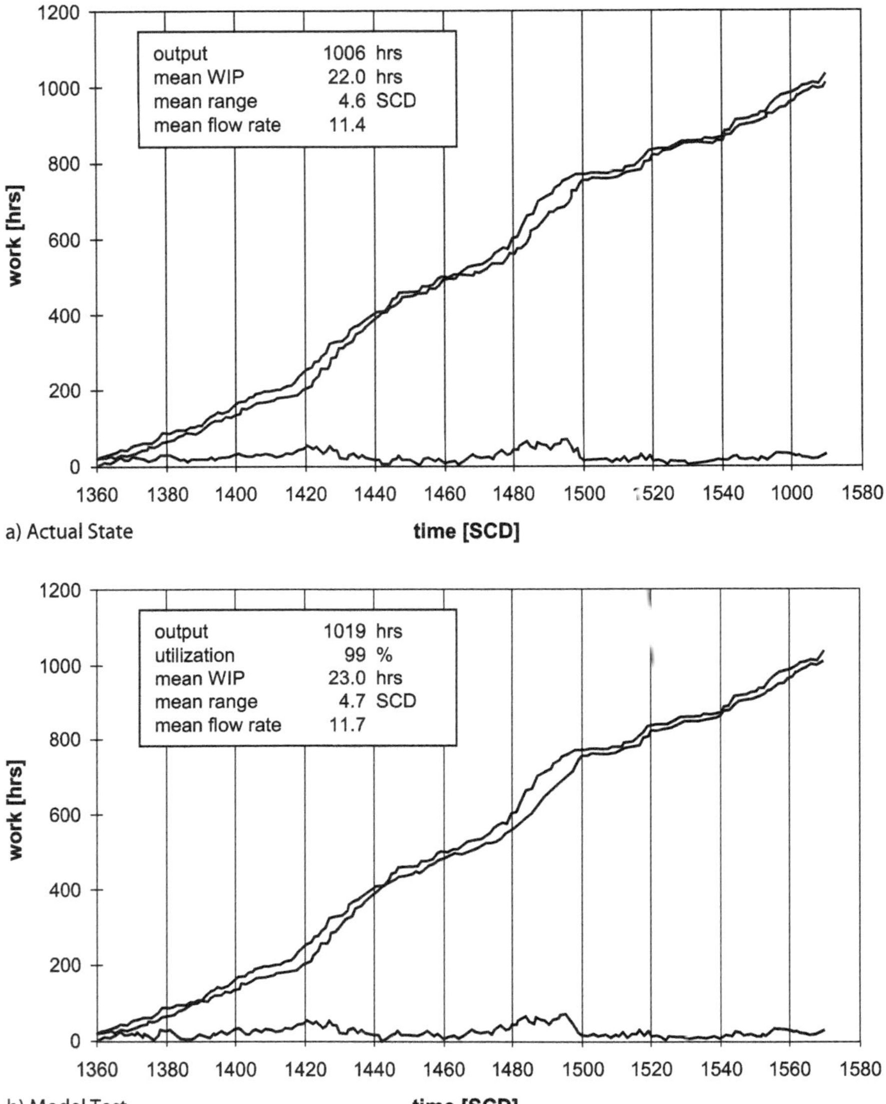

Fig. 3.6 Comparison of the Actual State and Simulation Model Test for Workstation 61570; Electronic Manufacturer

reacted to changes in the WIP, even when that reaction was to some extent delayed. It can be seen that the output rate grew shortly after the load was increased. It can therefore be inferred that there was room for additional capacity. Nonetheless, the time-related delays resulted in a temporary increase in the WIP.

The allocated capacities were then modeled carefully before the simulation was conducted using the compiled data and the PROSIM III simulation system. The minimal capacity that was to be made available in the shop per calendar week in order to process the waiting order volume was determined based on the output rate feedback from the production floor. At least one of the reasons for this limitation was that the causes for the decreased capacity were not analyzed or simulated. Following that, the manufacturing process was simulated by releasing the orders to the simulated workstation at the same point in time that they were also made available to the real workstation. The input pattern for both the simulated workstation and that on the plant floor were thus identical.

The result of the simulation is illustrated in Fig. 3.6b. It is evident in both the performance figures as well as the output rate and WIP curves that it was possible to achieve a strong correspondence between the field conditions and the results of the simulation. However, when interpreting this it should be taken into consideration that the opportunities for a detailed modeling of the resources and system load, which are present in this case are in particular, not available when planning for the future. Furthermore, since human decisions can also influence the results of the planning and control model, the planning and control model which the simulation is based on is generally speaking, just a more or less accurate approximation of reality. Due to these uncertainties, the individual parameters for the simulation cannot be accurately predicted and instead can only be estimated.

The consequence of such a procedure should be documented with the help of the previously illustrated example. It is assumed that the sum of the station's load (approx. 1000 hours of work content) during the simulation period (210 SCD) is known and not however, the chronological distribution of the input. Therefore, the capacity for both of the following analyses is allocated evenly over the entire simulation period.

When determining the dimensions for the system, it is initially assumed that the capacity is provided only in relation to the volume of orders to be processed. In the workstation observed here, the capacity is rounded-off to 5 hours/SCD. Based on this specification another simulation run was conducted in which the input pattern corresponded to the original conditions. The results are presented in Fig. 3.7. It can be seen that with the allotted constant capacity a bottleneck clearly arises. In the first quarter of the analysis period the arriving orders are processed without any delays. Following that, due to the (temporarily) increased load a queue develops which is not able to be processed during the simulation phase. The capacity, which at the beginning of the examination period was unused was lacking in the second interval. As a consequence of the increased WIP the mean range tripled in comparison to the actual state even though the utilization calculated over the entire period was only 91%.

In order to avoid such effects, the capacity during a simulation analysis is usually allocated very generously in order to circumvent such unwanted bottlenecks from temporarily arising (see [Zülc-96a]). Figure 3.7b illustrates the results of such an approach. In this case the allotted available capacity was increased to 6 hrs/SCD. With otherwise unchanged conditions it was possible to achieve a state in which the key performance figures for the output rate, WIP, range and flow rate corresponded well with the those from the shop floor. Nevertheless, the capacity can still only be utilized 81%. A capacity oriented system design based on such

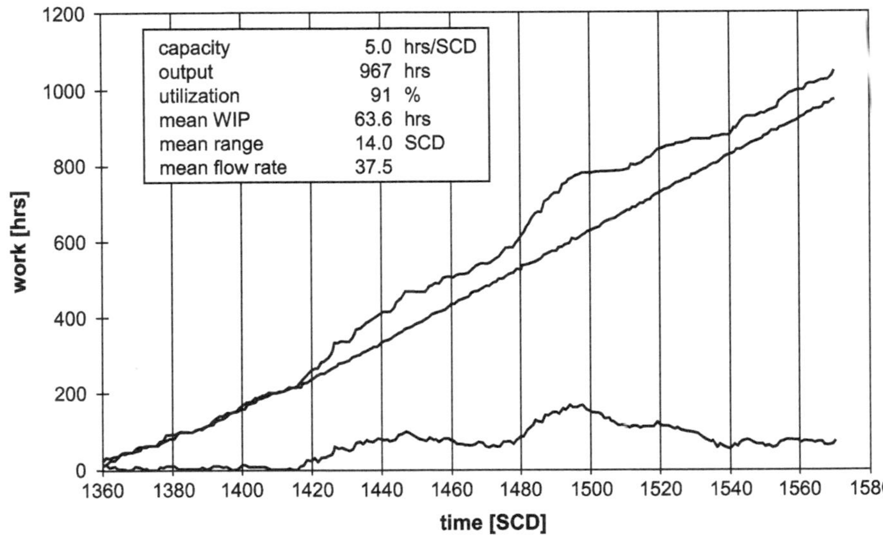

a) **Simulation Results for a Constant Capacity (assuming: 5.0 hrs/SCD)**

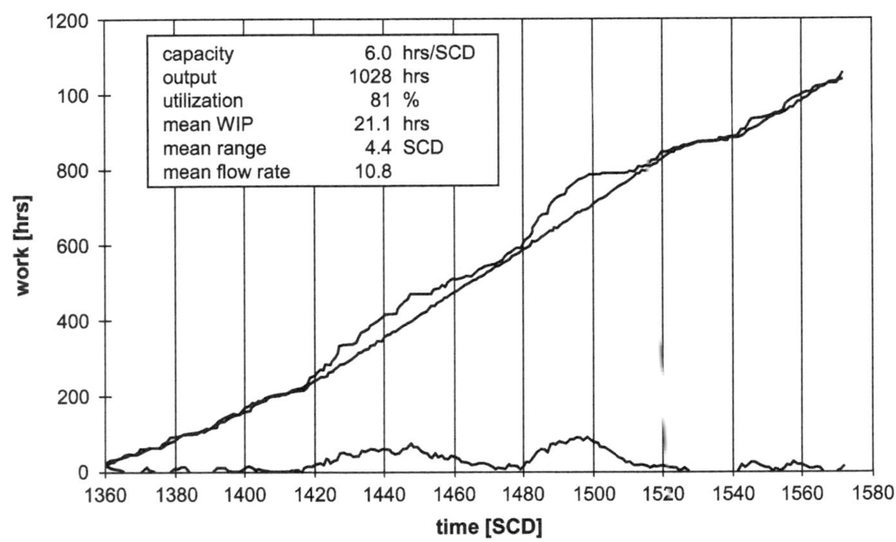

b) **Simulation Results for a Constant Capacity (assuming: 6.0 hrs/SCD)**

Fig. 3.7 Results of Selected Simulation Runs for Workstation 61570

simulation results is highly questionable. Moreover, the WIP's chronological behavior deviates considerably from the actual state's. Therefore, the time-dynamic behavior established in the simulation has no significant meaning.

Not only does this example demonstrate that the information derived from the simulation is only as good as the data that the model is based on, it also illustrates how well the processes inherent to the system can actually be reproduced from it ([Hess-88], [Schl-95], [Mich-97]).

This conclusion is supported by the results of a comparative analysis using various simulation tools: In his research on simulations, Krauth presented the results of a study in which different users simulated an identical assignment with different simulators [Krau-92*]. The task consisted of examining the output rate and throughput time behavior of an assembly system with varying WIP levels. The system had eight work stations, the assembly times and system runtimes as well as the placement and distance of the stations were specified in detail. Despite this relatively easy assignment, a comparison of the results demonstrated that no two system applications delivered the same results and that in some cases there were actually considerable differences.

These comments are not meant as an argument against simulation based analyses though. As long as all of the model components (load, planning, control and resource models) can be accurately described and are applied appropriately, simulation can deliver a good reproduction of the actual system.

However, one problem specific to simulations cannot be overlooked: The more accurate the computer simulation is (in particular the graphic representation of the processing sequence on the monitor), the more risk there is that the observer will not be able to differentiate between the emulation and the original, and is thus no longer able to objectively interpret and evaluate the results of the simulation. When this critical distance is no longer given there is a risk that the model's mistakes will remain unknown or that their significance will be underestimated. The simulation then merely becomes an end in itself [Birk-95].

Independent of the model's inaccuracy, it remains to be determined whether or not the simulation provides any clues about the direction in which the best solution can be found. Every simulation actually demonstrates how one specifically chosen model impacts a production system's performance. No general knowledge is gained, instead there are just individual results. The direction in which the best solution is to be found can only be determined through a multitude of simulations with different model variations (see e. g., [Dang-91], [Kull-87], [VDI-93]).

Existing simulation systems, however, offer no assistance in choosing and determining subsequent experiments. Choosing possible methods for designing them is therefore frequently accomplished according to the trial and error principle. Due to the combinatorial effect, even with a few parameters a large number of simulation runs are needed in order to find the best results. Thus, a number of authors ([Zülc-96b], [Rums-95] and [Scha-92]) suggest using a decision support system which assists the user in planning and conducting simulation experiments. Depending on the implemented optimization methods however, in order to find the best known results it can still require hundreds of simulation runs even with a problem which has 10 variable parameters [Rums-95].

The abovementioned problems explain why simulation is established in research but not in the field. Up until now, simulations have been applied particularly in detailed analyses of narrowly defined parts of a production system e. g., individual machines or plants. Employing simulations as a long-term decision support system for designing complex integrated production structures is in practice rarely considered [Günz-93] even though it seems to have particularly good potential especially in this area.

Chapter 4
Deriving the Logistic Operating Curves Theory

Due to the specific problems and application limits associated with both queuing theory and simulations, they are rarely applied in order to support the logistic oriented evaluation and design of complex production systems. In light of this and the growing demands on production logistics, a mathematical approach which could calculate Logistic Operating Curves for production processes using an approximation equation proved to be worthwhile. This equation, which in the meantime is also applied in the field (see Chap. 7), distinguishes itself from others in that it allows LOC to be calculated with less data and ensures more reliable outcomes.

The theory underlying the approximation equation was gained from a *deductive-experimental model design*. By combining basic modeling approaches the goal was to draw on their advantages and avoid their disadvantages. The model's fundamental structures are deductively determined. Therefore, the model is in principle universal and can be both transferred and generally interpreted within its range of validity. The developed model is then adapted to the real conditions through an experimental based parameterization. Here, simulations are systematically run and the individual parameters are gradually changed. When generalizable rules and laws are derived by analyzing the simulation results, the gained knowledge can be integrated into a mathematical model. This in turn extends the validity of the model's deductive component. Most importantly though, the generalized results from the experiments can be applied via a mathematical model without having to conduct further simulation trials in special cases.

Deriving the deductive components of the model is simplified considerably by idealizing the conditions on which they are based. The system behavior that is of interest can then often be quite easily described even quantitatively, whereby the range of validity is naturally limited by the idealized assumptions. In the *Logistic Operating Curves Theory* (LOC Theory) developed from this basis, the deductive model component consists by definition of both the ideal Logistic Operating Curves (introduced below) and the not yet parameterized, approximation equation used to calculate the LOC for the actual output rate and time parameters. In the following, *time parameters* is used as a general expression for the range, throughput time and inter-operation time.

4.1 Ideal Logistic Operating Curves

Deriving the ideal Logistic Operating Curves is based on analytically examining idealized production processes and the ideal processing figures which are determined from that. The following conditions describe the underlying ideal manufacturing process:

- There is exactly one order at every workstation at any point in time.
- An order is processed immediately after it enters the workstation; the inter-operation time between two operations is therefore zero.

When these conditions are true, the workstations do not have to wait for an order (no idle time) and the orders do not have to compete for capacities (no waiting time). Furthermore, in order to derive the ideal processing figures the following assumptions will also be made:

- Only one workstation will be considered.
- The upper limit of the workstation's output is given by the maximum possible output rate.
- The orders and their work content are given and cannot be changed on short notice.
- The production orders are transported in lots; overlapping productions are thus not allowed.
- The transport time between two operations is negligible.
- There is sufficiently precise planning data available in order to establish the work content.
- The examination period is long enough that the work content structure of the orders which are to be processed can be considered representative of the system.

4.1.1 Ideal Minimum WIP Level

A Throughput Diagram produced under the above conditions is presented in Fig. 4.1. Simplified, the time axis is indicated not in shop calendar days but instead in standard hours. The height and length of the individual throughput elements are thus determined exclusively by the orders' work content. Therefore, when the x and y axis are equally scaled, the throughput elements resemble squares.

Similarly, given the conditions mentioned above, the WIP on the workstation results directly from the work content of the orders to be processed (Fig. 4.1b).

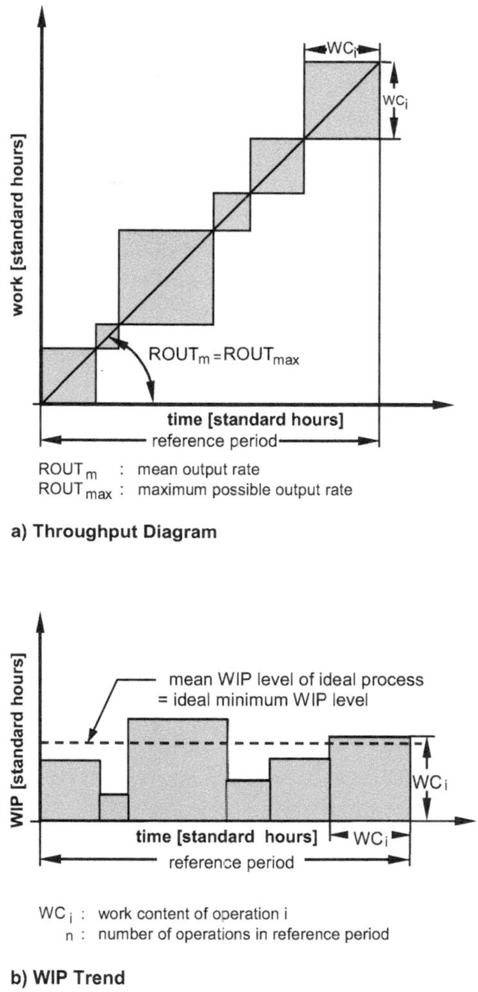

Fig. 4.1 Depiction of an Ideal Production Process in a Throughput Diagram

The mean value of this WIP, which in the following will be referred to as the *ideal minimal WIP level* ($WIPI_{min}$), is calculated by dividing the sum of the areas created from the work content by the sum of the individual work content values:

$$WIPI_{min} = \frac{\sum_{i=1}^{n}(WC_i \cdot WC_i)}{\sum_{i=1}^{n} WC_i} \tag{4.1}$$

where $WIPI_{min}$ ideal minimum WIP [hrs],
WC_i work content per operation [hrs],
n number of orders.

In calculating the ideal minimum WIP it is also taken into consideration that orders with a greater work content not only cause a greater WIP, but also correspondingly tie-up WIP longer. This is taken into account by including the areas in Eq. 4.1 when forming the means. Furthermore, there are additional equations for calculating the ideal minimum WIP with which the special influence of heterogeneous work content structures on this WIP value and – as will be shown – the logistical system behavior in general can be expressed.

Therefore, using the equations for the mean and standard deviation [Sach-84], Eq. 4.1 can be rewritten so that the ideal minimum WIP can also be calculated through this distribution parameter (see also [Erdl-84] and [Nyhu-91]):

$$WIPI_{min} = WC_m + \frac{WC_s^2}{WC_m} \tag{4.2}$$

where $WIPI_{min}$ ideal minimum WIP [hrs],
WC_m mean work content [hrs],
WC_s standard deviation of the work content [hrs].

Finally, based on Eq. 2.4 the following relation can be derived:

$$WIPI_{min} = WC_m \cdot \left(1 + WC_v^2\right) \tag{4.3}$$

where $WIPI_{min}$ ideal minimum WIP [hrs],
WC_m mean work content [hrs],
WC_v coefficient of variation for the work content [-].

Equations 4.2 and 4.3 show that the $WIPI_{min}$ is determined by the level of the work content (the mean) and by the work content's variance. It is important to note here that the standard deviation or the coefficient of variation for the work

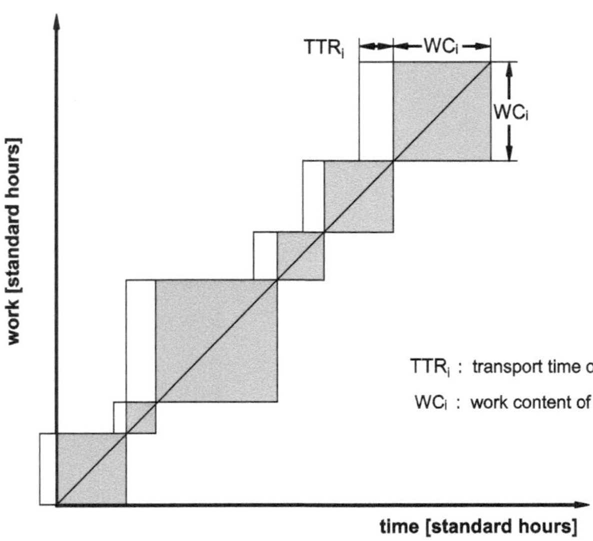

Fig. 4.2 Ideal Production Process in consideration of Transport Operations

content is squared in the calculation. That is, the inhomogeneity of the work content's variance in particular, decisively influences the ideal minimum WIP.

Up until now, it has been assumed that the transport time between two operations is negligible. However, if this is not the case and the transport is not considered an independent process, then the WIP that is bound to the transport process also has to be considered when calculating the $WIPI_{min}$. In Fig. 4.2 the ideal production process for this case is represented in a Throughput Diagram. In order to ensure the workstation's full utilization with the lowest possible WIP it is necessary that the corresponding transport process ends precisely when the preceding operation is completed. The additional WIP areas result from the product of the transport time and work content. Thus,

$$WIPI_{min} = \frac{\sum_{i=1}^{n}(WC_i \cdot WC_i)}{\sum_{i=1}^{n} WC_i} + \frac{\sum_{i=1}^{n}(TTR_i \cdot WC_i)}{\sum_{i=1}^{n} WC_i} \tag{4.4}$$

where $WIPI_{min}$ ideal minimum WIP [hrs],
WC_i individual work content per operation [hrs],
TTR_i individual transport time per operation [hrs].

When the transport times are constant or independent of the work content, Eq. 4.3 can be used to simplify Eq. 4.4:

$$WIPI_{min} = WC_m \cdot \left(1 + WC_v^2\right) + TTR_m \tag{4.5}$$

where $WIPI_{min}$ ideal minimum WIP [hrs],
WC_m mean work content [hrs],
WC_v coefficient of variation for the work content [-],
TTR_m transport time [hrs].

If in addition to the transport, other process dependent inter-operation times are required (e. g., when a product for technical reasons needs to be stored while it cools down or dries), they can be dealt with similarly to the transport time.

4.1.2 Maximum Possible Output Rate

In deriving the ideal minimum WIP it was a prerequisite that the upper limit for the workstation's output rate be given by the maximum possible output rate. As already explained in the section about calculating the operation time (see comments to Eq. 2.5), the *maximum possible output rate* basically results from the capacity with the upper output limit being determined by the corresponding *restrictive capacity factor* (machines or operators):

$$ROUT_{max} = \min\left\{CAP_m, CAP_o\right\} \tag{4.6}$$

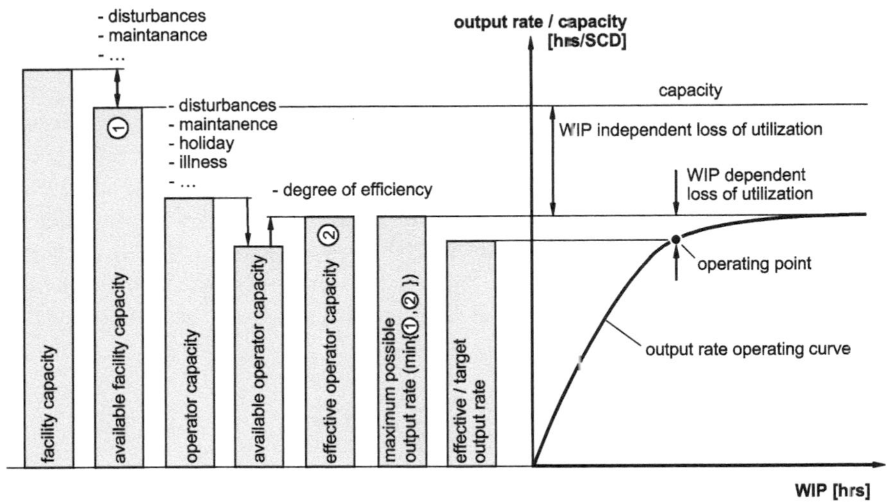

Fig. 4.3 Differentiation of the Capacity and Maximum Possible Output Rate

where $ROUT_{max}$ maximum possible output rate [hrs/SCD],
CAP_m available machine capacity [hrs/SCD],
CAP_o available operator capacity [hrs/SCD].

The correlation between the different capacity data and the maximum possible output rate parameter is represented in Fig. 4.3. The fact that the available capacities could be limited by disruptions (e. g., machine failure) is taken into consideration. If the capacity data is given in actual utilization hours and not in planned hours, then the available operator capacity has to be adjusted based on the loss of utilization.

On the right side of Fig. 4.3 the basic shape of an Output Rate Operating Curve is illustrated. We can see that OROC only depict the WIP dependent loss of utilization. All further parameters that influence the output rate or utilization of the workstation are already included in determining the WIP independent, maximum possible output rate.

4.1.3 Constructing Ideal Logistic Operating Curves for the Output Rate and Time Parameters

Based on a suggestion from Wedemeyer [Wede-89], the ideal Logistic Operating Curves pictured in Fig. 4.4 can be derived from the definition of a workstation's $WIPI_{min}$ and $ROUT_{max}$ (see also [Nyhu-91]). Thus, the upper limit of the workstation's output is at first determined by the maximum possible output rate which can be achieved with the ideal minimum WIP. If the abovementioned prerequisites are then partially neglected additional operating zones, which can be used to generally describe the system behavior when the WIP changes, can be derived.

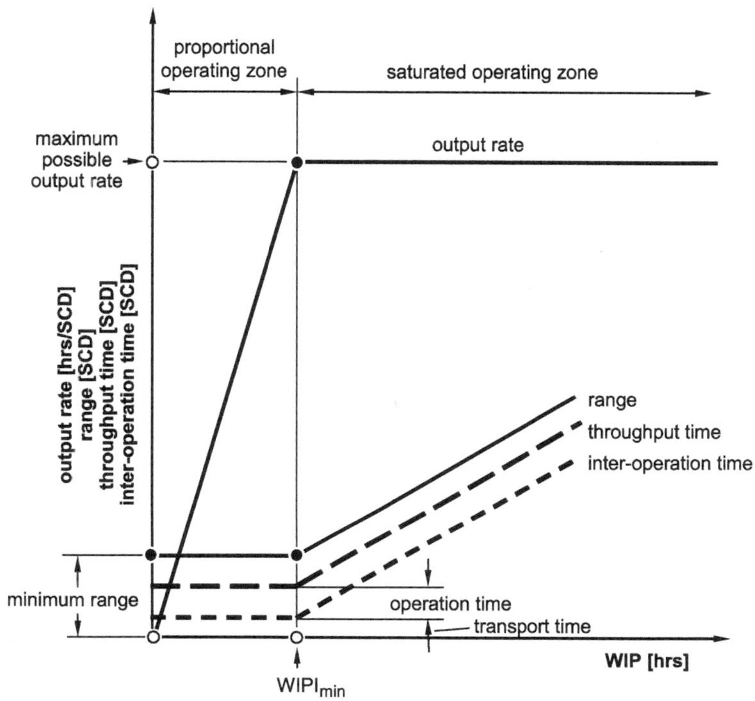

Fig. 4.4 Constructing the Ideal Logistic Operating Curves for the Output Rate and Time Parameters

The *ideal Output Rate Operating Curve (ideal OROC)* is based on the following consideration: If due to a break in material flow there are temporarily no orders on a workstation, then the mean WIP is lower than the ideal minimum WIP. At the same time, due to the idle time there is also a loss of output proportional to the reduction in WIP (*proportional operating zone* of the ideal OROC). If on the other hand, there are temporarily more orders on the workstation the mean WIP is increased, however, the output rate is not elevated, because the system is already operating at the threshold of the maximum possible output rate (*saturated operating zone* of the ideal OROC).

The *ideal Range Operating Curve (ideal ROC)* illustrated in Fig. 4.4 can be derived based on similar considerations. According to the Funnel Formula (Eq. 2.16) though, it is also the direct result of the WIP to output rate ratio. Therefore, in the saturated operating zone of the ideal OROC, the range behaves proportionally to a change in the WIP. Nevertheless if the WIP falls short of the ideal minimum then the output rate is reduced proportionally to the WIP. Consequently, the range remains at a constant level – the minimum range. Thus,

$$R_{min} = \frac{WIPI_{min}}{ROUT_{max}} \tag{4.7}$$

where R_{min} minimum range [SCD],
$WIPI_{min}$ ideal minimum WIP [hrs],
$ROUT_{max}$ maximum possible output rate [hrs/SCD].

By applying Eqs. 2.5, 2.6, 4.5 and 4.7 the *minimum range* can be expressed as:

$$R_{min} = TOP_m \cdot \left(1 + TOP_v^2\right) + \frac{TTR_m}{ROUT_{max}} \tag{4.8}$$

where R_{min} minimum range [SCD],
TOP_m mean operation time [SCD],
TOP_v coefficient of variation for the operation time [-],
TTR_m mean transport time [hrs],
$ROUT_{max}$ maximum possible output rate [hrs/SCD].

The throughput time can also not fall below a specific minimum defined by the orders' operation time and where applicable the related transport time:

$$TTP_{min} = TOP_m + \frac{TTR_m}{ROUT_{max}} \tag{4.9}$$

where TTP_{min} minimum throughput time [SCD],
TOP_m mean operation time [SCD],
TTR_m mean transport time [hrs],
$ROUT_{max}$ maximum possible output rate [hrs/SCD].

Finally, the minimum inter-operation time corresponds to the mean transport time, given in shop calendar days:

$$TIO_{min} = \frac{TTR_m}{ROUT_{max}} \tag{4.10}$$

where TIO_{min} inter-operation time [SCD],
TTR_m mean transport time [hrs],
$ROUT_{max}$ maximum possible output rate [hrs/SCD].

The ideal LOC for the throughput time and inter-operation time are plotted above the minimum WIP in Fig. 4.4 with the same slope as the Range Operating Curve. As will be shown in the following section (see Sect. 4.4), this is only permissible when the processing sequences are independent from the orders' work content. This is the case for example with the FIFO sequencing rule and in generally also with the slack time rule.

4.2 Deriving an Approximation Equation for Calculating an Output Rate Operating Curve

The conditions required for deriving the ideal LOC are not given on the shop floor. Particularly in complex manufacturing processes with omni-directional material flow structures (e. g., in job shop production) there are a number of factors such as how the order input is distributed or whether there are dynamic bottlenecks situations that cause the real operating points to deviate from the ideal Logistic Operating Curves.

If a number of operational points generated by a realistic simulation model are interpolated into a *simulated Output Rate Operating Curve*, it becomes evident that there is no longer a definite angulation point (Fig. 4.5, see also Fig. 3.5): The transition between the proportional operating zone and the saturation operating zone is fluid. In a real production there always has to be a *WIP buffer* in order to prevent output losses. Initially, the necessary size of the buffer is determined by the targeted utilization level. Evaluating numerous simulation trials has also shown that the required WIP buffer is dependent on the parameters that determine the *ideal Logistic Operating Curve's point of angulation*. Thus, the larger the mean work content is and the greater its variance or transport time are, the larger the WIP buffer has to be ([Erdl-84], [Nyhu-91]) and therefore the greater the $WIPI_{min}$ is. The available capacity flexibility as well as the load fluctuations in the order input however, also influence the required WIP level.

Nevertheless, it can still be determined that the ideal OROC sets the limiting curve for the real Output Rate Operating Curve. This suggests that the ideal OROC could be used as a reference coordinate system for a calculated Output

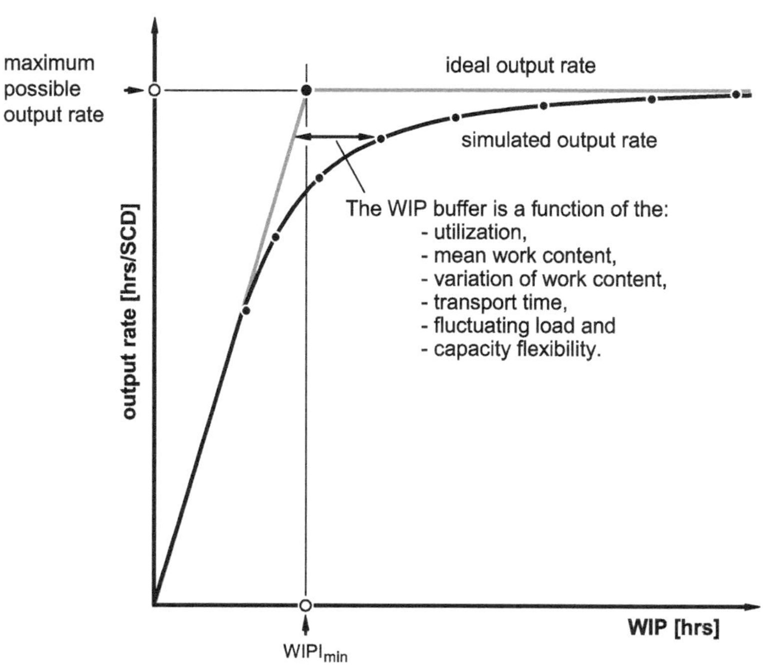

Fig. 4.5 Comparison of Ideal and Simulated Output Rate Operating Curves

Rate Operating Curve. This in turn creates the possibility to determine a mathematical function that meets the following requirements:

- The function must intersect or begin at the original coordinates system's point of origin (0,0).
- The workstation is fully utilized with a finite WIP level.
- The function must be able to be modeled through parameters in such a way that it approximates the ideal OROC in the range of the angulation point in the preferred form.

We will now describe the procedure for deriving and parametrizing such a function.

4.2.1 C_{norm} Function as the Basic Function for a Calculated Output Rate Operating Curve

In order to determine an appropriate function, we can first simplify our search by looking for an equation that in accordance with the requirements, can be modeled in a Cartesian coordinate system.

At this point, we can draw upon differential calculus, which uses the so-called C_{norm} *function* for specific characterizations of a vertex' neighborhood. Figure 4.6 shows this norm for a number of values from C. It is developed as a generalization of the circle. All points on a circle with the radius of 1 satisfy the equation of a circle:

$$1 = x^2 + y^2 \tag{4.11}$$

The idea behind generalizing it, is to replace the 2 with an open parameter – in this case 'C'. So that the function is valid for every value of C in all of the coordinate system's quadrants, the function for the absolute values has to be written in terms of x and y.

$$1 = |x|^C + |y|^C \tag{4.12}$$

The resulting function is illustrated for different values of C in Fig. 4.6. As the value of C becomes smaller, the curves approximate the axes more closely. As can be seen, a section of the curve in the second quadrant of the C_{norm} function for $C = 0.25$, which already strongly resembles an Output Rate Operating Curve, is bolded in the graph.

In order to mathematically describe this function it is necessary to parametrize it, in that the functions x(t) and y(t) are created (the parameter t corresponds with a running variable):

$$\left.\begin{aligned} x(t) &= t \\ y(t) &= -\sqrt[C]{1 - t^C} \end{aligned}\right\} \tag{4.13}$$

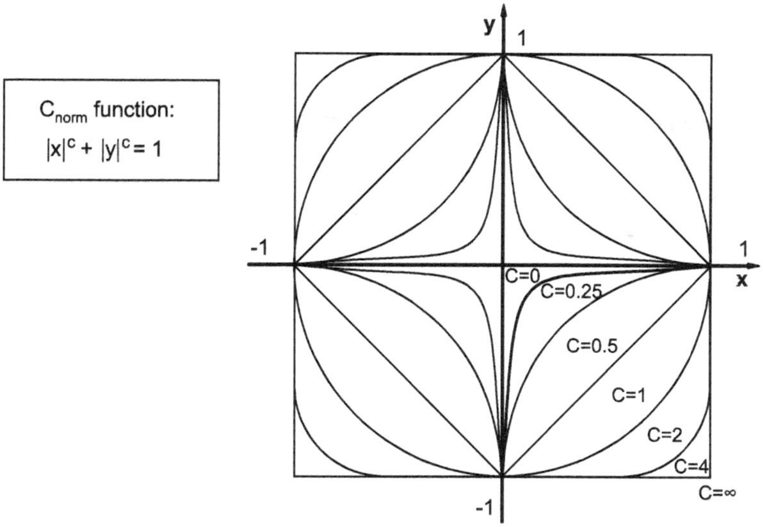

Fig. 4.6 C_{norm} Function for Various Values of C

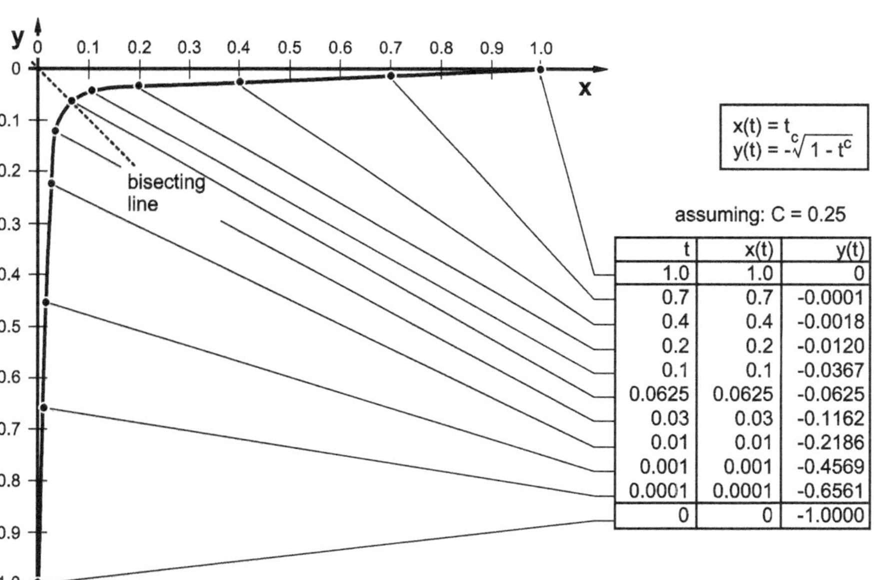

t	x(t)	y(t)
1.0	1.0	0
0.7	0.7	-0.0001
0.4	0.4	-0.0018
0.2	0.2	-0.0120
0.1	0.1	-0.0367
0.0625	0.0625	-0.0625
0.03	0.03	-0.1162
0.01	0.01	-0.2186
0.001	0.001	-0.4569
0.0001	0.0001	-0.6561
0	0	-1.0000

Fig. 4.7 Parametrizing the C_{Norm} Function with $c = 0.25$ (2nd quadrant)

The parameterization of the C_{norm} function for a chosen $C = 0.25$ is illustrated as an example in Fig. 4.7. In total 11 different values for x(t) and y(t) were calculated using Eq. 4.13, where $0 \leq t \leq 1$. These were then plotted and joined with one another in order to form the graph of the function. It should be noted here that no constant increment was chosen for the running variable t. Using the t-values which form the basis for Fig. 4.7, it is possible to evenly cover the graph of the function with relatively few points. However, this is valid only for C = 0.25. If a constant increment is to be used instead, then the increment must be very small in order to be able to describe the function with sufficient precision, particularly for smaller values of t and x. When the value of C is different other increments for t are necessary.

4.2.2 Transforming the C_{norm} Function

Once a basic function that resembles a Logistic Operating Curve is found, it has to be made more congruent with an actual Output Rate Operating Curve. An affine transformation (affinis [lat.]: related to) is thus used to stretch and shear the function until the C_{norm} curve finally corresponds to an OROC. This process can be broken down into four steps which can be followed in Fig. 4.8a to 4.8d.

First, the chosen section of the C_{norm} Function (i. e., the second quadrant) is shifted in the positive direction of the y-axis so that the initial value is equal to the coordinate system's point of origin (0,0). A displacement factor of 1 therefore, has to be added to each of the y-values. Whereas the curve was previously described with x(t) and y(t), it is now written:

$$\left.\begin{aligned} x &= x(t) \\ y &= y(t) + 1 \end{aligned}\right\} \quad (4.14)$$

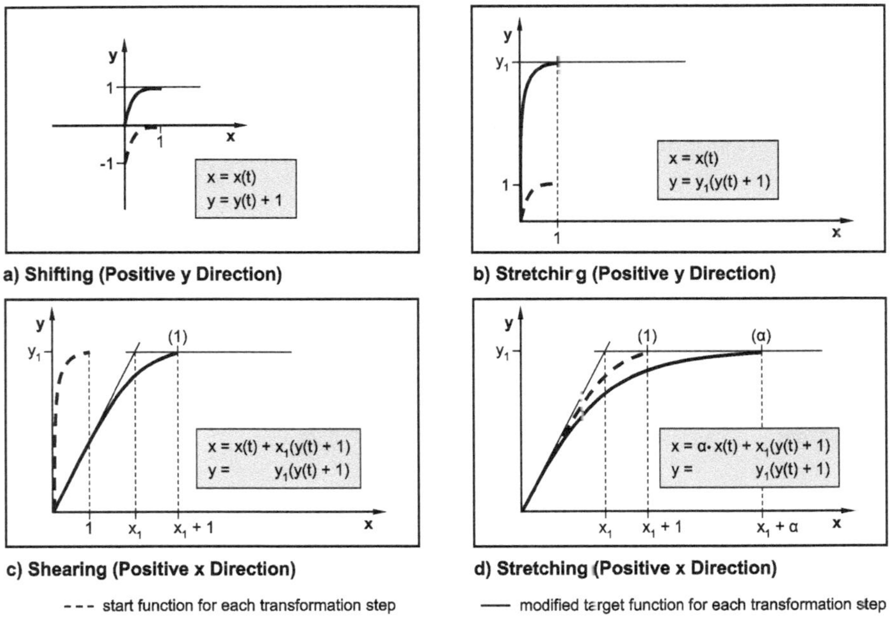

Fig. 4.8 Steps in the Affine Transformation in order to Mathematically Describe the Output Rate Operating Curve

In the second step, the curve is stretched in the direction of the positive y-axis, so that the maximum value of the function on the y-axis corresponds to the maximum possible output rate. In the following, this is generalized as y_1. Since the curve previously had a height of 1 and afterwards takes on that of y_1, the corresponding stretch factor is thus also y_1. Therefore,

$$\left.\begin{aligned} x &= x(t) \\ y &= y_1 \cdot \left(y(t)+1\right) \end{aligned}\right\} \tag{4.15}$$

The up till now ideal coordinate system is still formed by the positive y-axis and the maximum possible output rate. In the third step, this coordinate system has to be changed so that it is congruent to the ideal Output Rate Operating Curve. This ideal OROC is described in that the horizontal axis remains horizontal, however the point of angulation is displaced in the direction of the positive x-axis to the minimum WIP (also called x_1 in the following). The previously vertical y-axis thus opens up like a scissors' or shears' arm. In fact this transformation process is called shearing. The transformation is accomplished by shifting each point on the y-axis by a different value in the positive x-axis direction. Only the shears pivot point (0,0) has to remain the same. The value of the shear, for any y-value can be determined by applying the rule of three. At the height y_1 it has to be shifted by the value x_1 and therefore at the height y, by $y \cdot (x_1/y_1)$ in the positive x-axis direction:

$$\frac{x_1}{y_1} \cdot y = \frac{x_1}{y_1} \cdot y_1 \cdot \left(y(t)+1\right) = x_1 \cdot \left(y(t)+1\right)$$

Thus, the sought function becomes:

$$\left.\begin{aligned} x &= x_1 \cdot \left(y(t)+1\right) + x(t) \\ y &= y_1 \cdot \left(y(t)+1\right) \end{aligned}\right\} \tag{4.16}$$

In the fourth and last step, the curve itself is finally stretched along the positive x-axis. The required stretch factor can at this point not yet be clearly determined. Since it will be stretched to an as of yet unknown degree, the parameter α is introduced. Only the curve should be manipulated by this stretching in the x-direction and not the ideal coordinate system as the latter already has the preferred form. Therefore, only the second part of the equation which applies to x, is to be modified. The following function thus results:

$$\left.\begin{aligned} x &= x_1 \cdot \left(y(t)+1\right) + \alpha \cdot x(t) \\ y &= y_1 \cdot \left(y(t)+1\right) \end{aligned}\right\} \tag{4.17}$$

The stretch factor α takes into consideration the WIP buffer which in addition to the minimum WIP is required in order to ensure the workstations performance on the shop floor. Above, it was already determined that the larger the ideal minimum WIP (x_1) is the larger the WIP buffer needs to be. The stretch can therefore be expressed as a multiple of x_1. This in turn means that the scale of the curve, not the curve itself, is changed.

$$\alpha = \alpha_1 \cdot x_1 \tag{4.18}$$

By substituting the parametrized C_{norm} function (Eq. 4.13) for x(t) and y(t), the Output Rate Operating Curve is now described as

$$\left.\begin{aligned} x &= x_1 \cdot \left(1-\sqrt[C]{1-t^C}\right)+\alpha_1 \cdot x_1 \cdot t \\ y &= y_1 \cdot \left(1-\sqrt[C]{1-t^C}\right) \end{aligned}\right\} \tag{4.19}$$

This function is defined only within the limits $0 \le t \le 1$. When $t=0$ both x and y take on the value of zero as well. Therefore, where $t=0$ the function begins at the coordinate system's point of origin.

Thus, following Eq. 4.19 for $t=1$:

$$x = (\alpha_1 + 1) \cdot x_1$$

$$y = y_1$$

This point represents the maximum value of the function in the positive direction of both the x and y-axes. It follows from this that with Eq. 4.19 an Output Rate Operating Curve can only be calculated for up to $(\alpha_1 + 1)$ times the $WIPI_{min}$. At the resulting WIP level, the workstation has arithmetically reached full utilization. In Sect. 6.11 however, it will be shown that Logistic Operating Curves can also be developed for higher WIP levels.

4.2.3 Parametrizing the Logistic Operating Curves Equation

Equation 4.19 allows a function similar to a Logistic Operating Curve to be calculated using the four parameters: x_1 and y_1, which correspond to the minimum WIP and maximum possible output rate, and the function's parameters C and α_1. The precise values of these however, still have to be determined. Before this can be done we first need to discuss the parameters and their significance (Fig. 4.9).

When deriving a mathematical model for calculating the Output Rate Operating Curve the first point to be considered is the definition of an ideal minimum WIP. It influences both where the angulation point is located as well as the slope of the proportional operating zone for the ideal OROC. In deriving Eq. 4.19 $WIPI_{min}$ as parameter x_1 is incorporated into the model through the shearing and stretching in the positive x-direction. If the minimum WIP is varied, both the ideal and calculated OROC move similarly, parallel to the x-axis. The absolute distance of the two Output Rate Operating Curves changes. However, because the value of α_1 is assumed to be a constant, the relation between the two remains the same, because the stretching in the positive x-direction is expressed as a multiple of x_1. The two

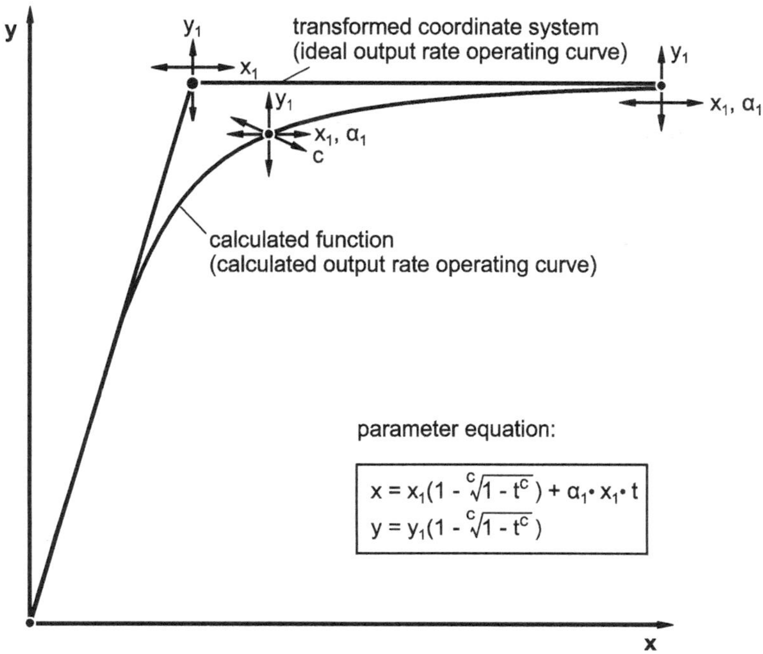

Fig. 4.9 Significance of Parameters within the Mathematical Description of the Output Rate Operating Curve

Logistic Operating Curves are translated vertically through changes in the maximum possible output rate y_1. The proportional operating zone of the OROC therefore also changes though. The ideal Output Rate Operating Curve is clearly determined by x_1 and y_1. The values of the two parameters result directly from the workstations' structural conditions and along with the ideal LOC determine the limits within which the calculated LOC can still be changed.

In contrast, the parameters α_1 and C cannot be derived from the workstation specific values, because the C_{norm} function used to calculate the Logistic Operating Curves is a freely chosen function and can be replaced by other suitable functions. Therefore, it is only possible to determine these undefined parameters and estimate the quality of the resulting function through empirical research. Nevertheless, a few general comments can be made here about the chosen function.

The value of C in the C_{norm} function determines how closely the calculated Output Rate Operating Curve approximates the ideal OROC's angulation point. The smaller C is, the more tightly the calculated OROC appears to approximate it. In the extreme case where $C=0$, the calculated curve is congruent to the ideal Output Rate Operating Curve.

In order to find a typical value for C, it is assumed that the point of intersection for the C_{norm} function and the angle bisector in the original coordinate system is the point of angulation (Fig. 4.10a). The coordinates (x_{AP}, y_{AP}) of this point can thus be determined generally. The relation $x_{AP}=-y_{AP}$, which results from this

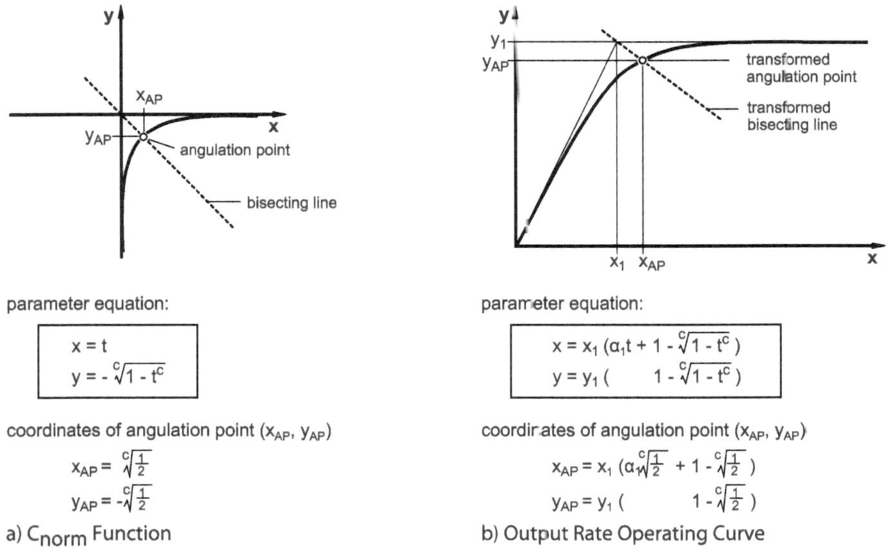

Fig. 4.10 Determining the Function Parameters

assumption is substituted into the parameter equation for the C_{norm} function in order to first determine the value t_{AP} which the running variable takes at this point:

$$\left.\begin{aligned} x_{AP} &= \quad t_{AP} \\ y_{AP} &= -\ t_{AP} = -\sqrt[C]{1 - t_{AP}^{C}} \end{aligned}\right\} \tag{4.20}$$

For t_{AB} it follows that

$$t_{AP} = \sqrt[C]{\frac{1}{2}} \tag{4.21}$$

Given the relation determined above for t_{AB}, the coordinates for the angulation point are:

$$\left.\begin{aligned} x_{AP} &= \quad \sqrt[C]{\frac{1}{2}} \\ y_{AP} &= -\sqrt[C]{\frac{1}{2}} \end{aligned}\right\} \tag{4.22}$$

The point of angulation can therefore also be established for the transformed function by substituting the relation for t_{AB} into the equation for the Output Rate Operating Curve (Eq. 4.19):

$$\left.\begin{aligned} x_{AP} &= x_1 \cdot \left(1 - \sqrt[C]{\frac{1}{2}}\right) + \alpha_1 \cdot x_1 \cdot \sqrt[C]{\frac{1}{2}} \\ y_{AP} &= y_1 \cdot \left(1 - \sqrt[C]{\frac{1}{2}}\right) \end{aligned}\right\} \tag{4.23}$$

The resulting angulation point rests on the line created by transforming the original coordinate system's bisecting angle.

This agreement however, is not yet sufficient for conclusively determining the parameters α_1 and C. Thus, based on numerous investigations of both real and simulated processes it was further identified that the point at which the WIP related loss of output is equal to approximately 6% should be considered the angulation point on a real Output Rate Operating Curve. If this is applied to the C_{norm} function, we find that the function of the bisecting angle must intersect at $y = -0.06$. For the preliminary determination of C it then follows (see Eq. 4.22) that

$$y_{AP} = -\sqrt[C]{\frac{1}{2}} = -0.06 \tag{4.24}$$

This equation can now be solved for C:

$$C = \frac{\ln\left(\frac{1}{2}\right)}{\ln(0.06)} \cong 0.25 \tag{4.25}$$

After determining C, the parameter α_1 still has to be established. In deriving Eq. 4.19 this parameter only appears when the curve is stretched in the direction of the x-axis. It thus also influences the location of the angulation point x_{AP} along the x-axis. If the equation for the defined angulation point's x-coordinate (Eq. 4.23) is solved for α_1, the following relationship results:

$$\alpha_1 \cong \frac{\frac{x_{AP}}{x_1} - 1 + \sqrt[C]{\frac{1}{2}}}{\sqrt[C]{\frac{1}{2}}} \tag{4.26}$$

In order to solve this equation it is first necessary to determine the WIP level (here, x_{AP}) at which the loss of utilization is approximately equal to 6%, that is where the WIP dependent utilization is 94%. This WIP level can then be set in relation to the ideal minimum WIP (here, x_1). In order to establish this relation, the results of numerous simulations gained through various research projects were drawn upon (see [Bech-84], [Erdl-84], [Nyhu-91]). In analyzing the Output Rate Operating Curves generated in these projects the above defined angulation point was found to occur mostly when the ratio of the measured WIP_m to the $WIPI_{min}$

is approximately 1.55. (The individual values for each of the simulated Logistic Operating Curves fluctuated between 1.4 und 1.7). Thus,

$$\frac{x_{AP}}{x_1} \cong 1.55 \tag{4.27}$$

By substituting the value of this ratio and the above determined value for C into Eq. 4.26, α_1 is given as:

$$\alpha_1 \cong \frac{1.55 - 1 + \left(\frac{1}{2}\right)^4}{\left(\frac{1}{2}\right)^4} \cong 10 \tag{4.28}$$

In principal, another point of angulation can also be defined for establishing the function's parameters. However, it can be shown that the calculated Output Rate Operating Curves are only negligibly different when the loss of utilization at the angulation point is assumed to be between 3% and 10%.

The OROC for three different parameter constellations are shown in Fig. 4.11. The parameters C and α_1 were defined according to the above described procedure (Eqs. 4.25 and 4.26). It can be seen that the definition of the angulation point determines how closely the function approximates the ideal OROC and at which WIP levels there are mathematically no longer WIP dependent losses of utilization. Given a 3% loss of utilization for the point of angulation, the resulting function is somewhat steeper with low WIP levels, however it has to be strongly stretched by the parameter ($\alpha_1 = 35$). In contrast, if a greater loss of utilization is assumed for the angulation point (here, 10%) the resulting function is not as steep

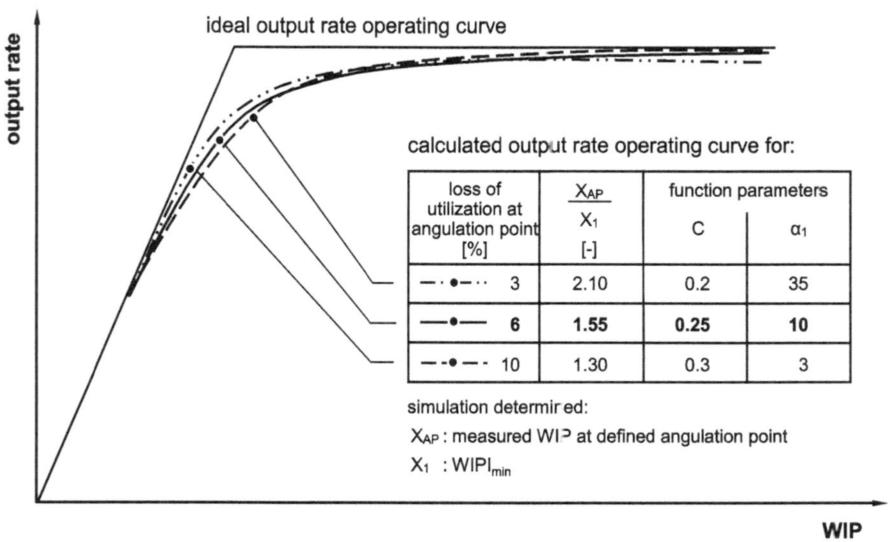

Fig. 4.11 Calculated Output Rate Operating Curves with Varied Function Parameters

with low WIP levels, but is also not as strongly stretched ($\alpha_1 = 3$). The mathematical full utilization is therefore achieved at relatively low WIP levels.

Despite the differences, it can still be established that the principle shape of the Output Rate Operating Curve is generally independent of how the angulation point is defined. Furthermore, it can also be determined that the absolute deviation between the three functions represented here is not very significant. The analyses conducted in [Nyhu-91] confirmed however, that the greatest conformance between the calculated OROC and the underlying simulation results could be attained with the parameter constellation $C = 0.25$ and $\alpha_1 = 10$. Nevertheless, in the following discussion, the parameter α_1 will remain variable so that we can continue to adapt and apply different conditions to the equation for the LOC and thus extend the validity of the Logistic Operating Curves Theory.

It should be noted at this point that although different simulation trials were drawn upon in order to determine the LOC parameters, all of them had a few common characteristics. Every simulation trial was conducted using either the PROSIM III simulator (see Sect. 3.2) or its predecessor. In each case, job shop productions were emulated and numerous model tests preceded the simulation runs. What this means in particular is that the capacities set in the simulations were adjusted to the production process' real behavior. It can generally be assumed here, that the real capacities were adjusted, at least within limits, to the required capacities. Furthermore, in the simulations the order input was determined according to the Load Oriented Order Release algorithm ([Bech-84], [Wien-95a*], [Wien-87]). This method of production control offers the possibility to specifically vary the WIP in a (simulated) manufacturing process. It was thus drawn upon within the trials for example, to simulatively develop Logistic Operating Curve points. Applying this control method however also meant that the load on the individual workstations was actively controlled and adjusted to the workstations' available capacity. The significance of this, especially with regards to determining α_1 and its range of validity is examined more closely in Sect. 4.7.

4.3 Calculating Output Rate Operating Curves

The approximation equation derived in the previous section provides a function with which an Output Rate Operating Curve can be calculated. By inserting the parameters developed in Sect. 4.2.3 into Eq. 4.19 the OROC is given as

$$\left.\begin{aligned} WIP_m(t) &= WIPI_{min} \cdot \left(1-\left(1-\sqrt[4]{t}\right)^4\right) + WIPI_{min} \cdot \alpha_1 \cdot t \\ ROUT_m(t) &= ROUT_{max} \cdot \left(1-\left(1-\sqrt[4]{t}\right)^4\right) \end{aligned}\right\} \tag{4.29}$$

The following substitutions were thus carried out in Eq. 4.19:

$x(t) \rightarrow$ $WIP_m(t)$: mean WIP (as a function of t) [hrs],
$y(t) \rightarrow$ $ROUT_m(t)$: mean output rate (as a function of t) [hrs/SCD],

$x_1 \rightarrow WIPI_{min}$: ideal minimum WIP [hrs],
$y_1 \rightarrow ROUT_{max}$: maximum possible output rate [hrs/SCD],
$C \rightarrow 1/4$: value of C in the C_{rorm} function [-].

Based on Eq. 4.29 and given $WIPI_{min}$, $ROUT_{max}$ and α_1, a pair of values can be calculated for the WIP and output rate for each t ($0 \leq t \leq 1$). By combining a number of these pairs ($WIP_m(t)$, $ROUT_m(t)$) an approximated Output Rate Operating Curve results point-by-point. We will not explain this calculation here however, as a concrete example can be found in Sect. 6.1.1.

In Fig. 4.12 a number of pairs of variates as well as the OROC that was approximated with them are shown. Based on this graph, we will now point out a few specific characteristics of this equation system.

First of all, by transforming Eq. 4.29 for the output rate it can be determined that the expression within the parentheses represents the ratio of the calculated output rate $ROUT_m(t)$ to the maximum possible output rate $ROUT_{max}$. The bracketed expression can therefore be interpreted as the mean WIP dependent utilization $U_m(t)$:

$$U_m(t) = \frac{ROUT_m(t)}{ROUT_{max}} \cdot 100 = \left(1 - \left(1 - \sqrt[4]{t}\right)^4\right) \cdot 100 \qquad (4.30)$$

where $U_m(t)$ mean WIP dependent utilization [%],
$ROUT_m(t)$ mean output rate [hrs/SCD],
$ROUT_{max}$ maximum possible output rate [hrs/SCD],
t running variable ($0 \leq t \leq 1$).

The WIP component of the Eq. 4.29, $WIP_m(t)$, can be broken down into two significant components $WIP_a(t)$and $WIP_b(t)$, which can be directly interpreted:

$$WIP_m(t) = WIP_a(t) + WIP_b(t) \qquad (4.31)$$

$$WIP_a(t) = WIPI_{min} \cdot \left(1 - \left(1 - \sqrt[4]{t}\right)^4\right) = \frac{WIPI_{min} \cdot U_m(t)}{100} \qquad (4.32)$$

$$WIP_b(t) = WIPI_{min} \cdot \alpha_1 \cdot t \qquad (4.33)$$

where $WIP_m(t)$ mean WIP [hrs],
$WIP_a(t)$ mean active WIP [hrs],
$WIP_b(t)$ mean buffer WIP [hrs],
$WIPI_{min}$ ideal minimum WIP [hrs],
$U_m(t)$ mean WIP dependent utilization [%],
α_1 stretch factor [-],
t running variable ($0 \leq t \leq 1$).

According to this description, one of the WIP components results directly from the orders being actively processed $WIP_a(t)$. This component can at the most be as large as the ideal minimum WIP, which is the case when there is an order being processed on the observed workstation at every point in time. However, if there is a loss of utilization due to a break in material flow then this part of the WIP is accordingly, proportionally reduced.

The second part of the work in process $WIP_b(t)$, represents the queued orders and can thus be interpreted as the WIP buffer. In Eq. 4.33 it can be seen that this component is also determined by the $WIPI_{min}$. Transferred to the production floor, this means that the WIP buffer required to ensure the utilization is dependent on the mean work content and in particular its variance. Furthermore, the absolute value for the WIP buffer is determined by the empirically developed stretch factor α_1. The visual representation in Fig. 4.12 shows however, that it does not behave proportionally to the utilization of the workstation. Instead, as the utilization increases, $WIP_b(t)$ grows above average. This corresponds with experiences in the field, where especially when a high utilization is to be ensured, a very high WIP level has to be maintained in order to be able to compensate for all the variance in the order input. However, if a small loss of utilization can be accepted it is possible to greatly reduce the WIP. In Eq. 4.33 this correlation is described by the running variable t (see also Eq. 4.30).

We can now develop an equation for the Output Rate Operating Curve, which is no longer a function of the running variable t. First, Eq. 4.30 has to be solved for t:

$$t = \left(1 - \sqrt[4]{1 - \frac{U_m}{100}}\right)^4 \tag{4.34}$$

where U_m mean WIP dependent utilization [%]
t running variable ($0 \leq t \leq 1$).

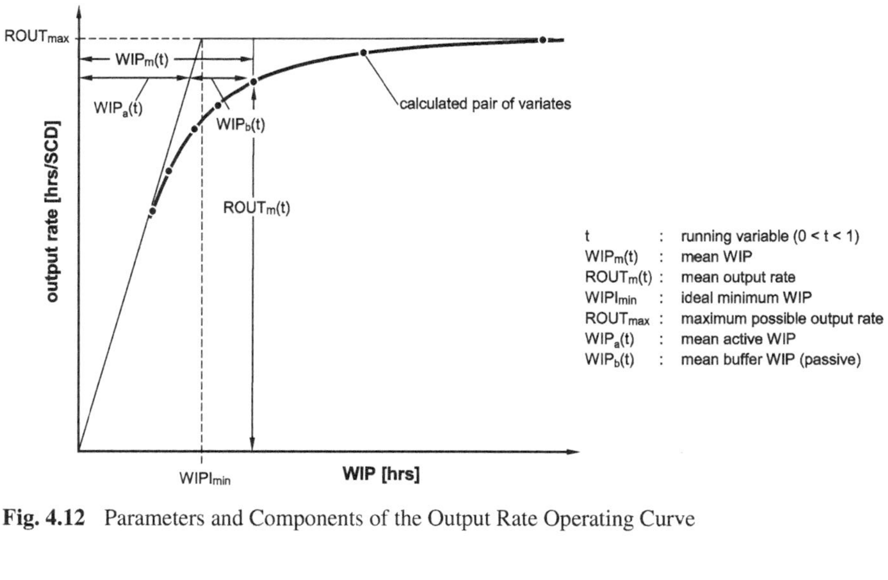

Fig. 4.12 Parameters and Components of the Output Rate Operating Curve

By substituting Eq. 4.34 in 4.29 the WIP can be written as a direct function of the utilization U_m (Eq. 4.35) or of the mean output rate $ROUT_m$ (Eq. 4.36):

$$WIP_m(U_m) = WIPI_{min} \cdot \frac{U_m}{100} + WIPI_{min} \cdot \alpha_1 \cdot \left(1 - \sqrt[4]{1 - \frac{U_m}{100}}\right)^4 \tag{4.35}$$

$$WIP_m(ROUT_m) = WIPI_{min} \cdot \frac{ROUT_m}{ROUT_{max}} + WIPI_{min} \cdot \alpha_1 \cdot \left(1 - \sqrt[4]{1 - \frac{ROUT_m}{ROUT_{max}}}\right)^4 \tag{4.36}$$

where WIP_m mean WIP [hrs],
$WIPI_{min}$ ideal minimum WIP [hrs],
α_1 stretch factor [-],
U_m mean utilization $(0 < U_m < 100)$ [%],
$ROUT_m$ mean output rate $(0 < ROUT_m < ROUT_{max})$ [hrs/SCD],
$ROUT_{max}$ maximum possible output rate [hrs/SCD].

In the following however, the notation will continue to be applied according to Eq. 4.29 in order to make clear that the variables are calculated based on the Logistic Operating Curves Theory.

4.4 Calculating Operating Curves for the Time Parameters

Approximation equations can also be derived for the Range Operating Curve (ROC), Throughput Time Operating Curve (TTOC), and Inter-Operation Time Operating Curve (ITOC) by affine transforming the C_{norm} function or by other appropriate functions. The procedure for doing so corresponds extensively with that used for deriving the Output Rate Operating Curve. The differences essentially result when determining another coordinate system: Instead of the ideal Logistic Operating Curve, the transformation here is based on the ideal Logistic Operating Curves for the various throughput time parameters.

However, the Throughput Time Operating Curves can also be derived from the Output Rate Operating Curve and the fundamentals of production logistics described in Chap. 2. We will describe this second method in the following.

The range, according to the Funnel Formula (Eq. 2.16) results from the ratio of the WIP to the output rate. This relation also obviously applies to the calculated values $WIP_m(t)$ und $ROUT_m(t)$ (Fig. 4.13). Thus for the range,

$$R_m(t) = \frac{WIP_m(t)}{ROUT_m(t)} \tag{4.37}$$

where $R_m(t)$ mean range [SCD],
$WIP_m(t)$ mean WIP [hrs],
$ROUT_m(t)$ mean output rate [hrs/SCD].

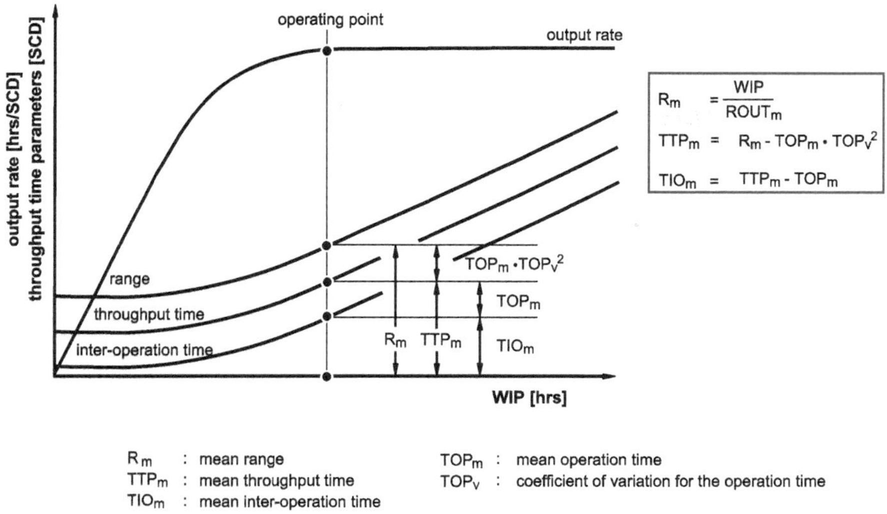

Fig. 4.13 Mathematical Correlation Between Chosen Time Parameters (Work Content Independent Sequencing Rules)

The Logistic Operating Curves for the unweighted throughput time cannot be directly described because – due to the different input dimensions – Little's Law (see Sect. 2.3) cannot be applied for the WIP (in Little's Law the WIP is given in number of orders). Furthermore as mentioned in Sect. 2.3, it is not permissible to express the WIP in standard hours through the product of the mean work content WC_m and mean WIP in number of orders.

$$WIP_m \neq WIPO_m \cdot WC_m \qquad (\text{when } WC_v > 0) \tag{4.38}$$

where WIP_m mean WIP (in standard hours) [hrs],
$WIPO_m$ mean WIP (in number of orders) [-],
WC_m mean work content [hrs],
WC_v coefficient of variation for the work content [-].

The evidence for this inequality is easily found in the definition of the ideal minimum WIP: When deriving $WIPI_{min}$ it is assumed that there is always exactly one order on the system, therefore $WIPO_m$ is equal to 1. According to Eq. 4.3 though, when $WC_v > 0$ the mean WIP is greater than the mean work content. Therefore, at least for this operating state, it is proven that directly converting the two WIP values through the mean work content is not allowable.

In order to then be able to derive an equation for calculating the Throughput Time Operating Curve the relationships and equations which were explained while developing the ideal Logistic Operating Curve can be drawn upon. There it was assumed that when the processing sequence was independent of the work content (e. g., with FIFO processing) the Throughput Time and Inter-Operation Time Operating Curves run parallel to the Range Operating Curve. This correlation was also proven empirically in ([Ludw-92] and [Ludw-93]. Moreover, the minimum values for the range R_{min} and the throughput time TTP_{min} could also be expressed (Eqs. 4.8 and 4.9). The difference between these two values describes the distance between the two Logistic Operating Curves:

$$R_{min} - TTP_{min} = TOP_m \cdot TOP_v^2 \tag{4.39}$$

where R_{min} mean range [SCD],
TTP_{min} minimum throughput time [SCD],
TOP_m mean operation time [SCD],
TOP_v coefficient of variation for the operation time [-].

Therefore, for the Throughput Time Operating Curve it follows:

$$TTP_m(t) = R_m(t) - TOP_m \cdot TOP_v^2 \tag{4.40}$$

where $TTP_m(t)$ mean throughput time [SCD],
$R_m(t)$ mean range [SCD],
TOP_m mean operation time [SCD],
TOP_v coefficient of variation for the operation time [-].

When the sequencing of the order processing is independent of the work content and the work content of the orders are non-uniform, the mean throughput time is always smaller than the mean range. Furthermore, the size of the difference is essentially determined by the operation time's variance.

Since the throughput time consists of the operation time and the inter-operation time, it can be said for the Inter-Operation Time Operating Curve that

$$TIO_m(t) = TTP_m(t) - TOP_m \tag{4.41}$$

where $TIO_m(t)$ mean inter-operation time [SCD],
$TTP_m(t)$ mean throughput time [SCD],
TOP_m mean operation time [SCD].

The Logistic Operating Curves for the throughput time and the inter-operation time could be derived in this manner, only under the condition that the sequencing is independent of the work content. Numerous studies from different authors have however, proven that sequencing rules influence the unweighted throughput time (see e. g., [Berr-70], [Hold-86]). As already explained in Sect. 2.2, the throughput time is shorter when the SPT sequencing rule is applied, than when the FIFO principle is followed and longer when the LPT rule is applied. The size of the possible deviation is influenced among other things by the WIP level: Figure 4.14 [Ulfe-86] shows that the impact of an order sequence interchange is more prominent with longer queues than with lower WIP levels Lower WIP levels enforce manufacturing according to the FIFO principle to a continuously greater degree; FIFO is thus referred to as the natural rule of production.

Since neither the operation time nor the mean transport time (where applicable) can be influenced by the sequencing rules, only the pre- and post-processing waiting times (see Fig. 2.3) remain as manipulable time components. The sum of the

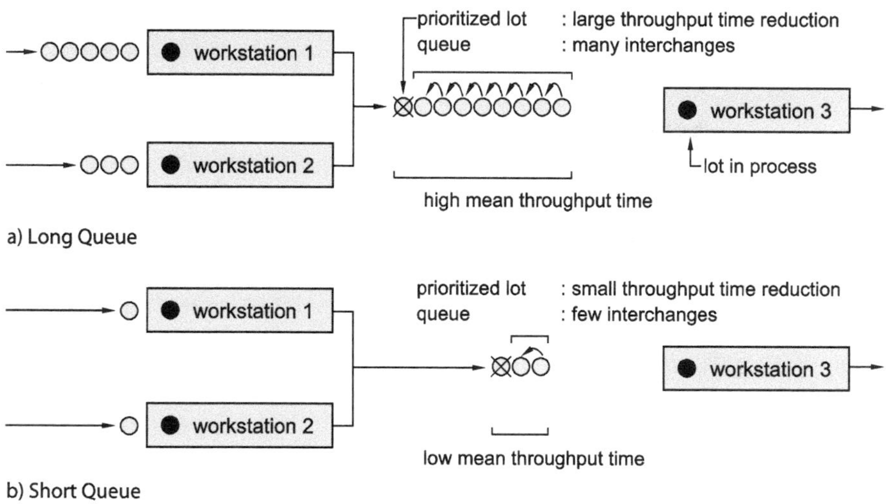

Fig. 4.14 Influence of Order Sequence Interchanges on Differently Sized Queues (Ulfers, Siemens AG)

waiting time for a queue which is conducted according to the FIFO principle can be calculated as:

$$TW_{FIFO}(t) = TTP_m(t) - TOP_m - TTR_m \tag{4.42}$$

where $TW_{FIFO}(t)$ mean waiting time with the FIFO sequencing rule [SCD],
$TTP_m(t)$ mean throughput time [SCD],
TOP_m mean operation time [SCD],
TTR_m mean transport time [SCD].

When WIP levels are low the waiting time is approaching zero, sequencing rules therefore have no effect. With increasing WIP levels however, there are growing opportunities to prioritize specific orders and to therefore influence the waiting time. The greater the work content's variance is, the greater the impact is. Analyses of specific simulation trials have shown that when LPT or SPT sequencing is drawn upon, the waiting time for individual workstations can be described with close approximation using Eq. 4.43 or 4.44:

$$TW_{LPT}(t) \approx TW_{FIFO}(t) \cdot \left(1 + TOP_v^2\right) \tag{4.43}$$

$$TW_{SPT}(t) \approx TW_{FIFO}(t) \cdot \frac{1}{\left(1 + TOP_v^2\right)} \tag{4.44}$$

where $TW_{LPT}(t)$ mean waiting time with the LPT sequencing rule [SCD],
$TW_{SPT}(t)$ mean waiting time with the SPT sequencing rule [SCD],
$TW_{FIFO}(t)$ mean waiting time with the FIFO sequencing rule [SCD],
TOP_v coefficient of variation for the throughput time [-].

Thus, for the throughput time

$$TTP_{LPT}(t) \approx TW_{FIFO}(t) \cdot \left(1 + TOP_v^2\right) + TOP_m + TTR_m \tag{4.45}$$

and

$$TTP_{SPT}(t) \approx TW_{FIFO}(t) \cdot \frac{1}{\left(1 + TOP_v^2\right)} + TOP_m + TTR_m \tag{4.46}$$

where TTP_{LPT} mean throughput time with the LPT sequencing rule [SCD],
TTP_{SPT} mean throughput time with the SPT sequencing rule [SCD],
$TW_{FIFO}(t)$ mean waiting time with the FIFO sequencing rule [SCD],
TOP_v coefficient of variation for the operation time [-],
TOP_m mean operation time [SCD],
TTR_m mean transport time [SCD].

The resulting Logistic Operating Curves are presented in Fig. 4.15. For the Throughput Time Operating Curve we can summarize the following basic characteristics:

- The range is expressed solely through the WIP to output rate ratio and is thus independent from the orders' sequencing rules.
- The (unweighted) throughput time is always smaller than the range when a sequencing rule which is independent of the work content is applied and the work content is non-uniform. The difference of the two Logistic Operating Curves is determined by the extent of the work content's variance.
- The lower the WIP level, the less influence sequencing rules have on the throughput time. Above the ideal minimum WIP however, the slope of the TTOC is influenced by the sequencing rules which are applied. The impact is then greater, the greater the variance of the work content is.

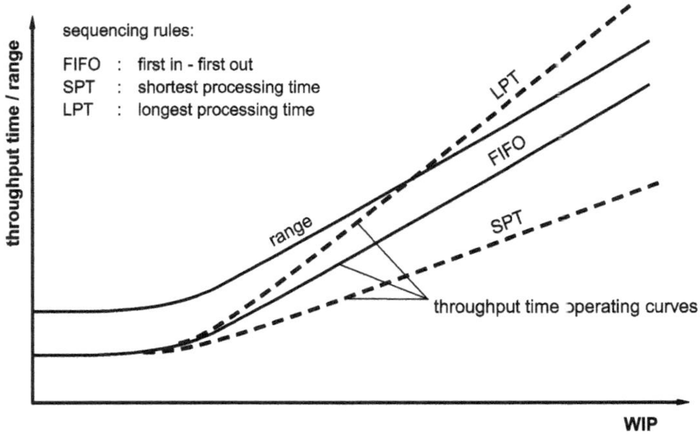

Fig. 4.15 Throughput Time Operating Curves for Different Sequencing Rules

Finally, it should also be noted that a relaxed operating state is required in order to calculate Throughput Time Operating Curves. In particular, the mean range must correspond with the mean weighted throughput time i. e., the WIP cannot have a tendency to change during the investigation period. Furthermore it should be pointed out that Eqs. 4.43 to 4.46 only apply to individual workstations. In workshops where the workstations can substitute each other, sequencing rules have to some extent a much stronger effect on the throughput time.

4.5 Normalized Logistic Operating Curves

For a number of problems it is helpful to normalize reference parameters in order to be able to draw conclusions that are independent of the system specific conditions. In order to do so it is necessary to determine appropriate reference values for such normalizations. It seems obvious that for the output rate and WIP level they can be based on the ideal operating state and thus expressed as a relative parameter. In our discussion about Output Rate Operating Curves we already conducted a similar type of normalization, using the definition of the mean WIP dependent utilization U_m (as a ratio of $ROUT_m$ to $ROUT_{max}$, see Sect. 4.3). In order to describe a *relative WIP level*, the mean WIP is set in relation to $WIPI_{min}$. Thus:

$$WIP_{rel}(t) = \frac{WIP_m(t)}{WIPI_{min}} \cdot 100 \tag{4.47}$$

where $WIP_{rel}(t)$ mean relative WIP level [%],
$WIP_m(t)$ mean WIP level [hrs],
$WIPI_{min}$ ideal minimum WIP [hrs].

If, for the purpose of normalizing, the ideal minimum WIP and the maximum possible utilization are each set at 100%, then it follows that for the *normalized Output Rate Operating Curve*:

$$\left.\begin{aligned} WIP_{rel}(t) &= 100 \cdot \left(1 - \left(1 - \sqrt[4]{t}\right)^4\right) + 100 \cdot \alpha_1 \cdot t \\ U_m(t) &= 100 \cdot \left(1 - \left(1 - \sqrt[4]{t}\right)^4\right) \end{aligned}\right\} \tag{4.48}$$

where $WIP_{rel}(t)$ mean relative WIP level [%],
$U_m(t)$ mean WIP dependent utilization [%],
α_1 stretch factor [-],
t running variable ($0 \le t \le 1$).

The Utilization Operating Curve (UOC) calculated using this equation is depicted in Fig. 4.16, where $\alpha 1 = 10$. The graph describes how a change in the WIP impacts the utilization of the workstation, independent of the existing work content structures and the workstation's capacity. It shows, for example that the WIP

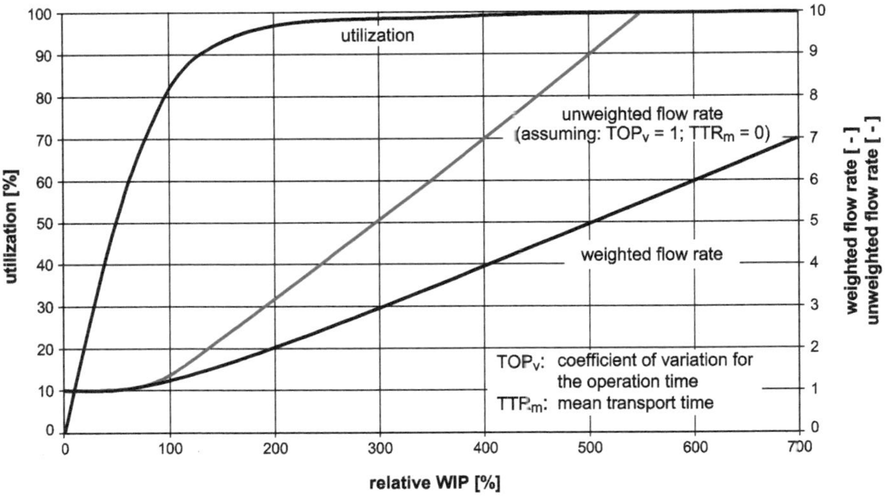

Fig. 4.16 Normalized Logistic Operating Curves

dependent loss of utilization is approximately 17% when the mean WIP corresponds to WIP_{min}. If the WIP is doubled the loss of utilization is reduced to approximately 3% and if tripled to approx. 1%.

A relative measure of the throughput time is the *flow rate*. In the following, we will differentiate between the weighted and unweighted flow rate. If, in analogy to the Funnel Formula the relative WIP is set in relation to the utilization we obtain the mean weighted flow rate RF_{mw} as a normalized parameter for the range. This can also be calculated through the ratio of the range to minimum range:

$$RF_{mw}(t) = \frac{WIP_{rel}(t)}{U_m(t)} = \frac{R_m(t)}{R_{min}} \tag{4.49}$$

where $RF_{mw}(t)$ mean weighted flow rate [-],
$WIP_{rel}(t)$ mean relative WIP [%],
$U_m(t)$ mean WIP dependent utilization [%],
$R_m(t)$ mean range [SCD],
R_{min} minimum range [SCD].

The *mean unweighted flow rate* RF_m results from the ratio of the mean unweighted throughput time to minimum throughput time (see Eq. 2.11):

$$RF_m(t) = \frac{TTP_m(t)}{TTP_{min}} \tag{4.50}$$

where $RF_m(t)$ mean unweighted flow rate [-],
$TTP_m(t)$ mean throughput time [SCD],
TTP_{min} minimum throughput time [SCD].

With a few conversions the equations for the weighted flow rate $RF_{mw}(t)$ and the unweighted flow $RF_m(t)$ can also be expressed as:

$$RF_{mw}(t) = 1 + \frac{100 \cdot \alpha_1 \cdot t}{U_m(t)} \tag{4.51}$$

$$RF_m(t) = 1 + \frac{100 \cdot \alpha_1 \cdot t}{U_m(t)} \cdot \left(1 + \frac{TOP_m \cdot TOP_v^2}{TOP_m + TTR_m}\right) \tag{4.52}$$

or when the transportation time is negligible,

$$RF_m(t) = 1 + \frac{100 \cdot \alpha_1 \cdot t}{U_m(t)} \cdot \left(1 + TOP_v^2\right) \tag{4.53}$$

where $U_m(t)$ mean WIP dependent utilization [%],
α_1 stretch factor [-],
TOP_m mean operation time [SCD],
TOP_v coefficient of variation for the operation time [-],
TTR_m mean transport time [SCD],
t running variable ($0 \le t \le 1$).

Finally, the relationship between the unweighted and weighted flow rate is expressed as

$$RF_m(t) = RF_{mw}(t) \cdot \left(1 + TOP_v^2\right) - TOP_v^2 \tag{4.54}$$

where $RF_{mw}(t)$ mean weighted flow rate [-],
$RF_m(t)$ mean unweighted flow rate [-],
TOP_v coefficient of variation for the operation time [-].

In Fig. 4.16, in addition to the Utilization Operating Curve the Logistic Operating Curves for the weighted and unweighted flow rate are also plotted. In the latter case it is assumed that the transport time is insignificant and that the coefficient of variation for the operation time is 1. As can be seen with the help of the preceding equations the more inhomogeneous the orders' work content is, the steeper the Logistic Operating Curve for the unweighted flow rate is. The above graph shows that already when the coefficient of variation is equal to 1 (which was used here and which is frequently found in the field), an unweighted flow rate of >5 is required in order to be able to ensure that the workstation is fully utilized.

The fundamental differences between the workstation specific (individual) and normalized Logistic Operating Curves are comparatively summarized above in Fig. 4.17 [Ludw-94]. In particular, it needs to be emphasized that due to their independence from individual output rate limits and work content structures, the normalized LOC support a strategic Logistic Positioning as well as a consistent parameter adaptation during the basic functions of production planning and control. However, should the absolute performance figures for the logistic objectives be required for setting a specific task, the workstation based Logistic Operating Curves have an advantage.

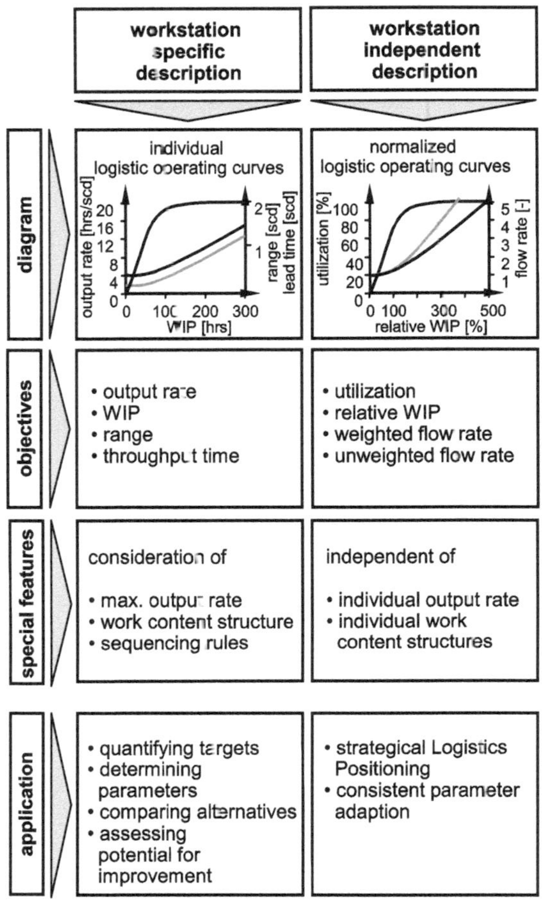

Fig. 4.17 Logistic Operating Curves as an Explanation Model for the Basic Laws of Production Logistics (Ludwig)

4.6 Logistic Operating Curves Theory and Little's Law – a Model Synthesis

One of the prerequisites which permits the Logistic Operating Curves Theory to be derived as discussed above, is that the WIP be given in standard hours. This is particularly useful when the workstation's load and utilization are being observed, because the capacity and output rate are generally also given in hours per unit of time. A higher WIP level is however only an expression of a large amount of available work and does not inevitably mean a large number of orders and long queues. Thus the level of the WIP on a workstation – given in standard hours – is not only dependent on the number of orders, but also on their work content.

In some cases, it is preferable to also be able to express the WIP in the number of orders. Directly converting between the two WIP specifications through the mean work content is nevertheless, not possible (see Sect. 4.4).

Here, it is possible to combine the LOC Theory and Little's Law in that the values for the throughput time and output rate, calculated through the LOC Theory, can be substituted in Eq. 2.22:

$$WIPO_m(t) = \frac{TTP_m(t) \cdot ROUT_m(t)}{WC_m} \tag{4.55}$$

where $WIPO_m(t)$ mean WIP (in number of orders) [-],
$TTP_m(t)$ mean throughput time [SCD],
$ROUT_m$ mean output rate [hrs/SCD],
WC_m mean work content [hrs].

It is now possible to determine the number of orders in the WIP, as a function of the Logistic Operating Curves equation's running variable t and to present it in combination with one of the previously calculated performance figures. As an example the WIP in mean number of orders is plotted over the relative WIP for different throughput time variation coefficients in Fig. 4.18. The graph underscores the significance of the work content structure's inhomogeneity for the logistic potential of a production process. Thus, with a relative WIP of 200% and a homogenous operation time ($TOP_v = 0$) there is an average of two orders on the system, whereas with $TOP_v = 1$ the WIP at this operating point already consists of an average of three orders and with $TOP_v = 2$, approx. six orders. Here, large orders increasingly block the workstation and a queue results due to new incoming orders, without the certainty of the utilization being increased.

By substituting Eqs. 2.5, 2.13 and 4.40 in 4.55 both WIP values can be converted into one another:

$$WIPO_m(t) = \frac{WIP_m(t)}{WC_m} - \frac{U_m(t) \cdot WC_v^2}{100} \tag{4.56}$$

$$WIP_m(t) = \left(WIPO_m(t) + \frac{U_m(t) \cdot WC_v^2}{100} \right) \cdot WC_m \tag{4.57}$$

where $WIPO_m(t)$ mean WIP in number of orders [-],
$WIP_m(t)$ mean WIP in standard hours [hrs],
$U_m(t)$ mean WIP dependent utilization [%],
WC_m mean work content [hrs],
WC_v coefficient of variation for the work content [-].

Since Eq. 4.40 is only applicable when the sequencing rules are independent of the work content, the range of validity for Eqs. 4.56 and 4.57 is also limited accordingly.

With the help of the above equation, it is also possible to establish Logistic Operating Curves based on unweighted parameters. In Fig. 4.19 a curve of this type is presented for $TOP_v = 1$. It can be seen here that ideal LOC can also be described based on the definition of the ideal processing state even for unweighted parameters (see Sect. 4.1). Thus, the $WIPI_{min}$ for an individual workstation is one order and the minimal flow rate one. How the calculated LOC deviates from the ideal

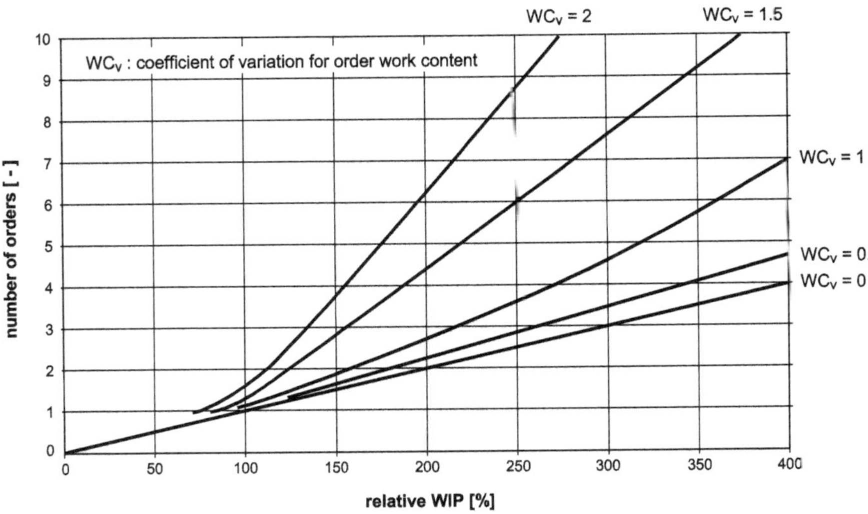

Fig. 4.18 Number of Orders as a Function of the Relative WIP (Work Content Independent Sequencing Rules)

LOC though, no longer depends only on the parameter α_1, but instead also depends on the coefficient of variation for the throughput time and the sequencing rules.

At this point, it is interesting to compare these results with those from queuing theory. A similar graph was developed using the M/G/1 Model in Sect. 3.1.2

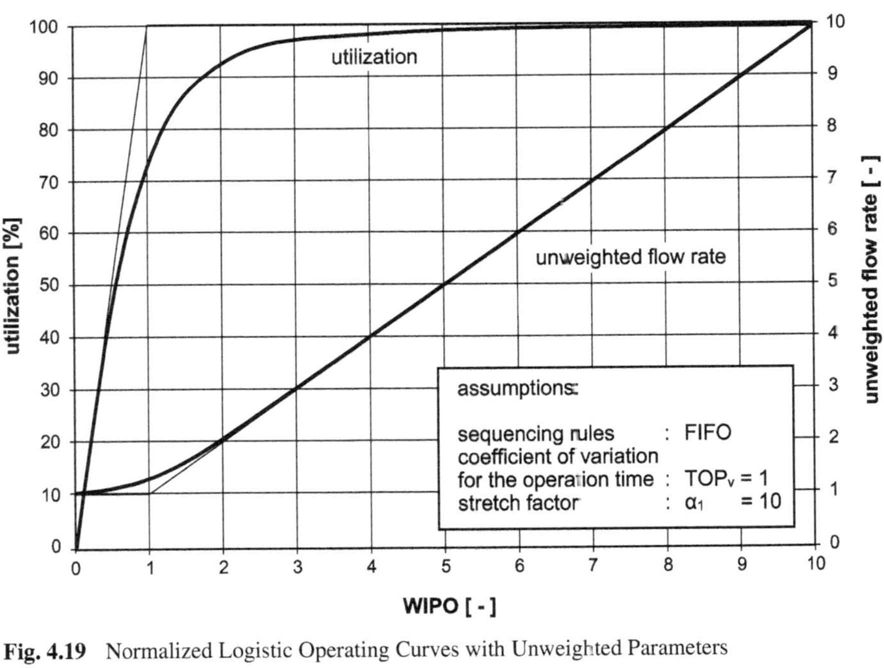

Fig. 4.19 Normalized Logistic Operating Curves with Unweighted Parameters

(Fig. 3.3). According to it, the underlying queuing model says that even with a relative homogenous work content ($TOP_v = 0.5$) the loss of utilization caused by the WIP is approximately 15% when there is an average of four orders on the workstation. On the other hand, the LOC Theory states that with an average of four orders only a very minimal loss of utilization is to be expected – even when there is greater variance in the work content ($TOP_v = 1.0$) as was assumed when creating Fig. 4.18.

4.7 Verifying the Logistic Operating Curves Theory

Once a theory for calculating Logistic Operating Curves is established, the next step is to validate the model. Ideally, a workstation's key performance figures for various operating states should be compared with the calculated LOC. However it has to be a prerequisite that the basic conditions such as the structure of the work content and capacity are identical and that only the WIP differs in the individual operating states. Circumstances such as these can generally only be generated with the help of simulation experiments. The LOC Theory is thus first verified by comparing the calculated Logistic Operating Curves with the results of various simulation applications (Sect. 4.7.1).

The goal in deriving the Logistic Operating Curves Theory was to replace simulations for chosen problems in practical applications. At least, in the model utilization there are no simulation results available for validating the model. Therefore, in Sect. 4.7.2. we have also included a practical example of how a model can be validated based on analysis of the production process.

4.7.1 Simulation Based Model Validation

The simulation based model validation is based upon simulation trials that were conducted using the PROSIM III simulation system described in Sect. 3.2. The manufacturing of a parts spectrum from a company involved in the automobile industry (organized according to the job shop principle) for example, was emulated on a computer and subsequently operated under specifically changed conditions. The development of both the simulation and the individual model components (see Sect. 3.2) as well as the ensuing model test were oriented on a previously conducted, comprehensive analysis of the throughput time and WIP. A detailed description of the production, which consisted of 55 workstations, and the model tests can be found in [Nyhu-91].

During the simulation experiments, the modeled workshop (the resource model) and the production schedule that was to be processed within a simulated nine month time frame remained unchanged. By applying different methods for determining the lot size, the production order data and thus in particular the work content structure of the orders to be processed was varied. Afterwards, a series of simulations were run for each of the generated order data during which the level of WIP was changed using the Load Oriented Order Release algorithm [Wien-95a*].

As an example, the results for a single workstation are reproduced in Fig. 4.20. The simulated operating points as well as the LOC that were calculated for the output rate, range and throughput time using Eqs. 4.29, 4.37 and 4.40 are compared for three different sets of orders. These samples were generated using different methods of lot size determination (Wagner Whitin Method, Throughput Oriented Lot Size Determination and the firm's original order file). Above all, the data differ from one another in their composition and in particular with regards to the performance figures for the work content structure (mean and variance of the work content).

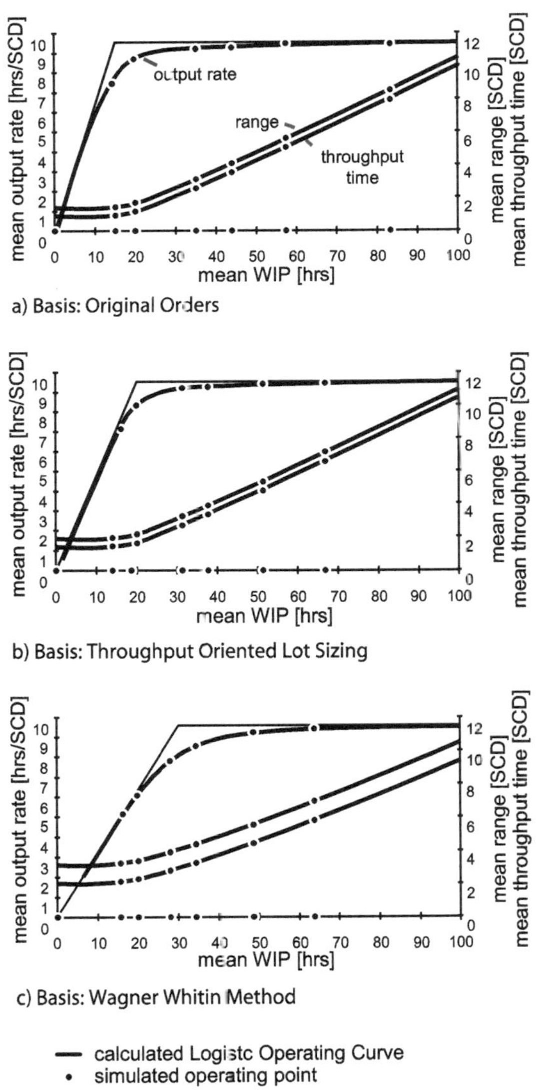

Fig. 4.20 Simulation Based Validation of Calculated Logistic Operating Curves (Workstation 511033; Slack Sequencing Rule)

In order to calculate the LOC, the stretch factor α_1 was set at 10 for each case. In the simulation, the orders were completed according to the urgency of the due-date (slack sequencing rule) and were thus processed independent of the work content. The Throughput Time Operating Curve could therefore be calculated using Eq. 4.40.

It should be noted that in all three cases there was a good match between the simulation results and the calculated operating curves (OROC, ROC and TTOC). This could also be confirmed with the other workstations. Extensively analyzing the deviation [Nyhu-91] showed that the mean deviation was less than 2%. Finally, for further comparison, the simulations conducted by Bechte [Bech-84] and Erdlenbruch [Erdl-84] regarding model validation were also drawn upon. Also here it was possible to determine that there was a strong conformance between the simulations and the calculations. It can therefore be concluded that despite what at first seems quite a strong simplification, the mathematical model successfully describes the logistical system behavior of the simulated production with a sufficient accuracy even without simulation.

Nevertheless, the underlying simulation trials point out a few significant characteristics, which need to be taken into consideration when interpreting and generalizing the results. In particular it can be determined that the simulation trials were preceded by a model test. In the model test the workstations' capacities were set to correspond with the output rate reported back in the real process (see also Sect. 3.2.3). Since on the shop floor, the capacities are adjusted at least to the demanded capacities over the medium term, this is also the case in the simulation model in which the capacities are emulated corresponding to the output rate feedback.

Furthermore, due to the use of the Load Oriented Order Release algorithm the WIP on the workstation is actively controlled. As a result, the production processes in the individual simulation runs are steady and have an almost constant level of WIP.

It is now conceivable that a strong deviation from these conditions would influence the system's logistic behavior. Thus, the simulation trials presented in Sect. 3.2.3 demonstrate that even with a generally high WIP level, a greater loss of utilization is to be expected when the order input is greatly varied and the capacities are fixed. In contrast, with flexible capacities it is possible to directly inhibit the WIP from increasing by increasing the capacities and in turn when the load is reduced the capacities can be quickly decreased. Flexible capacities therefore permit the necessary WIP buffer to be lowered accordingly.

Both of the states outlined here, can also be expressed with the Logistic Operating Curves Theory. However in order to do so, the stretch factor α_1 has to be adjusted in order to take into account the changed conditions. Additional simulation experiments were therefore conducted in which the distribution of the load and/or the capacity's flexibility were changed. The results of these experiments, which were also confirmed in similar studies from Burmeister [Burm-97], are summarized in Fig. 4.21. The graph shows that the stretch factor α_1 is always equal to 10 when in reaction to a stronger mid to long term load fluctuation the capacity is adjusted (essentially through an increase or decrease in operator capacity; high capacity flexibility) or when the load is balanced by implementing corresponding PPC methods (low load variability). In other words $\alpha_1 = 10$, when there is a *real* correlation between the workstation's input and output (see Sect. 3.1.3). If in contrast – as is the case in many simulation applications – the capacities are fixed (no capacity flexibility), then greater variance in the load results in the need to increase the value of α_1 in order for the calculated LOC to coincide with the simulation results.

Incidentally, the maximum value plotted for α_1 ($\alpha_1 = 120$) in Fig. 4.21 was determined not only by analyzing the simulation results, but also by comparing the LOC Theory with the M/G/1 queuing model. Among other things, the M/G/1 model is characterized by rigid (both load and WIP independent) capacities and exponentially distributed interarrival times (see Sect. 3.1.1). It is thus assumed that there could be a very long time span between two consecutive input events on a workstation and that there is great variance in the load. In order to now be able to reproduce one of the LOC generated by the M/G/1 model (see Fig. 3.3) with the Logistic Operating Curves Theory the abovementioned α_1 value is required.

The value range $30 < \alpha_1 < 120$ could only be simulatively established in cases where individual workstations were depicted. When simulating larger production areas, no α_1 value > 30 was found even when there was greater variance in the order input and despite the rigid capacities (except for those workstations that were purely initial ones). The reason for this is that the load fluctuations on the individual workstations are limited for different reasons. First of all, the input on one workstation results from the output of the previous station. Due to the limited capacities, the load fluctuations are inevitably smoothed out for the following workstations. Furthermore, the workstations within the material flow are often supplied with orders from a number of upstream workstations which allows the load fluctuations to reciprocally compensate for one another. Even with a directed

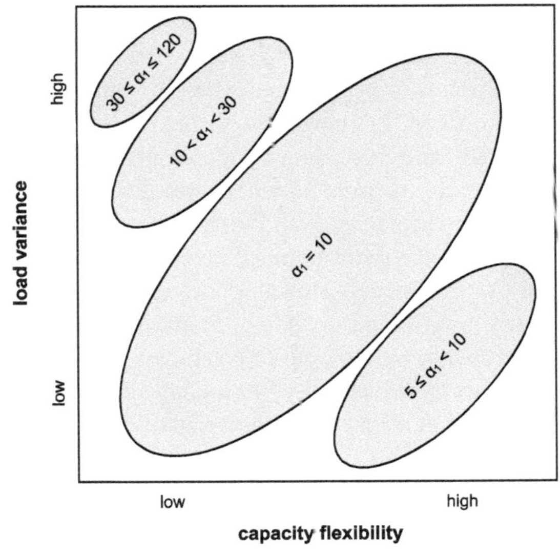

Fig. 4.21 Influence of the Load Variance and Capacity Flexibility on the Stretch Factor α_1

material flow the load fluctuations caused by the order input or disruptions in processing for example, are generally strongly reduced because the individual system capacities are better aligned to one another than in a networked job shop.

In cases where the capacities are extremely flexible it is possible to very closely approximate the ideal production process, which is expressed in the LOC Theory by a low α_1 value. Burmeister's studies [Burm-97] showed that with a low load variability and highly flexible capacities (in his experiments the capacities could be quickly changed by $\pm 50\%$) the α_1 value could be reduced to approximately 5.

The principle validity of the LOC Theory has also been confirmed in the research of other authors as well. The studies that were conducted under totally different conditions than those which apply to job shop productions should be emphasized here in particular. Thus, a simulation of an automatic assembly conducted by Winkelhake [Wink-88] demonstrated that the performance figures for the WIP and output rate also behave similarly in linked production processes and that they can be described in the form of Logistic Operating Curves. The stochastic influences in such facilities are not due to the work content's variability or the resulting order input variability, but instead are primarily created by processing disruptions. The effects nevertheless are similar: In order to prevent a loss of utilization, WIP buffers need to be maintained at the plants.

In another study using a dynamic workstation model – emulated and operated by a continuous simulation system – Pritschow [Prit-95*] analyzed the input/output behavior of a workstation with non-uniform work content and variability in the order input. In Fig. 4.22, the model it was based on and the results generated by this trial are represented in comparison to the LOC Theory; the agreement between the two is obvious.

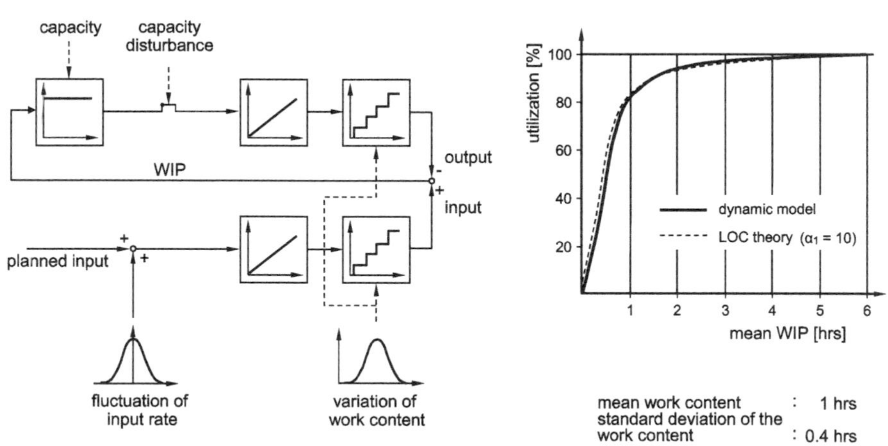

a) Dynamic Model   b) Comparison of Results

Fig. 4.22 Comparing the Results from a Dynamic Workstation Model with those of the Logistic Operating Curves Theory (ISW, Stuttgart)

4.7.2 Validating the Model Based on Field Analyses

Since the goal of the Logistic Operating Curves Theory is to replace the simulation with a mathematical model for practical applications, simulations are not always available for validating the model. Furthermore, simulation based model validation also needs to be critically discussed: Ultimately it is not the accuracy of a model that is verified by such procedures, but rather the extent to which the results of one model (here, the LOC Theory) correspond to the results of another (in this case simulation). Thus, if a simulation model has an unidentified failure, it is automatically transferred to the LOC method when adjusting the parameters according to the simulation results. Moreover, the primary goal is to implement the Logistic Operating Curves in the field, therefore the final validation has to be the shop floor itself.

It is helpful here to draw upon another technique for validating models: In process analyses, it is possible to evaluate individually acquired operating states in detail and to compare the results based on this analysis with those determined through the LOC (Fig. 4.23). Should there be a strong agreement, it can generally be assumed that the production process can be sufficiently described with the calculated LOC. One of the advantages of the following outlined procedure for validating models is that it can be conducted based directly on the processing data from the shop floor and thus supports a continuous operational and system specific verification. The model only needs to be adapted (when necessary by using simu-

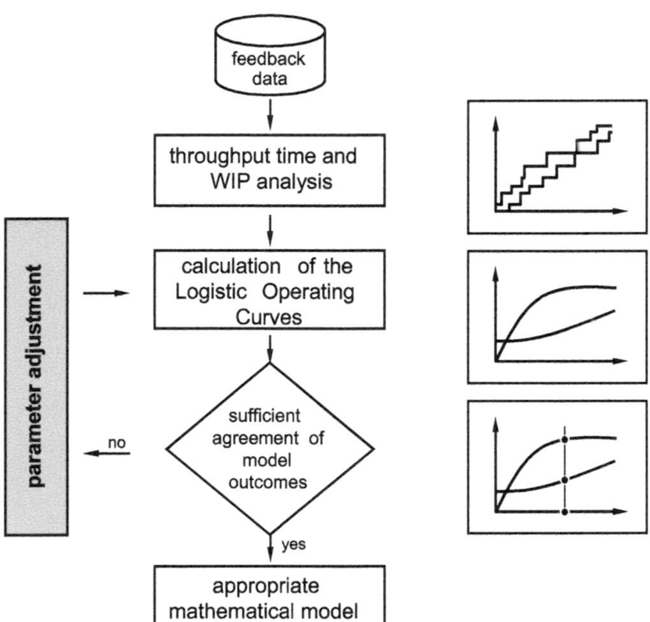

Fig. 4.23 Procedure for a Production Data Based Model Validation

lation) if it is determined that there are sustained, significant deviations between the model and the production floor.

In order to validate the model, workstations which can be said to typically represent operating states in the *underload*, *overload*, and *transitional zones* need to be chosen from the process analysis. The following examples are from a study at a medical equipment manufacturer. Feedback data was collected at the company's mechanical production plant (24 workstations) over a period of 18 weeks and evaluated with the help of the performance figures and methods described in Chap. 2.

4.7.2.1 Underload Operating Zone

As an example for this operating zone we chose Workstation 1034 (copy turning). The measured production process from this system is reproduced as a Throughput Diagram in the upper part of Fig. 4.24. In addition to the input, output and WIP behavior, the throughput elements of the individually executed operations as well

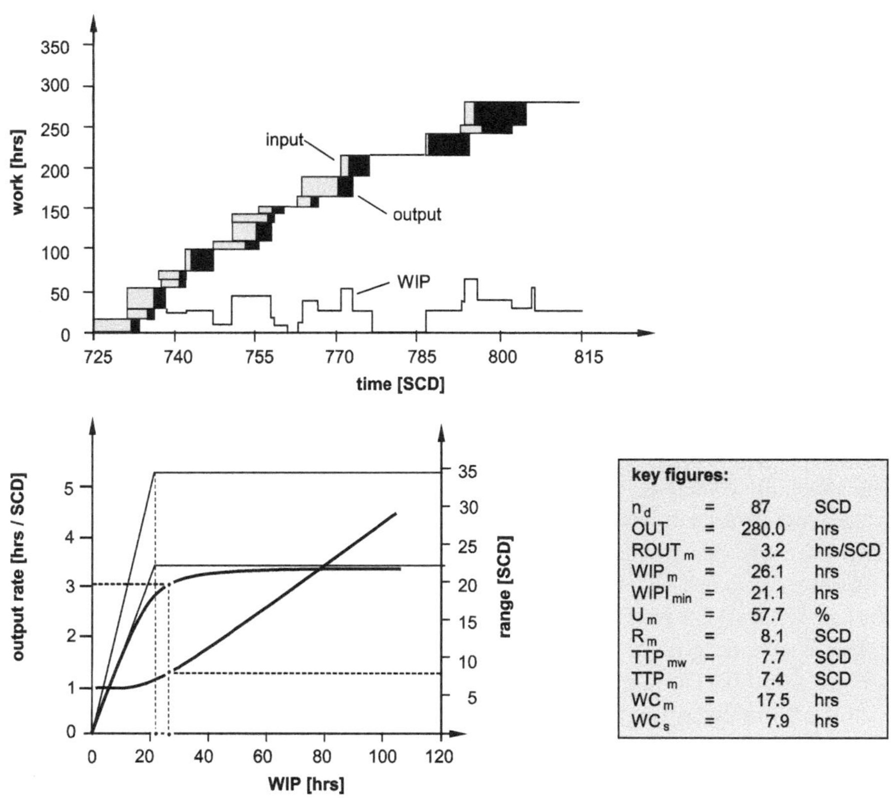

Fig. 4.24 Validating the Logistic Operating Curves Theory based on Actual Data (Underload Operating Zone, Workstation 1034 – Copy Turning)

as the corresponding operation elements (calculated based on the planned capacities) are also plotted. A selection of the calculated performance figures are provided in the lower portion of the figure, next to the calculated LOC.

Based on the Throughput Diagram, the following statements can initially be made:

- There were basically no order sequence interchanges made on this station during the evaluation period.
- During the first 10 weeks there was WIP on the workstation almost continuously. After this period, the rate of utilization calculated based on the planned capacity is very low ($U_m = 57.7\%$). It can thus be concluded that the planned capacities were not available, otherwise the WIP could not have occurred as it has here.
- On a total of 10 (from 87) work days, there was no WIP on the workstation. This corresponds to just under 11.5% of the investigation period. Therefore, there was an 11.5% WIP dependent loss of utilization.

The Logistic Operating Curves were then calculated using the determined performance figures. Since developing LOC is extensively discussed in Chap. 6 (also for when the data provided for the capacities is erroneous), a detailed description of the calculation process will not be given at this point. Instead, only the relevant key data and results will be interpreted here.

In determining the ideal minimum work in progress $WIPI_{min}$ it is first assumed that the transport time is negligible. The $WIPI_{min}$ can therefore be calculated through the mean work content ($WC_m = 17.5$ hrs) and its standard deviation ($WC_s = 7.9$ hrs). Thus, according to Eq. 4.2 the $WIPI_{min}$ is 21.1 hrs.

The previous explanation of the Throughput Diagram indicates that the given planned capacities cannot be drawn upon in order to determine the maximum possible output rate $ROUT_{max}$. We will thus refer to the normalized Logistic Operating Curves (Fig. 4.16) which allow us to draw a conclusion about the calculable utilization with the given relative WIP. If the measured mean WIP of 26.1 hours is set in relation to the $WIPI_{min}$ (Eq. 4.47), a relative WIP of approximately 123% results. Given the assumed stretch factor $\alpha_1 = 10$, the normalized Output Rate Operating Curve states that the loss of utilization is approximately 10.7%. Using this value, the $ROUT_{max}$ and the Logistic Operating Curves for the output rate and range were then calculated. These are represented in Fig. 4.24 alongside two ideal OROC (one based on the planned capacity, the other using $ROUT_{max}$).

Comparing the LOC with the results of the process analysis shows that they strongly correspond with each other. It is also clear in the calculated LOC that the station is operating in the underload zone. The WIP dependent losses of utilization derived from the Throughput Diagram are negligibly greater than those calculated with the normalized Logistic Operating Curves. However, due to the fact that the WIP is calculated using feedback data provided only on a daily basis, there is a level of uncertainty about the measures. These deviations should therefore be considered insignificant.

Furthermore, the diagram clearly indicates that the planned capacities could not have been fully available. Due to the low load the operator may have been employed at other workstations and thus only able to work sporadically at the observed station when needed.

4.7.2.2 Transitional Operating Zone

Workstation 1043 (CNC lathe) serves as an example of a workstation operating in the transitional zone. It can be seen in the workstation's corresponding Throughput Diagram (Fig. 4.25) that there was a low and almost steady level of WIP during the evaluation period. Only on 3 workdays, in the transition between intervals 15 and 16, was there no WIP. The WIP dependent loss of utilization was therefore approximately 3.5%. This conclusion is generally found again in the calculated LOC: The measured operating state is in the transitional zone of the Output Rate Operating Curve. Therefore in view of the WIP levels only negligible output losses are to be expected. Whereas, reducing the WIP further would lead directly

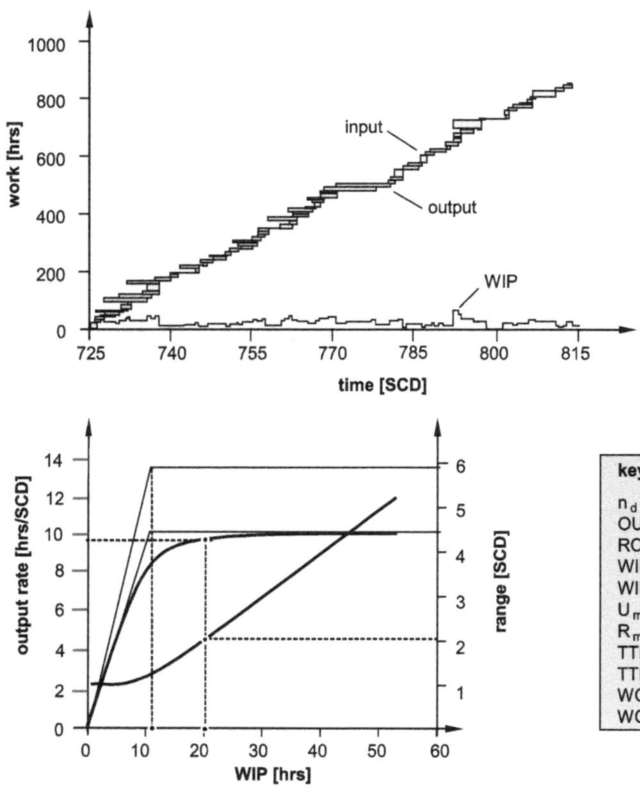

key figures:			
n_d	=	87	SCD
OUT	=	848.0	hrs
$ROUT_m$	=	9.7	hrs/SCD
WIP_m	=	20.1	hrs
$WIPI_{min}$	=	10.6	hrs
U_m	=	72.3	%
R_m	=	2.1	SCD
TTP_{mw}	=	2.2	SCD
TTP_m	=	2.0	SCD
WC_m	=	7.5	hrs
WC_s	=	4.8	hrs

Fig. 4.25 Validating the Logistic Operating Curves Theory Based on Real Data (Transitional Operating Zone, Workstation 1043 – CNC Lathe)

to disruptions in the material flow, increasing it would have no significant influence on improving the output rate as long as the capacity requirements and therefore the load, are not simultaneously increased through additional orders.

4.7.2.3 Overload Operating Zone

Workstation 1032 (plastic component lathe) is clearly in the overload zone of the LOC. The Throughput Diagram for this workstation is presented in Fig. 4.26. It illustrates that during the evaluation period there were strong WIP fluctuations i. e., the input and output were poorly coordinated.

The mean range R_m was 13 SCD, whereas the mean throughput time TTP_m was 11.5 SCD. There were frequent sequence order interchanges which were however, obviously not influenced by the orders' work content (nothing was processed according to the SPT or LPT principles). The high mean WIP level ($WIP_m = 73.1$ hrs with an average of approximately 10 orders) indicates that the WIP can be significantly reduced on the workstation.

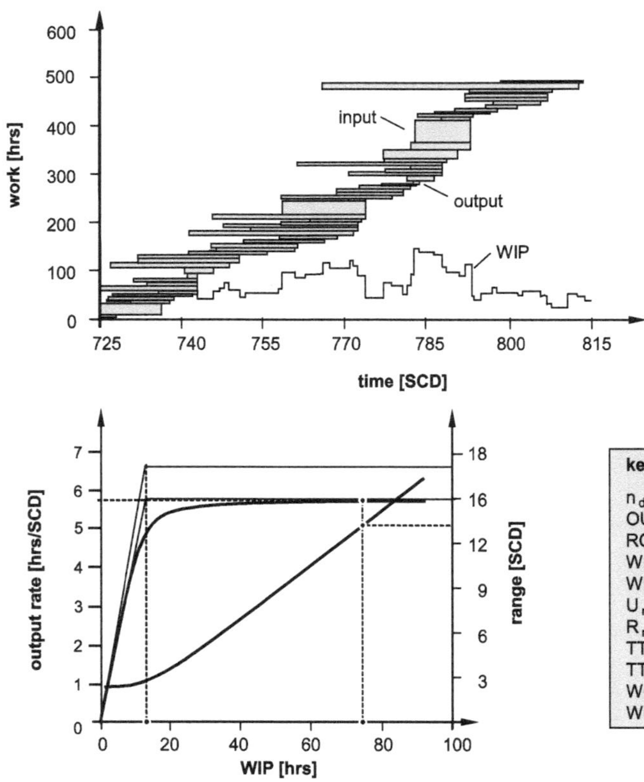

key figures:			
n_d	=	87	SCD
OUT	=	491.0	hrs
$ROUT_m$	=	5.6	hrs/SCD
WIP_m	=	73.1	hrs
$WIPI_{min}$	=	12.5	hrs
U_m	=	87.4	%
R_m	=	13.0	SCD
TTP_{mw}	=	12.8	SCD
TTP_m	=	11.5	SCD
WC_m	=	6.4	hrs
WC_s	=	6.2	hrs

Fig. 4.26 Validating the Logistic Operating Curves Theory Based on Real Data (Overload Operating Zone, Workstation 1032 – Plastic Component Lathe)

The calculated Logistic Operating Curves and the plotted operating point confirm that the workstation is clearly in the overload zone. Other information derived through the LOC Theory also apply. Thus with Eq. 4.40 it can be derived from the difference of the range and throughput time, that work content independent sequence rule must have been used most of the time.

These three examples confirm that this mathematical model can describe the production process with sufficient accuracy. In particular, the underload and transition operating zones make it possible to validate the LOC Theory and especially the stretch factor α_1 by comparing the Throughput Diagram and calculated LOC. In the field examples analyzed here, the WIP dependent loss of utilization that can be read from the Throughput Diagram, could be calculated with a good approximation using the base value $\alpha_1 = 10$.

This method of validating the model can however, only be conducted with individual workstations. When a Throughput Diagram is established for a capacity group, it can generally no longer be recognized if and to what extent 'zero stock phases' exist on the individual workstations. The comparisons based on that can therefore also not be conducted.

4.8 Extending the Logistic Operating Curves Theory

The Logistic Operating Curves Theory is derived based on a number of conditions which at the same time limit the model's range of validity. However, as we will show in the following, the prerequisites for the model that were described in Sect. 4.1 can be suspended for the most part, without having to fundamentally change the model.

4.8.1 Hierarchically Aggregating Logistic Operating Curves

The Logistic Operating Curves Theory is developed from the idealized description of the processing flow on a workstation. However, for many problems it is beneficial to be able to concisely describe the logistic potential of more complex production systems not only through a multitude of single analyzes, but also for a super-ordinate system. As will be shown below this is also possible using the LOC Theory.

The Funnel Model and the Throughput Diagram form the starting point for the LOC Theory and are also usually not only applied to individual workstations. The basic idea is to instead, understand the workstations that participate in the order processing as a network of interconnected funnels and to model them. As an example, Fig. 4.27 illustrates the funnel model for a workshop. The illustrated material flow between the workstations is derived from the production orders that need to be processed and their work-flow. This flow oriented mapping documents the

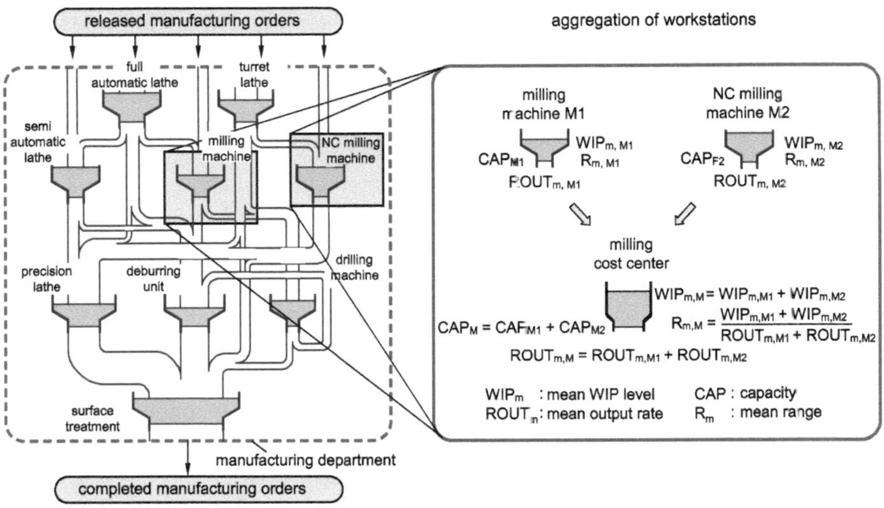

Fig. 4.27 Hierarchical Aggregation of Workstations in the Funnel Model

integration of the individual workstations in the operational manufacturing flow. Furthermore, it is also possible to summarize the individual workstations corresponding to the organizational structure and to thus determine performance figures and flow charts for basically any level of aggregation.

The right side of Fig. 4.27 depicts an example of the aggregated performance figures based on a milling cost center. It can be seen that the performance figures described in Chap. 2 deal partly with parameters that only need to be added in order to determine the corresponding performance figures for the entire system. (It is of course also still possible to calculate this with the feedback data according to the formula described in Chap. 2.) Thus, the mean output rate for the cost center results from the sum of the individual workstation's output values. This also applies to the capacity as well as the WIP. However, this is not the case for other parameters. In order to calculate the mean range, the Funnel Formula still has to be applied, in this case with the sums of the WIP and output rate. The range thus calculated can be understood as the mean of an imaginary average workstation. This also applies in a similar form for all of the other throughput time and schedule deviation figures.

The hierarchical aggregation can now be transferred to the Logistic Operating Curves. In a super-ordinate system the ideal minimum WIP and the maximum possible output rate are performance figures which result from the sum of the individual values:

$$WIPI_{min} = \sum_{j=1}^{w} WIPI_{min,j} \tag{4.58}$$

$$ROUT_{max} = \sum_{j=1}^{w} ROUT_{max,j} \tag{4.59}$$

where $WIPI_{min}$ ideal minimum WIP [hrs],
$WIPI_{min,j}$ ideal minimum WIP on workstation j [hrs],
$ROUT_{max}$ maximum possible output rate [hrs/SCD],
$ROUT_{max,j}$ maximum possible output rate on workstation j [hrs/SCD],
w number of workstations [-].

The ideal Output Rate Operating Curve for the super-ordinate system is determined using these values. The fact that under ideal assumptions, the ideal minimum WIP cannot be exceeded without a loss of utilization also applies to hierarchically aggregated workstations.

The two WIP components that are described in the LOC (the active WIP and WIP buffer) also deal with additive parameters. Nevertheless, the WIP buffer is dependent upon the stretch factor α_1 as well. If this should vary between the individual workstations, then the OROC is calculated:

$$\left.\begin{aligned} WIP_m(t) &= \left(1-\left(1-\sqrt[4]{t}\right)^4\right)\cdot\sum_{j=1}^{w} WIPI_{min,j} + t\cdot\sum_{j=1}^{w}\left(WIPI_{min,j}\cdot\alpha_{1,j}\right) \\ ROUT_m(t) &= \left(1-\left(1-\sqrt[4]{t}\right)^4\right)\cdot\sum_{j=1}^{w} ROUT_{max,j} \end{aligned}\right\} \quad (4.60)$$

where $WIP_m(t)$ mean WIP [hrs],
$WIPI_{min,j}$ ideal minimum WIP on workstation j [hrs],
$\alpha_{1,j}$ stretch factor for workstation j [-],
$ROUT_m(t)$ mean output rate [hrs/SCD],
$ROUT_{max,j}$ maximum possible output rate on workstation j [hrs/SCD],
w number of workstations [-],
t running variable ($0 \le t \le 1$).

If α_1 is the same for all of the workstations, then Eq. 4.29 can also be used to calculate the super-ordinate system's OROC by applying the sums of the $WIPI_{min}$ and $ROUT_{max}$.

However, there is a peculiarity about the measured operating states in consolidated workstations: Due to the non-linearity of the Output Rate Operating Curve, the measured operating state is no longer positioned on the LOC (see Fig. 4.28). That is to say, if there is a significant reduction in the output rate on an individual workstation, it is reflected in the super-ordinate system even when there is a relatively high WIP level on the other workstations (and therefore also on the entire system). Thus, a difference between the calculated OROC and the measured operating state indicates that there is an irregular (relative) WIP situation on the individual workstations.

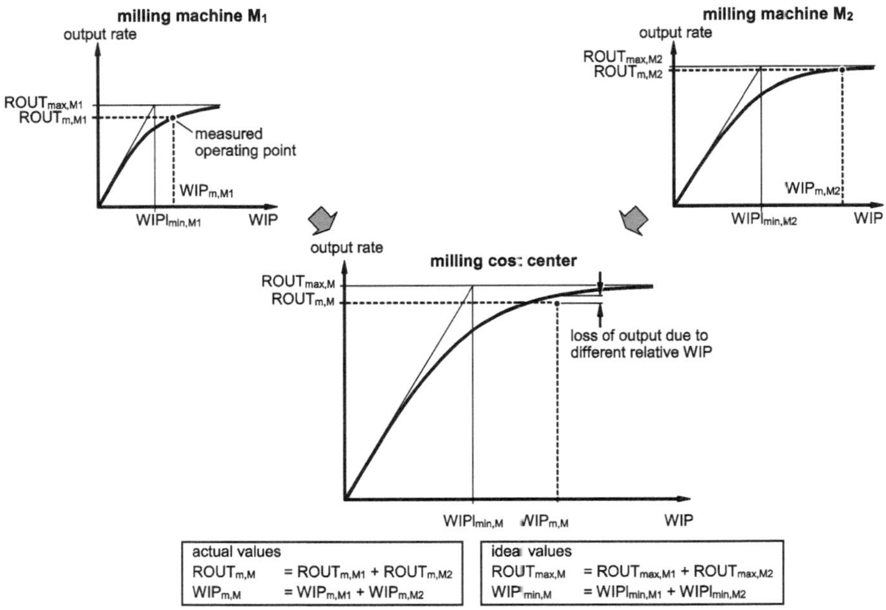

Fig. 4.28 Hierarchical Aggregation of Output Rate Operating Curves (Example with Two Workstations)

4.8.2 Manufacturing System Operating Curves (Author: Dipl.-Ing. Michael Schneider)

In addition to hierarchically aggregating Logistic Operating Curves for parallel workstations, consecutively positioned or randomly connected ones play a large role on the manufacturing floor, because a workstation is generally a part of a more or less complex production structure e. g., a processing chain or a manufacturing system. The interactions between the workstations and in particular how the capacity constraints of an individual workstation impacts the output rate of an entire production system cannot be represented with individual Logistic Operating Curves. In order to analyze and evaluate these types of production systems it is beneficial to be able to map the output rate behavior with a single Logistic Operating Curve.

The following method links the Logistic Operating Curves for individual workstations together to form a *Manufacturing System Operating Curve (MSOC)* for an arbitrary system of networked workstations (Fig. 4.29). The basic idea behind the MSOC is to model the workstations which are part of the system processing the orders as a network of funnels connected together through the material flow and to aggregate these into one comprehensive funnel. It is thus possible to determine the WIP, throughput time and output rate for the entire system. Bottlenecks become visible, and the impact of measures on the resources can be demonstrated e. g., how changing the WIP or capacity levels impact the order's throughput.

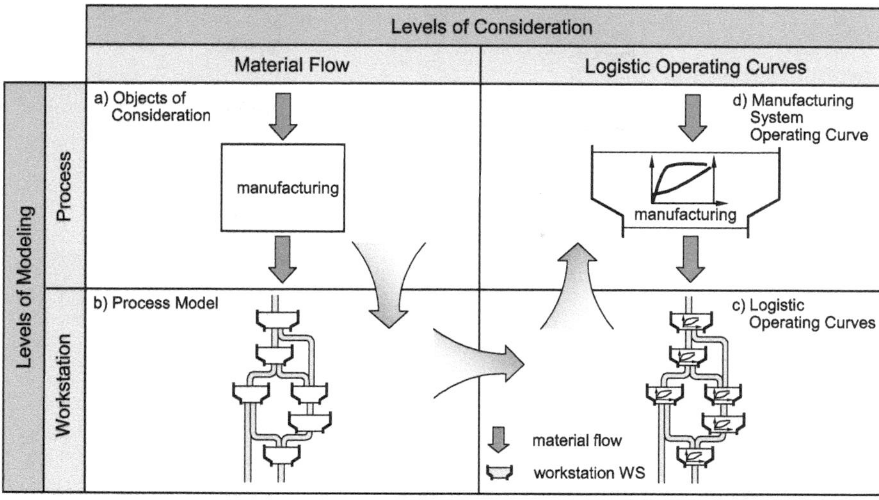

Fig. 4.29 Steps for Developing a Manufacturing System Operating Curve

The key to the Manufacturing System Operating Curve approach is defining an appropriate parameter for the flow in order to map the connections and thus the interactions between two workstations. Since the input and therefore the load of one workstation results directly from the output of the one upstream from it, the output of the previous workstation can be transferred to the input of the following one. However, this cannot be done by evaluating the output with a work content based parameter such as the mean work content WC_m: Since the content being processed on the workstations differ, the output rate based on the work content of the previous station will generally deviate from the input rate on the following workstation.

Instead, a parameter based on the flow of the orders can be used here to link the various workstations; the number of incoming and outgoing orders is well suited for this. As the flow parameter, the output rate is accordingly defined as the number of orders per time unit. However, since the WIP as the input parameter for the Output Rate Operating Curve is measured in standard hours, the workstation is still viewed with regards to the work content.

Based on the example of two directly coupled workstations, Fig. 4.30 illustrates how the LOC for individual workstations can be thus linked together as a comprehensive Manufacturing System Operating Curve. MSOC can be generated in three steps. In the first step, the individual Logistic Operating Curves are calculated for each of the workstations depending on their operating states. The WIP and output rate are measured in standard hours. Whereas, the operating state of Workstation 1 is in the overload zone of the LOC, Workstation 2 is in the underload zone. In the illustrated example, this is due to the fact that there is a greater mean work content on Workstation 1 with a comparable daily capacity.

In the second step, the dimension of the individual LOC is changed from standard hours into number of orders. The mean WIP_m is converted into the number of

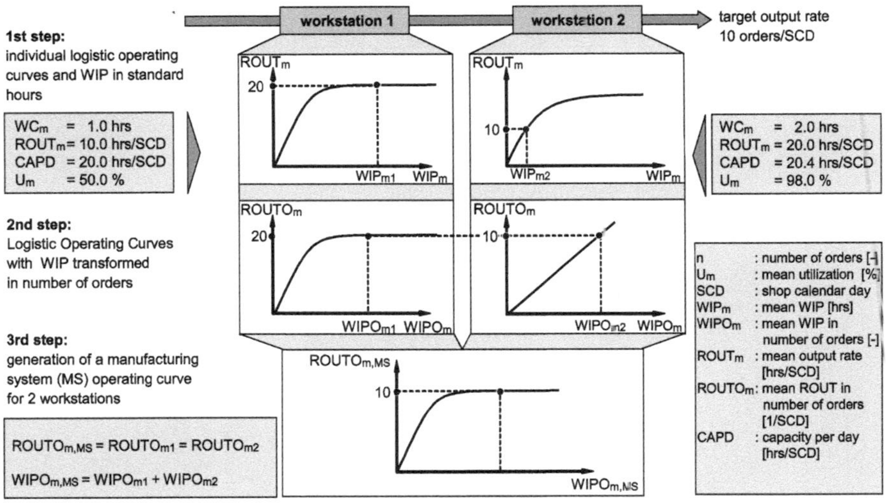

Fig. 4.30 Procedure for Developing a Manufacturing System Operating Curve for Two Workstations

orders $WIPO_m$ with the help of Eq. 4.56. Similarly, the mean output rate $ROUT_m$ is converted into the number of orders per shop calendar day $ROUTO_m$ with Eq. 2.23. The output rate of the workstations can then be directly compared. In the example, the output rate (measured in the number of completed orders) of the two workstations correspond with one another because each of the orders from Workstation 1 are subsequently processed by Workstation 2, and there is no additional input for Workstation 2.

Based on this information, the operating state for the entire system can be determined. The output rate corresponds to the output rates of the individual workstations and the WIP results from the sum of the individual WIP levels. Finally, the mean throughput time is calculated according to Eq. 2.24 from the quotient of the mean WIP and mean output rate:

$$TTP_{m,MS} = \frac{WIPO_{m,MS}}{ROUTO_{m,MS}} \tag{4.61}$$

where $TTP_{m,MS}$ mean unweighted throughput time of the manufacturing system [SCD],

$WIPO_{m,MS}$ mean WIP of the manufacturing system in number of orders [-],

$ROUTO_{m,MS}$ mean output rate of the manufacturing system in number of orders [1/SCD].

The throughput rate calculated using this method is an unweighted time parameter, because the WIP and output rate used to determine it are measured in number of orders.

In the third step, the shape of the MSOC can now be determined. However, because the equation for calculating it is valid not only for directly coupled worksta-

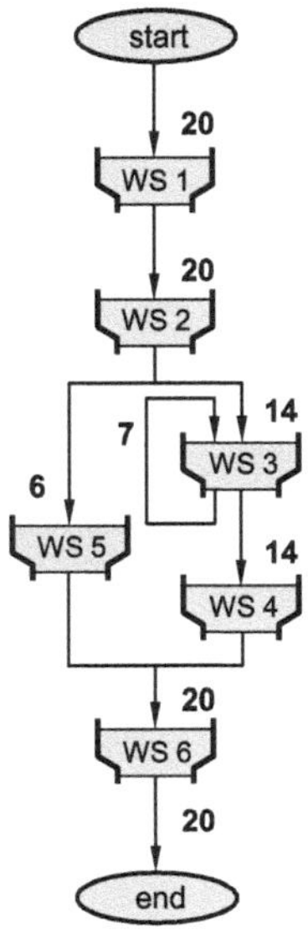

a) Flow Chart

from WS_i / to WS_j	WS 1	WS 2	WS 3	WS 4	WS 5	WS 6
start	20	0	0	0	0	0
WS 1	0	20	0	0	0	0
WS 2	0	0	14	0	6	0
WS 3	0	0	0	14	0	0
WS 4	0	0	0	0	0	14
WS 5	0	0	0	0	0	6
WS 6	0	0	0	0	0	0
total	20	20	14	14	6	20

b) Material Flow Matrix

from WS_i / to WS_j	WS 1	WS 2	WS 3	WS 4	WS 5	WS 6
start	1	0	0	0	0	0
WS 1	0	1	0	0	0	0
WS 2	0	0	0.7	0	0.3	0
WS 3	0	0	0	0.7	0	0
WS 4	0	0	0	0	0	0.7
WS 5	0	0	0	0	0	0.3
WS 6	0	0	0	0	0	0
MFC_i	1	1	0.7	0.7	0.3	1

c) Material Flow Coefficient Matrix

Fig. 4.31 Determining the Material Flow Coefficient from the Material Flow Matrix

tions, but also for omni-directional material flow we will first have to consider how they are mapped.

In order to be able to map omni-directional material flows, the corresponding material flow coefficients of the workstations are first determined. The *material flow coefficient* MFC_i describes the workstation's relative proportion of the manufacturing system's material flow. The coefficient is determined from the material flow matrix of the system being examined. The material flow matrix states the number of orders that are transported from a specific workstation to one that follows it, within the reference period. In order to calculate the material flow coefficient, the number of orders that are reported back on a workstation is divided by the total number of orders processed during the reference period. Figure 4.31 illustrates the procedure for this calculation in a simple example.

In order to calculate a Manufacturing System Operating Curve, we need additional information about the ideal minimum WIP and maximum possible output rate. The ideal process is characterized here in that no production orders on a bottleneck workstation have to wait for other orders to be processed, instead they can all be processed immediately after their arrival. At the same time, there can be no idle time due to lack of orders in the bottleneck workstation of the observed manufacturing system. The ideal minimum WIP is thus the lowest possible WIP on the observed manufacturing system or processing chain that still ensures the full utilization of the bottleneck workstation. The maximum possible output rate of the MSOC is the yielded maximum output rate of the manufacturing system in its ideal state.

The ideal $WIPIO_{min,MS}$, is calculated as

$$WIPIO_{min,MS} = \sum_{i=1}^{w} \frac{\frac{ROUTO_{max,BN}}{MFC_{BN}}}{\frac{ROUTO_{max,i}}{MFC_i}} \cdot WIPIO_{min,i} \tag{4.62}$$

where $WIPIO_{min,MS}$ ideal minimum WIP of the manufacturing system in number of orders [-],
$WIPIO_{min,i}$ ideal minimum WIP of the workstation i in number of orders [-],
$ROUTO_{max,BN}$ maximum output rate of the bottleneck workstations in number of orders [1/SCD].
$ROUTO_{max,i}$ maximum output rate of the workstation i in number of orders [1/SCD],
MFC_{BN} material flow coefficient of the bottleneck workstation [-],
MFC_i material flow coefficient of the workstation i [-],
w number of workstations i [-].

According to Eq. 4.62 $WIPIO_{min,MS}$ essentially results from the sum of the $WIPIO_{min,i}$ on the individual workstations in the manufacturing system. Furthermore, the different capacities of the workstations as well as how they are integrated into the material flow are taken into consideration. Thus the ideal minimum WIP of each workstation is weighted with a coefficient that sets the maximum possible output rate as well as the material flow coefficient of the workstation in relation to the bottleneck. Therefore, if a workstation occupies only a small portion of the production system's material flow, its ideal minimum WIP is not fully considered because this station can structurally not be fully utilized. In contrast if a workstation is a bottleneck then its ideal minimum WIP is fully taken into consideration in calculating the manufacturing system's minimum WIP in order to target a WIP dependent full utilization.

The maximum possible output rate of the manufacturing system $ROUT_{max,MS}$ is based on the maximum possible output rate of the bottleneck workstation $ROUTO_{max,BN}$:

$$ROUTO_{max,MS} = \frac{ROUTO_{max,BN}}{MFC_{BN}} \tag{4.63}$$

where $ROUTO_{max,MS}$ maximum output rate of the manufacturing system in number of orders [1/SCD],
$ROUTO_{max,BN}$ maximum output rate of the bottleneck system in number of orders [1/SCD],
MFC_{BN} material flow coefficient of bottleneck workstation [-].

If all of the orders pass through the bottleneck workstation, then the maximum possible output rate corresponds to that of the bottleneck workstation. The correlation between the logistic objectives output rate, WIP and throughput time can therefore clearly be described for the ideal manufacturing system.

On the manufacturing floor though the processes deviate considerably from these ideal conditions. WIP buffers are therefore needed in order to avoid a loss of utilization on the bottleneck workstation. By inserting the equations for the ideal minimum WIP (Eq. 4.62) and the maximum possible output rate (Eq. 4.63) in Eq. 4.29 it follows that for the output rate of the MSOC

$$\left.\begin{aligned} WIPO_{m,MS}(t) &= WIPIO_{min,MS} \cdot \left(1-\left(1-\sqrt[4]{t}\right)^4\right) + WIPIO_{min,MS} \cdot \alpha_1 \cdot t \\ ROUTO_{m,MS}(t) &= ROUTO_{max,MS} \cdot \left(1-\left(1-\sqrt[4]{t}\right)^4\right) \end{aligned}\right\} \tag{4.64}$$

where $WIPO_{m,MS}(t)$ mean WIP of the manufacturing system [-],
$WIPIO_{min,MS}$ ideal minimum WIP of the manufacturing system [-],
$ROUTO_{m,MS}(t)$ mean output rate of the manufacturing system [1/SCD],
$ROUTO_{max,MS}$ maximum possible output rate of the manufacturing system [1/SCD],
t running variable ($0 \le t \le 1$),
α_1 stretch factor (standard value: 10).

In order to validate the Manufacturing System Operating Curve numerous simulation tests were conducted. Different simulation models, based on data from actual enterprises, were integrated into the simulation system *ProSim III* (see Sect. 3.2) and used for this purpose (see [Wien-01*] and [Schn-02*]).

Figure 4.32 illustrates the procedure used to develop a simulated MSOC based on the example of a supplier in the transporter industry which had a total of 16

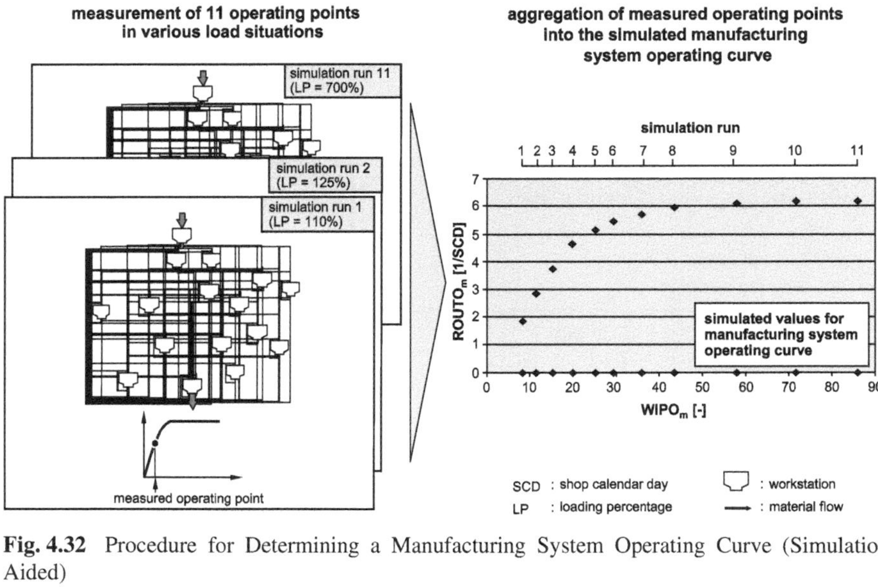

Fig. 4.32 Procedure for Determining a Manufacturing System Operating Curve (Simulation Aided)

different workstations. In eleven simulation trials, the WIP in the production was gradually increased. For each simulation, the mean WIP and the mean output rate of the production were measured and the values were aggregated into a Manufacturing System Operating Curve.

The simulated Manufacturing System Operating Curve has the typical shape of a LOC: Close to the manufacturer's maximum output rate, increases in the WIP no longer lead to a significant increase in the output rate. Generally, the simulated and calculated MSOC correspond sufficiently well with one another. The mean relative deviation in the example is approximately 4.5% and is acceptable for solving practical problems.

4.8.3 Workstations with Common WIP Buffers

When a workshop consists of a number of identical workstations, they can access a common WIP buffer. Since the risk of a disruption in the matrial flow is thus reduced for the individual workstations, the necessary WIP buffer can also generally be reduced.

The following example should clarify this as well as the consequences that it has for the Logistic Operating Curves equation (Fig. 4.33):

It first has to be assumed that in order to fully utilize a workstation with its own buffer, it is sufficient enough to maintain an average of one order in its WIP

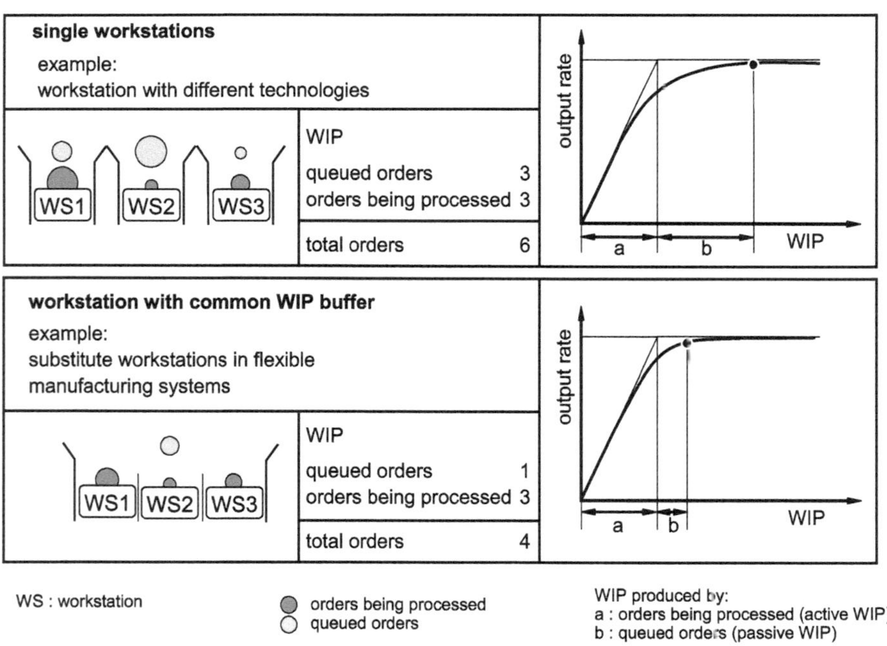

Fig. 4.33 Reducing the WIP by Generating a Common Buffer

buffer. Furthermore, since there is one order being processed on the workstation, the mean total WIP obviously consists of two orders per individual workstation. Thus for three workstations, the WIP is equal to 6 orders.

However should a number of workstations have access to a common WIP buffer, the situation can change fundamentally (Fig. 4.33, lower part). When it is ensured that there is always one order in the common WIP buffer, than it is also ensured that all of the concerned workstations can be utilized. This flexible method of allocating the WIP means that while the active WIP remains the same, the WIP buffer and therefore also α_1 can be reciprocally reduced proportional to the number of workstations involved. This was confirmed with the help of simulations. Nevertheless, analyses in the field showed that the impact of common WIP buffers is not as pronounced as was theoretically expected. That is to say, the flexibility of allocating the WIP to the individual workstations is limited by both specially equipped workstations and in particular by the operators' different qualifications. Since certain orders are thus reserved for certain workstations, the described effect does not in practice arise or at least not to the full extent.

The influence of the common WIP buffer can in principle be mathematically expressed by inserting an additional parameter into the LOC equation (in the WIP component of the equation for $WIP_b(t)$). However, we will refrain from doing so for two reasons: First of all, a parameter such as this can be determined, at least in the field, only empirically. Secondly, the calculated Logistic Operating Curves can also be changed into the preferred form by varying the stretch factor α_1. Nevertheless in special cases, it has to be verified whether or not the stretch factor α_1 can be adjusted (see also Sect. 6.3.3).

4.8.4 Considering Overlapping Production

One of the conditions for deriving the Logistic Operating Curves Theory, was that the orders have to be transported in traditional lots. As a result of increasing efforts to segment the manufacturing process, this condition is frequently no longer given. However, once the required adjustments have been made the approximation equation can in principle, still be applied.

Figure 4.34 shows two different fictional operating sequences as a Funnel Model, a Throughput Diagram and as an Output Rate Operating Curve. In the first case (Fig. 4.34a), the orders are processed and transported in traditional lots. It is assumed that in addition to the order being processed, there is a WIP buffer of one order available. Similarly in the second case (Fig. 4.34b), the orders are also processed according to traditional lots, and there is always one in the buffer. However, the orders here should be able to be delivered and removed on a split lot basis.

The representation of these processes in a Throughput Diagram, illustrates that through this measure, the WIP levels can be greatly reduced. From the perspective of the LOC Theory the possibility of reducing the WIP can be traced back to a

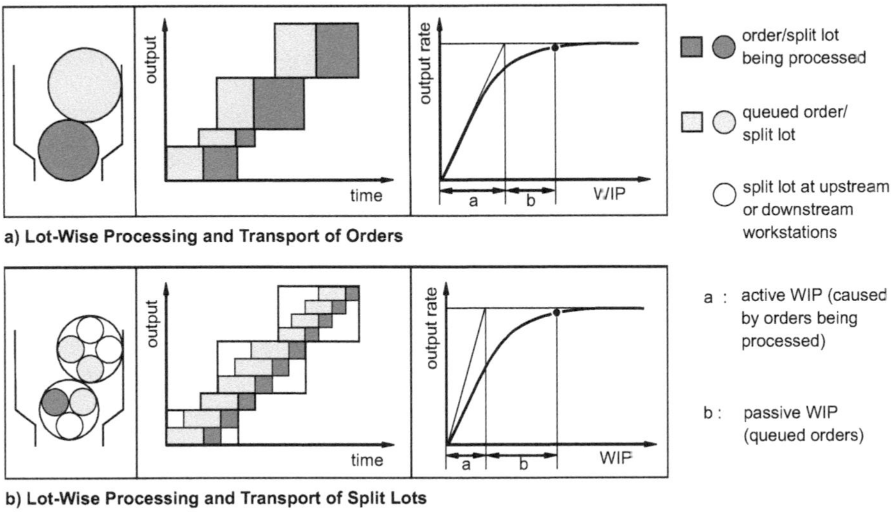

Fig. 4.34 Reducing the WIP through Overlapping Production

reduction in the ideal minimum WIP, which according to Eq. 4.5 results from the work content of the transport lots:

$$WIPI_{min,SPL} = WC_{m,SPL} \cdot \left(1 + WC_{v,SPL}^2\right) + TTR_m \tag{4.65}$$

where $WIPI_{min,SPL}$ ideal minimum WIP with split lot transports [hrs],
$WC_{m,SPL}$ mean work content of the split lots [hrs],
$WC_{v,SPL}$ coefficient of variation for the work content of the split lots [-],
TTR_m mean transport time [hrs].

The ratio of $WIPI_{min}$ to $WIPI_{min,SPL}$ is defined as the *degree of overlapping*:

$$OD = \frac{WIPI_{min}}{WIPI_{min,SPL}} \tag{4.66}$$

where OD degree of overlapping [-],
$WIPI_{min}$ ideal minimum WIP [hrs],
$WIPI_{min,SPL}$ ideal minimum WIP with split lot transports [hrs].

Incidentally, given the stated conditions the WIP buffer is not changed, because it is assumed that there still has to be one queued order. Therefore, it follows that for the OROC approximation equation:

$$\left.\begin{aligned} WIP_m(t) &= \frac{WIPI_{min}}{OD} \cdot \left(1 - \left(1 - \sqrt[4]{t}\right)^4\right) + WIPI_{min} \cdot \alpha_1 \cdot t \\ ROUT_m(t) &= ROUT_{max} \cdot \left(1 - \left(1 - \sqrt[4]{t}\right)^4\right) \end{aligned}\right\} \tag{4.67}$$

where $WIP_m(t)$ mean WIP [hrs],
$ROUT_m(t)$ mean output rate [hrs/SCD],
$WIPI_{min}$ ideal minimum WIP [hrs],
OD degree of overlapping [-],
α_1 stretch factor for the workstation [-],
$ROUT_{max}$ maximum possible output rate on a workstation [hrs/SCD].

The shifting of the ratio between the WIP buffer (passive WIP) and the WIP in progress (active WIP) in an overlapping production (which is noticeable in Eq. 4.67 and is hinted at in Fig. 4.34) is not unknown on the production floor. Therefore, in flexible manufacturing systems (FMS) and manufacturing segments, the smallest possible unit – a lot size of one – is frequently transported. Consequently, the WIP within a system can be strongly restricted. However, since the order lots are also generated and controlled as a whole within the production by the PPC in these systems, a WIP buffer develops in front of the FFS or manufacturing segments due to the work content of the order lot.

4.9 Prerequisites for Applying Calculated Logistic Operating Curves

Developing and applying calculated Logistic Operating Curves is bound to a number of prerequisites, which can be derived from the model's structure and parameters.

First, it is necessary to be able to establish ideal LOC for the observed production process. In order to do this, it has to be possible to describe an idealized production process. All of the parameters required to calculate the ideal minimum WIP and the maximum possible output rate not only have to be available or determinable but also have to be sufficiently accurate. An overview of the parameters which play a primary role in the logistic performance of a production system is provided in Fig. 4.35. Those highlighted with a gray background, must be available at the level of the individual workstations. Based on the correlations described in Sect. 4.8.1 and 4.8.2, the data for the super-ordinate systems can be hierarchically aggregated.

The Logistic Operating Curves Theory is a method that assumes a stable operating state. The chosen investigation period therefore has to be long enough that the individual parameters can be seen as representative of the workstation being examined. At the same time, it also has to be ensured that in particular there is no tendency for the work content structure to change during this time.

The processing state has to also be stable for the second part of the modeling – the describing of the WIP buffer through the stretch factor α_1. This condition is met when the WIP remains at an almost constant level. Generally, the input and output processes (measured in standard hours per time unit) have to run parallel to each other. The stretch factor α_1 is set depending on the variance, or expected variance of the WIP. Currently, this parameter can only be determined experimen-

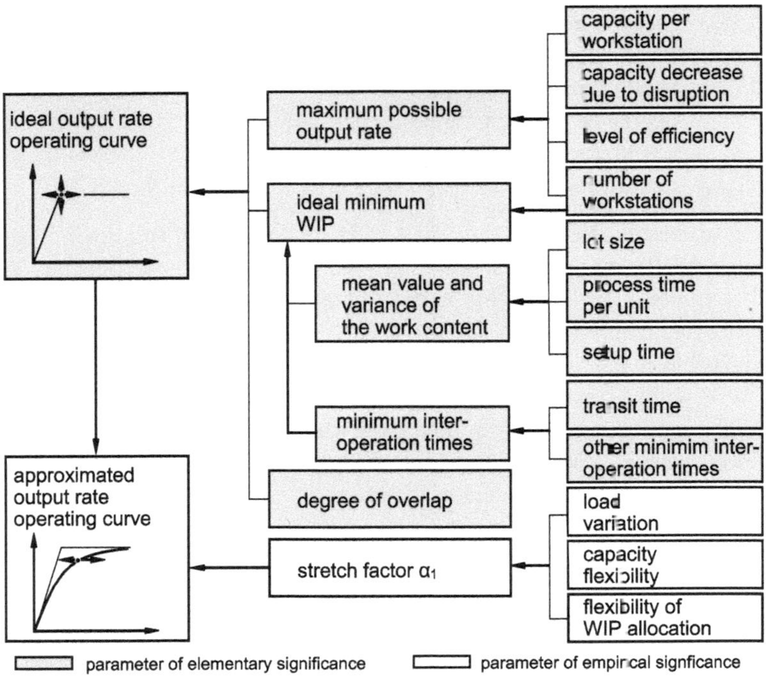

Fig. 4.35 Parameters for Approximated and Ideal Output Rate Operating Curves

tally and thus is empirically significant within the model. Numerous studies however have shown that the standard value which the work here is based on ($\alpha_1 = 10$), generally permits a near to real description of the interdependencies between the logistic objectives (see also Sect. 4.7). Nevertheless, the validity of this value should be verified especially when there are rigid capacities and a strongly fluctuating load as it will then be greater than 10.

Since the Logistic Operating Curves for the time parameters are derived from the Output Rate Operating Curve, the same conditions apply to them as well. However, for these there is also an additional requirement for the processing stability: The equations for calculating the throughput time and inter-operation time as well as all of the descriptions based on them, can only be applied failure-free when the mean range R_m corresponds to the mean weighted throughput time TTP_{mw}, even when the processing occurs according to the FIFO principle. This is always the case during long periods of investigation as long as there is not a general increase or decrease in the WIP. It is not the WIP fluctuation that is significant in this context, but rather the difference between the reference period's initial and end WIP levels (see also Sect. 2.2.3).

When some of the required parameters are not available, can only be approximately determined, or violate the requirements of a stable processing state it does not necessarily exclude the possibility of determining LOC. Depending on the type of error, the application possibilities and to some degree the quality of the conclusions will be more or less limited though. In Chap. 6 however, we will show that the LOC Theory is a quite tolerant and robust model.

4.10 Schedule Reliability Operating Curves (Author: Dr.-Ing. Kwok-Wai Yu)

One of the key challenges of a production firm is ensuring scheduling reliability. Knowledge about the factors which influence it as well as the interactions between the objectives described by the Logistic Operating Curves greatly contributes to mastering these logistic performance issues, and therefore to ensuring the enterprise's competitiveness.

Yu [Yu-01] developed a pragmatic method for calculating a *Schedule Reliability Operating Curve (SROC)*. The starting point for this method is a descriptive model of the output lateness.

4.10.1 Mean Relative Lateness Operating Curve

In the manufacturing area, output schedule deviations have two causes which to some degree occur separately but which can also overlap (see Sect. 2.1.3). On the one hand, orders are often started too late (input lateness) and on the other hand, the planned throughput time can differ from the actual throughput time (throughput time deviations or *relative lateness*). Since the completion of one operation is required before the next can be started, a delay in the output date causes a delay in the start date of the next order. This overlaps with a possibly longer operation duration and thus forms the (total) output lateness of the order.

Figure 4.36a illustrates a state often found on the production floor. It is characterized by a positive mean output lateness (delay) and a large variance among the individual values. The goal is to improve this situation. One the one hand, using appropriate measures the mean output lateness should be reduced to zero (Fig. 4.36b) and on the other hand, the variance should be reduced (Fig. 4.36c). The best result and greatest process reliability is achieved when both of these values are simultaneously reduced to a minimum (Fig. 4.36d).

Due to the diverse causes which can also overlap, input lateness is difficult to predetermine. The orders are either moved suddenly and unexpectedly (unplanned) e. g., through required delays, lack of material/production resources, or – due to planning – deferred or moved up e. g., in order to balance the load.

Calculating the input lateness using basic parameters is therefore not possible. As an alternative, they have to be measured over a longer time period. By doing so they can be extrapolated and superimposed onto the throughput time deviations in order to finally be able to derive the estimated output lateness.

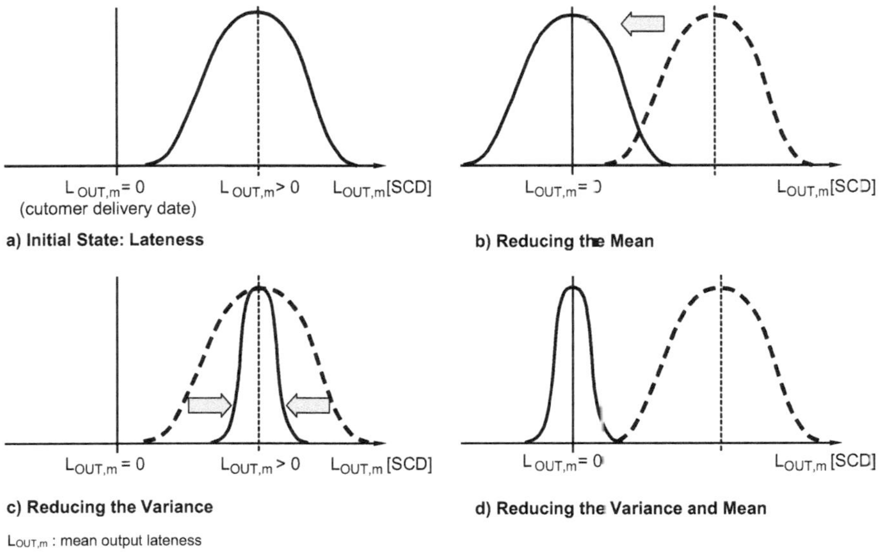

Fig. 4.36 Increasing the Process Reliability by Controlling the Output Lateness' Mean and Variance

In order to derive the Schedule Reliability Operating Curve, it is therefore assumed that there is no input lateness. The output lateness and thus the schedule reliability or adherence is determined exclusively by the relative lateness.

The relative lateness can be used to represent whether the scheduling situation in the output has improved or worsened in comparison to the input (see Sect. 2.1.2). The cause for this can for example be that despite PPC support, the planned WIP and throughput time levels can frequently not be met. Fluctuating WIP levels and throughput times can be due to processing disruptions but can also be traced back to planning mistakes (incorrect or unrealistic planning data) or uncontrolled order input. Finally, throughput times and therefore throughput deviations are also influenced by the sequencing rules that are applied (see Sect. 4.4).

The mean relative lateness can be visualized in a Throughput Time Operating Curve and broken down into its major components. Generally, the mean relative lateness can be caused by the WIP and/or by prioritizing orders. In Fig. 4.37 – Case 1, no orders are prioritized during the actual state. The mean relative lateness can thus be interpreted as a result of the actual WIP deviating from the planned WIP for that workstation. However, if the orders are also prioritized as in Fig. 4.37 – Case 2, then there is an additional throughput time deviation. Depending on what sequencing rule is applied, the throughput deviations can either increase or decrease the mean relative lateness (see also Sect. 4.4 and Fig. 4.15).

When this train of thought is transferred to all of the possible actual operating points, then a *Mean Relative Lateness Operating Curve* (RLOC) is derived by subtracting the planned throughput time from the values of the Throughput Time Operating Curve. The RLOC illustrates the dependence of the mean relative late-

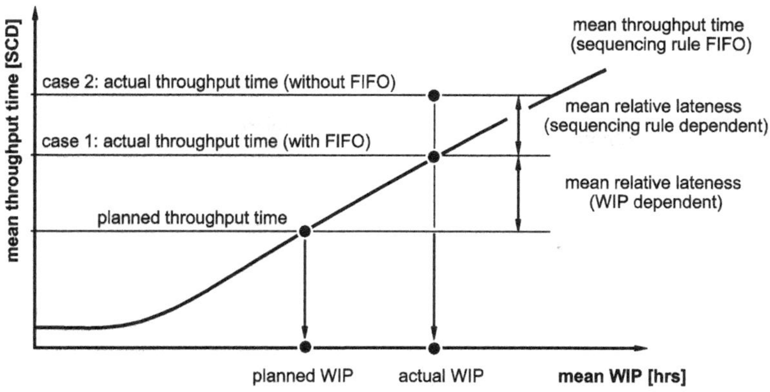

Fig. 4.37 WIP Dependent and Sequencing Rule Dependent Lateness

ness on the mean actual WIP for a mean planned throughput time (see Fig. 4.38). It results from a parallel, vertical shift of the Throughput Time Operating Curve by the value of the mean planned throughput time, which then intersects the x-axis at the point of the planned WIP. With the help of the actual WIP, the mean relative lateness can therefore be determined. When the actual WIP is smaller than the planned WIP then the mean relative lateness L_{rel} is negative (too early). Should the actual WIP be the same as the planned WIP then L_{rel} is equal to zero, and when it is greater than the planned WIP, L_{rel} is positive (too late).

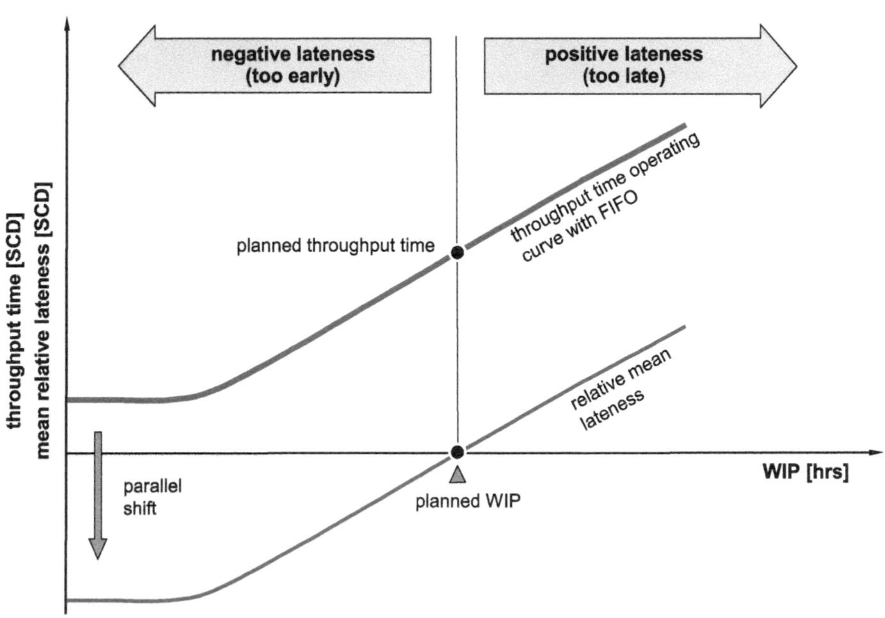

Fig. 4.38 Mean Relative Lateness Operating Curve

4.10.2 *Deriving an Operating Curve for Describing the Schedule Reliability*

The behavior of a production system's schedule is not sufficiently expressed by the mean lateness alone. Instead, it is of far greater importance whether or not the orders will be completed within an agreeable due-date tolerance i. e., whether or not the schedule reliability can be ensured. For our purpose, the individual terms relevant for scheduling will be differentiated as follows (Fig. 4.39):

- The due-date tolerance defines the range of a scheduling plan within which an order is considered punctually delivered.
- The schedule adherence describes whether or not an order will be delivered according to the planned schedule. It reflects the market's perspective. Generally, only the delayed orders decrease the schedule adherence.
- The schedule reliability describes whether or not an order was delivered within the due-date tolerance. It reflects the manufacturing perspective. An enterprise generally strives to avoid completing orders too early as well as too late.

Similar to the Logistic Operating Curves theory, the approximation equation used to calculate the schedule reliability is based on a deductive-experimental processing model. The modeling was based on the following steps [Yu-01]:

- Deriving the ideal Schedule Reliability Operating Curve from an ideal production process.
- Expanding the model in order to reflect the process flow found in the field.
- Adjusting the parameters by analyzing the simulation results.

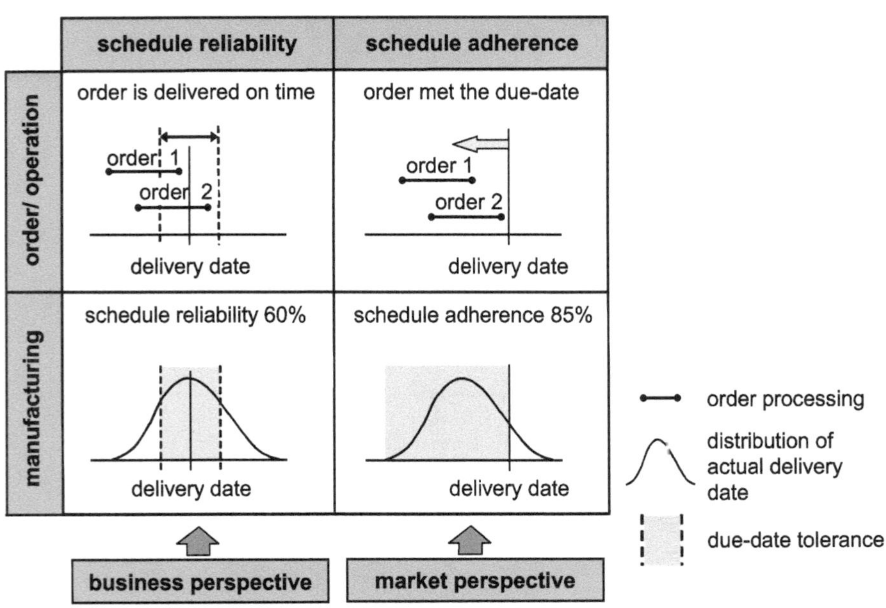

Fig. 4.39 Schedule Reliability and Adherence

The following conditions characterize the ideal state:

- There is no input lateness.
- Orders are completed according to the FIFO principle.
- There is no variance of the actual or planned throughout time.

Under these conditions, the schedule reliability is only dependent on the relative lateness. The ideal *Schedule Reliability Operating Curve* (SROC) thus represents the correlation between the schedule reliability, the relative lateness and the WIP. If the planned WIP corresponds to the actual WIP, under the aforementioned conditions, the planned throughput time then also corresponds to the actual throughput time and all of the orders will be completed on time.

When these prerequisites are met, the Schedule Reliability Operating Curve illustrated in Fig. 4.40a can be established. At the planned WIP level the schedule reliability is 100%, and at all other WIP values 0%. By implementing a time interval (a due-date tolerance within which the orders are considered on time) a range of WIP is created in which the schedule reliability is 100% (Fig. 4.40b). Outside of this range the schedule reliability is 0%.

The assumptions made for deriving ideal Schedule Reliability Operating Curves are not achievable on the shop floor. This is made particularly clear, for example, by the large variance usually found in the lateness. In Fig. 4.41a, three simplified, but typical situations of output lateness, are shown for different operating states. These are characterized by the workstation's actual WIP levels. Operating state I is defined by a lower than planned WIP level. In operating state II the actual WIP corresponds to the planned WIP and finally, in operating state III the actual WIP is greater than the planned WIP. The different operating states can

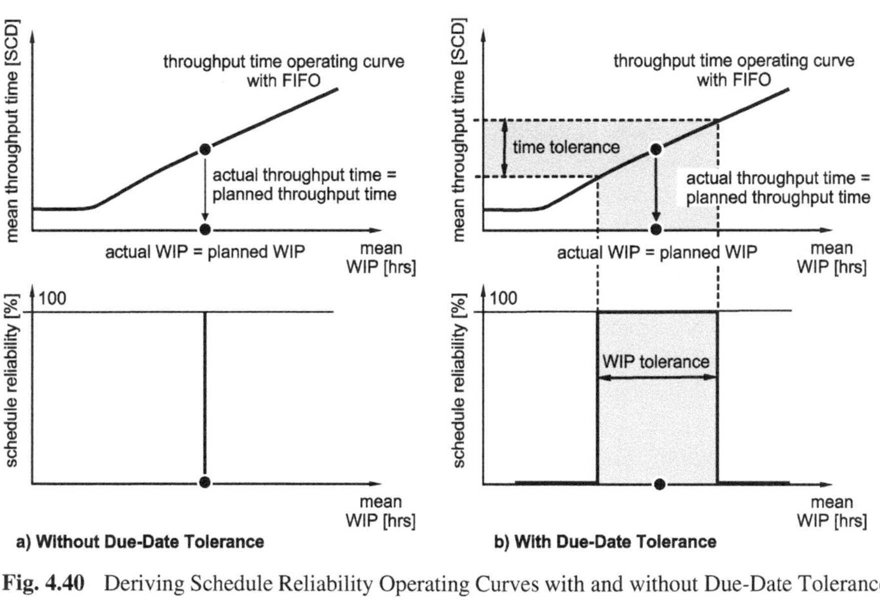

Fig. 4.40 Deriving Schedule Reliability Operating Curves with and without Due-Date Tolerance

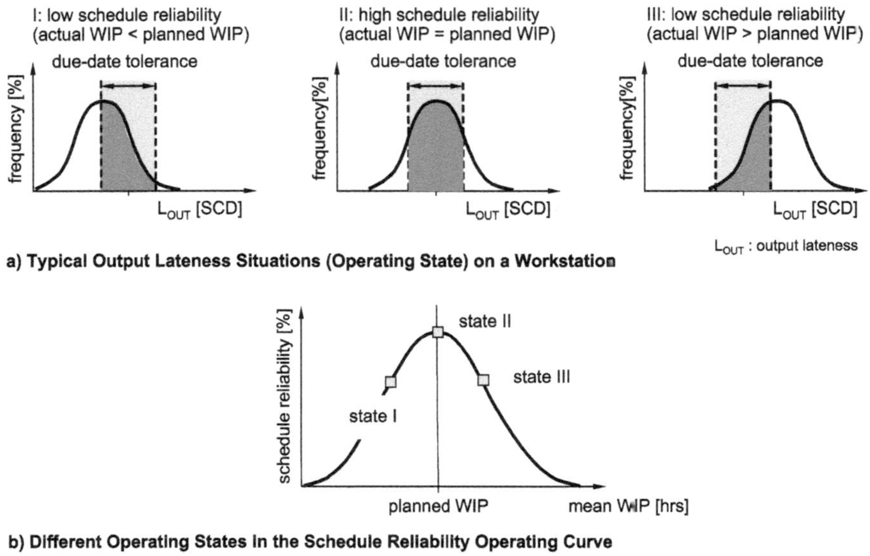

Fig. 4.41 Representation of Different Operating States in the Schedule Reliability Operating Curve

now be transferred to a Schedule Reliability Operating Curve. The values are determined by calculating the amount of orders which are within the due-date tolerance (grey area). The SROC illustrates that with increasing WIP values, the schedule reliability increases up to the planned WIP. If the WIP level continues to grow beyond this point, the scheduling reliability decreases. The more the WIP level climbs, the closer the scheduling reliability approaches zero.

Figure 4.42 depicts an example of a Schedule Reliability Operating Curve created simulatively. For the ideal SROC the due-date tolerance was converted into a WIP tolerance. The SROC was approximated through 15 simulation trials in which the mean WIP was gradually varied. It can be seen that the schedule reliability climbs as the WIP increases, up to a point at which the actual WIP corresponds to the planned WIP. The schedule reliability achieved at this point, is referred to as the practical maximum schedule reliability. It defines the schedule reliability value that a system, with the given logistic behavior and set due-date tolerance cannot exceed. Should the WIP value surpass this optimal point, the schedule reliability is then reduced again and with increasing or decreasing WIP approaches zero.

By analyzing these and other simulations it was possible to derive an approximation equation which, assuming that the throughput time is normally distributed, allows the scheduling reliability to be calculated as a function of the WIP. This equation, which is extensively derived in [Yu-01], is based on the distribution function $\phi(u)$ for the standard normal distribution and is written as:

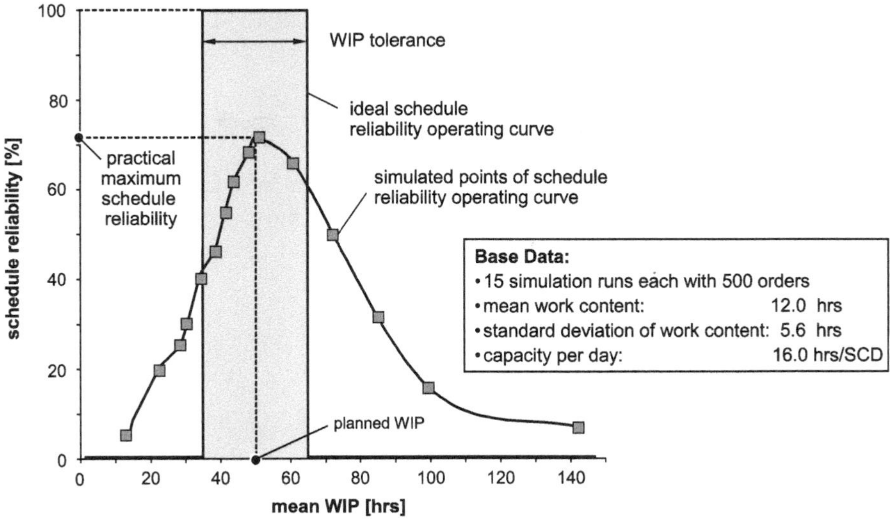

Fig. 4.42 A Simulated Schedule Reliability Operating Curve

$$ReS(WIP(t)) = \phi\left(\frac{b - TIO_m(WIP(t))}{TIO_s}\right) - \phi\left(\frac{a - TIO_m(WIP(t))}{TIO_s}\right) \qquad (4.68)$$

where ReS schedule reliability [%],
a lower bound [SCD],
b upper bound [SCD],
TIOs standard deviation of the inter-operation time [SCD],
TIOm(WIP(t)) mean inter-operation time [SCD],
WIP(t) WIP (as a function of t) [hrs],
t running variable [-].

The limits within which an order is considered to have been delivered reliably are determined by a and b. When the planned inter-operation time is 4 SCD and the due-date tolerance is ± 1 SCD the interval's limits are a = 3 SCD and b = 5 SCD. The standard deviation of the inter-operation time TIO_s is determined by a random distribution of the actual inter-operation times. It is assumed that the standard deviation of the inter-operation time is almost independent from the WIP when the FIFO principle is applied (see also [Ludw-93]). The mean inter-operation time TIO_m (t) can be derived from the Logistic Operating Curves and the related equations (Eqs. 4.29, 4.37, 4.40 and 4.41).

Equation 4.68 can be used to calculate the practical maximum schedule reliability. In the following we will explain this procedure based on an inter-operation time distribution where $TIO_m = 4.5$ SCD, $TIO_s = 1.12$ SCD and there is a due-date tolerance of ± 1 SCD:

Once the due-date tolerance is set, the proportion of values within the interval $3.5 \leq X \leq 5.5$ SCD has to be determined. The interval bounds a = 3.5 and b = 5.5 are converted according to Eq. 4.68:

$$\text{lower bound}: \quad \frac{a - TIO_m(WIP(t))}{TIO_s} = \frac{3.5 - 4.5}{1.12} = -0.89$$

$$\text{upper bound}: \quad \frac{b - TIO_m(WIP(t))}{TIO_s} = \frac{5.5 - 4.5}{1.12} = 0.89$$

By applying the distribution function $\phi(u)$ for the standard deviation (Table 4.1) we obtain the following values for the scheduling reliability:

$$\begin{aligned} ReS(WIP(t)) &= \phi(0.89) - \phi(-0.89) \\ &= \phi(0.89) + \phi(0.89) - 1 \\ &= 0.8133 + 0.8133 - 1 \\ &= 0.6266 \end{aligned}$$

Thus, based on the above parameters the schedule reliability would be approximately 63%.

In Fig. 4.43 schedule reliability values that were calculated based on the approximation equation are shown in comparison to simulated values. It can be seen that the two Logistic Operating Curves correspond well with another. The approximation equation therefore offers the possibility to describe the correlation between the schedule reliability, due-date tolerance and throughput time as a function of the planned and actual WIP with minimal effort.

It should be noted that in order to calculate the Schedule Reliability Operating Curve the processing state has to be balanced and there has to be no tendency for the WIP to change during the investigation period.

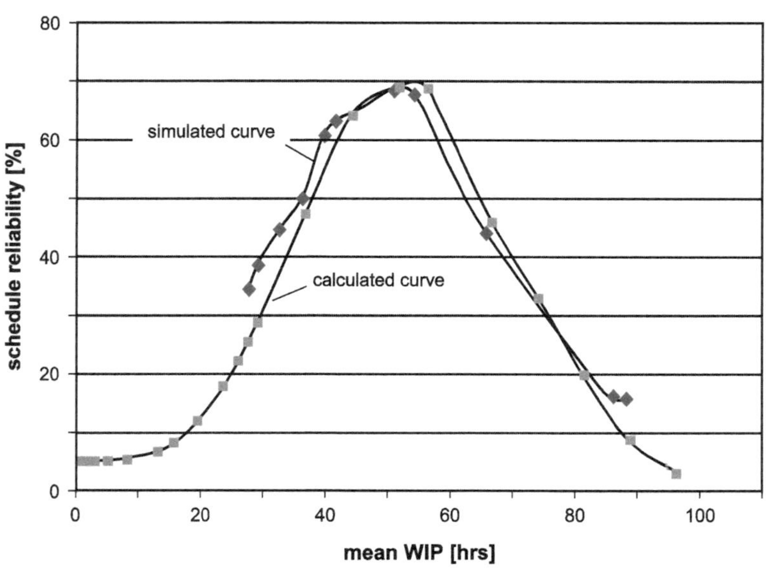

Fig. 4.43 Comparison of a Simulated Schedule Reliability Operating Curve and a Schedule Reliability Operating Curve Calculated Using the Distribution Density Function $\phi(u)$

Furthermore, in order to derive the SROC it also has to be a prerequisite that there is no input lateness. Strictly speaking the focus is thus on the schedule reliability of the relative lateness. Regardless of these prerequisites, the Schedule Reliability Operating Curve provides an important extension for supporting and monitoring the reliability and capability of a company's logistic process. The SROC are thus suitable for evaluating the manufacturing process within the context of a production control. They also indicate what level of schedule reliability can be reached with the existing structural conditions. Finally, in conjunction with the Logistic Operating Curves, they make it possible to comprehensively model the logistic objectives.

4.11 Summarizing the Derivation of the Logistic Operating Curves Theory

The Logistic Operating Curves Theory introduced here is a deductive-experimental processing model that integrates the essential advantages of both the underlying modeling methods.

In order to calculate the Logistic Operating Curves, a deductive approach – the ideal Logistic Operating Curve – was derived in a first modeling step. This step was based on a description of the order's flow through the production's individual workstations using the Funnel Model and the resulting Throughput Diagrams (Fig. 4.44). The ideal Logistic Operating Curves describe the theoretical limits of the key performance figures and the underlying interdependencies. In order to transfer the parameters into real processes, the deductive model was then expanded. This was accomplished by integrating a basic mathematical function (the

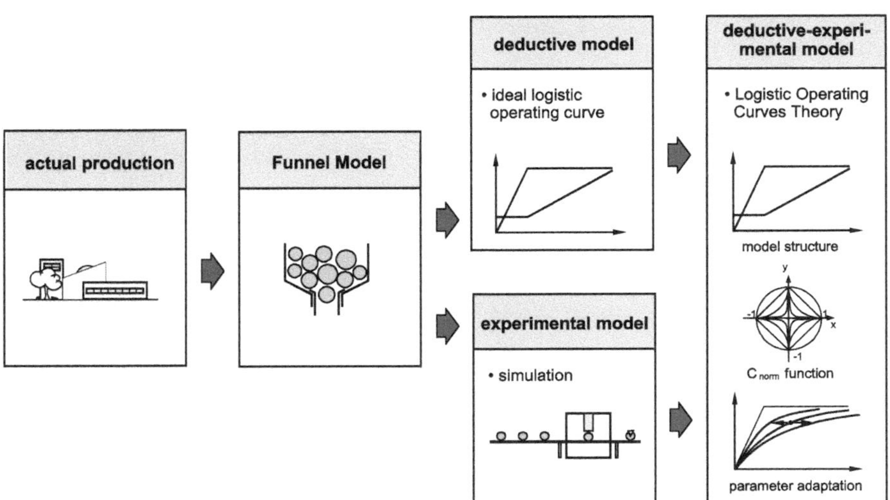

Fig. 4.44 Logistic Operating Curves as a Deductive-Experimental Model

Table 4.1 Distribution Function ϕ(u) for the Standard Deviation [Papu-94]

u	0	1	2	3	4	5	6	7	8	9
0	0.5000	0.5040	0.5080	0.5120	0.5160	0.5199	0.5239	0.5279	0.5319	0.5359
0.1	0.5398	0.5438	0.5478	0.5517	0.5557	0.5596	0.5639	0.5675	0.5714	0.5754
0.2	0.5793	0.5832	0.5871	0.5910	0.5948	0.5987	0.6026	0.6064	0.6103	0.6141
0.3	0.6179	0.6217	0.6255	0.6293	0.6331	0.6368	0.6406	0.6443	0.6480	0.6517
0.4	0.6554	0.6591	0.6628	0.6664	0.6700	0.6736	0.6772	0.6808	0.6844	0.6879
0.5	0.6915	0.6950	0.6985	0.7019	0.7054	0.7088	0.7123	0.7157	0.7190	0.7224
0.6	0.7258	0.7291	0.7324	0.7357	0.7389	0.7422	0.7545	0.7486	0.7518	0.7549
0.7	0.7580	0.7612	0.7642	0.7673	0.7704	0.7734	0.7764	0.7794	0.7823	0.7852
0.8	0.7881	0.7910	0.7939	0.7967	0.7996	0.8023	0.8051	0.8078	0.8106	0.8133
0.9	0.8159	0.8186	0.8212	0.8238	0.8264	0.8289	0.8315	0.8340	0.8365	0.8398
1	0.8413	0.8438	0.8461	0.8485	0.8508	0.8531	0.8554	0.8577	0.8599	0.8621
1.1	0.8643	0.8665	0.8686	0.8708	0.8729	0.8749	0.8770	0.8790	0.8810	0.8830
1.2	0.8849	0.8869	0.8888	0.8907	0.8925	0.8944	0.8962	0.8980	0.8997	0.9015
1.3	0.9032	0.9049	0.9066	0.9082	0.9099	0.9115	0.9131	0.9147	0.9162	0.9177
1.4	0.9192	0.9207	0.9222	0.9236	0.9251	0.9265	0.9279	0.9292	0.9306	0.9319
1.5	0.9332	0.9345	0.9357	0.9370	0.9382	0.9394	0.9406	0.9418	0.9429	0.9441
1.6	0.9452	0.9463	0.9474	0.9484	0.9495	0.9505	0.9515	0.9525	0.9535	0.9545
1.7	0.9554	0.9564	0.9573	0.9582	0.9591	0.9599	0.9603	0.9616	0.9625	0.9633
1.8	0.9641	0.9649	0.9656	0.9664	0.9671	0.9678	0.9686	0.9693	0.9699	0.9706
1.9	0.9713	0.9719	0.9726	0.9732	0.9738	0.9744	0.9750	0.9756	0.9761	0.9767
2	0.9772	0.9778	0.9783	0.9788	0.9793	0.9798	0.9803	0.9808	0.9812	0.9817
2.1	0.9821	0.9826	0.9830	0.9834	0.9838	0.9842	0.9846	0.9850	0.9854	0.9857
2.2	0.9861	0.9864	0.9868	0.9871	0.9875	0.9878	0.9881	0.9884	0.9887	0.9890
2.3	0.9893	0.9896	0.9898	0.9901	0.9904	0.9906	0.9909	0.9911	0.9913	0.9916
2.4	0.9918	0.9920	0.9922	0.9925	0.9927	0.9929	0.9931	0.9932	0.9934	0.9936
2.5	0.9938	0.9940	0.9941	0.9943	0.9945	0.9946	0.9948	0.9949	0.9951	0.9952
2.6	0.9953	0.9955	0.9956	0.9957	0.9959	0.9960	0.9961	0.9962	0.9963	0.9964
2.7	0.9965	0.9966	0.9967	0.9968	0.9969	0.9970	0.9971	0.9972	0.9973	0.9974
2.8	0.9974	0.9975	0.9976	0.9977	0.9977	0.9978	0.9979	0.9979	0.9980	0.9981
2.9	0.9981	0.9982	0.9982	0.9983	0.9984	0.9984	0.9985	0.9985	0.9986	0.9986
3	0.9987	0.9987	0.9987	0.9988	0.9988	0.9989	0.9989	0.9989	0.9990	0.9990
3.1	0.9990	0.9991	0.9991	0.9991	0.9992	0.9992	0.9992	0.9992	0.9993	0.9993
3.2	0.9993	0.9993	0.9994	0.9994	0.9994	0.9994	0.9994	0.9995	0.9995	0.9995
3.3	0.9995	0.9995	0.9995	0.9996	0.9996	0.9996	0.9996	0.9996	0.9996	0.9997
3.4	0.9997	0.9997	0.9997	0.9997	0.9997	0.9997	0.9997	0.9997	0.9997	0.9998
3.5	0.9998	0.9998	0.9998	0.9998	0.9998	0.9998	0.9998	0.9998	0.9998	0.9998
3.6	0.9998	0.9998	0.9999	0.9999	0.9999	0.9999	0.9999	0.9999	0.9999	0.9999
3.7	0.9999	0.9999	0.9999	0.9999	0.9999	0.9999	0.9999	0.9999	0.9999	0.9999
3.8	0.9999	0.9999	0.9999	0.9999	0.9999	0.9999	0.9999	0.9999	0.9999	0.9999
3.9	1.0000	1.0000	1.0000	1.0000	1.0000	1.0000	1.0000	1.0000	1.0000	1.0000

C_{norm} function) and a systematic analysis of simulation results in order to adjust parameters. Since the underlying experimental model was also ultimately based on the Funnel Model, the conditions for applying the deductive model could be fulfilled or gradually removed. This in turn strongly supported the transferability of the knowledge gained from the experimental model into the Logistic Operating Curves Theory.

The procedure for the deductive-experimental modeling demonstrates a number of fundamental advantages especially for the problems considered here: Based on the structure of the model, the Logistic Operating Curves Theory can be adjusted quite easily to changes in the basic conditions. It is therefore imaginable that this mathematical model could be transferred to entirely different processes. In order to do so though, it has to be possible to describe an ideal process, which can then be used as a basis for building the deductive model components. The Schedule Reliability Operating Curve is mentioned here as an example as it too is generated based on a deductive-experimental model.

Since the model structure and model parameters can for the most part be derived from basic laws, the interdependencies between the logistic objectives can be easily described. The model can therefore also be considered an *impact model*. In general, as long as the prerequisites for applying the model are met, it is also possible to derive conclusions from the model that can be directly transferred to other production processes. They can thus contribute to gaining a fundamental understanding of the system's stationary and time-dependent behavior. We will discuss this further in the following chapter.

Chapter 5
Basic Laws of Production Logistics

In the previous chapter, it was shown that despite what seems to be erratic circumstances it is possible to mathematically describe the behavior of a manufacturing system measured through the key performance figures for the logistic objectives.

Based on these mathematical descriptions, we will now develop basic interdependencies that can be universally applied and formulate them as the *Basic Laws of Production Logistics*. These laws are primarily meant to assist in understanding the logistic processes and how they can be potentially influenced in a production. As a starting point we will use the basic laws originally formulated by Wiendahl [Wien-89]. These will then be substantially expanded and to some degree reformulated. These laws can for the most part be traced back to the equations introduced in Chaps. 2 to 4. Only, the last two laws are based on empirically gained knowledge which at this point cannot yet be mathematically described.

5.1 First Basic Law of Production Logistics

The first basic law of production logistics is:

The input rate and output rate of a workstation have to be balanced over the long term.

The output rate – independent of the WIP level – must correspond to the load, over the long-term (Fig. 5.1). Here, it is assumed that the parameters used to determine the load and output rate cannot as a result of the process be changed. This condition is met when the input and output are measured in either planned hours or number of orders.

This law can be verified with a few simple considerations: If, in the long-term, the load is greater than the available capacity, then the WIP will theoretically grow indefinitely. Such a situation is inconceivable in the operational reality: A com-

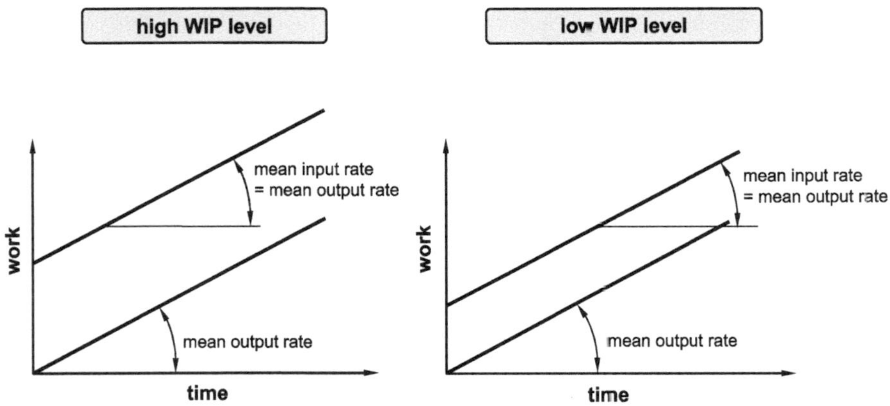

Fig. 5.1 First Basic Law of Production Logistics

pany reacts to a load that is continuing to increase either by correspondingly adapting the capacities or – in case that is not possible – by adjusting the load i. e., either by refusing orders or by allocating orders to other firms. If on the other hand, the load is less than the available capacity, then the output rate will inevitably follow the load, since only as many orders can be processed as those which have entered the system.

Based on this law, it follows that the WIP should not be reduced through capital expenditures aimed at long-term capacity increases. Although an increased capacity would initially help to reduce the WIP and throughput time, the output would not increase in the medium to long-term with these measures. In order for this to occur, there has to be a corresponding increase in demand and therefore also a long-term increase in the load.

5.2 Second Basic Law of Production Logistics

The second basic law of production logistics describes the correlation between the logistic objectives throughput time, range, WIP and output rate:

The throughput time and range of a workstation result from the ratio of the WIP and output rate.

By a given load and output rate, the throughput time can only be reduced by decreasing the WIP (Fig. 5.2). This law is based on the Funnel Formula as well as Little's Law (Eqs. 2.16 and 2.24). The mean range calculated using the Funnel Formula expresses the time required to process the WIP currently on a workstation at a constant output rate. In contrast, Little's Law describes how long on average a newly arriving order will have to stay at a workstation until it is completed. An extensive comparison of these two equations can be found in Sect. 2.3.

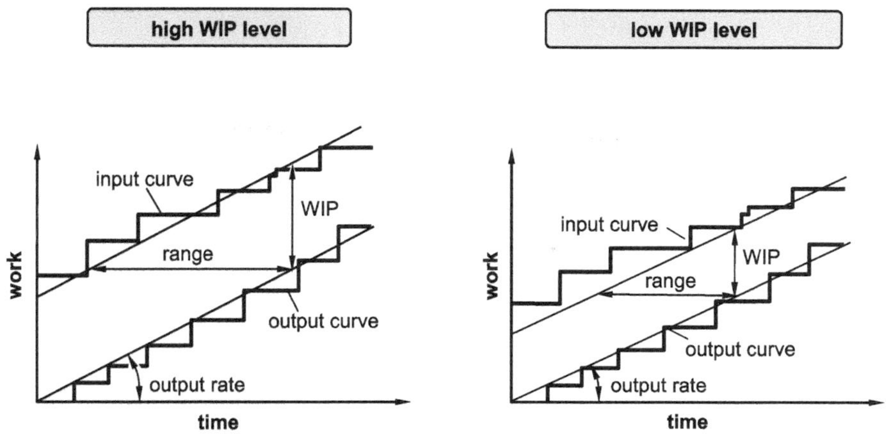

Fig. 5.2 Second Basic Law of Production Logistics

In order to reduce the WIP there are basically only two possibilities, either temporarily restrict the input or temporarily increase the capacity and therefore the output rate. The emphasis here is on temporarily. The extent to which these measures should be implemented both time-wise and quantitatively, results directly from the targeted WIP reduction. If the WIP on a workstation is to be reduced by 20 hours then the capacity can be increased for example by 4 hrs/SCD of overtime for 5 days or by 2 hrs/SCD of overtime for 10 days. Alternatively, a number of orders with a corresponding work content could be transferred to another workstation or placed externally. Finally, it is also possible to postpone orders, which have not yet entered into the manufacturing process (in the order of magnitude of the targeted throughput time reduction). Once the measures have been ended, it has to be ensured that the load and output rate once again run parallel to another.

5.3 Third Basic Law of Production Logistics

The WIP and throughput time cannot always be reduced with the given work content structure, without impacting the utilization. Depending on the targeted utilization, a certain level of WIP has to be maintained on a workstation. The third Basic Law of Production Logistics describes the correlation between the utilization, WIP and throughput time:

Decreasing the utilization of a workstation allows the WIP and throughput time to be disproportionately reduced.

When WIP dependent utilization losses are to be avoided on a workstation, a higher level of WIP has to be maintained on it in order to compensate for the stochastic influences, especially, the fluctuations in the order input. However if a minimal loss of utilization can be accepted, then the WIP can be greatly reduced.

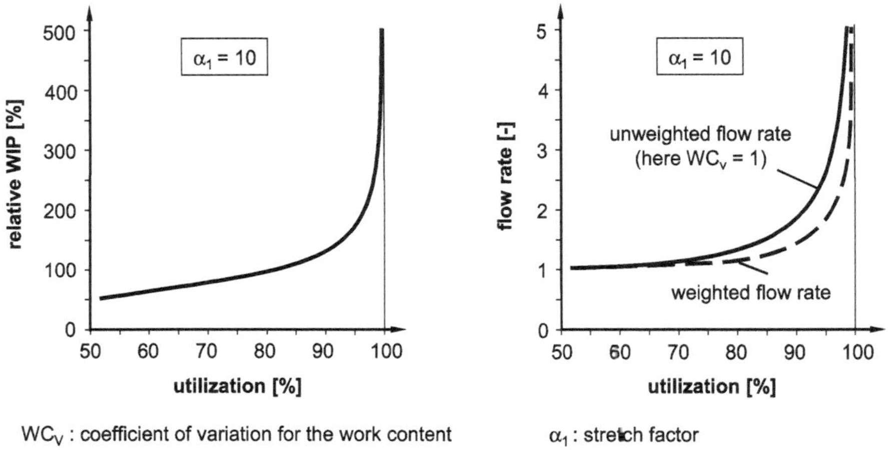

Fig. 5.3 Third Basic Law of Production Logistics

According to the Logistic Operating Curves Theory, with a relative WIP of 200% (the work in process is thus twice that of the ideal minimum WIP) the loss of utilization is approximately 3.5% (assuming: $\alpha_1 = 10$). By clearly decreasing the utilization above and beyond that the arriving orders can be processed on a workstation with almost no queuing time. As a result, there is to a large extent a proportional correlation between the WIP and the utilization (Fig. 5.3, left).

Similar statements can be made for the flow rate and – since it is defined as the ratio of throughput time to operation time – for the throughput time as well. When the utilization value is low, there are no or minimal queuing times and the flow rate therefore is close to one. Since the operation time theoretically represents the minimum throughput time, the flow rate can generally not be smaller than one. Should a higher system utilization be desired, the flow rate will increase greatly as a result (Fig. 5.3, right). The more inhomogeneous the work content of the orders to be processed is, the more the flow rate and therefore the throughput time will escalate when the utilization increases.

5.4 Fourth Basic Law of Production Logistics

The amount of work which has to be available on a workstation in order to avoid output losses due to disruptions in the material flow is primarily determined by the ideal minimum WIP. With a given capacity, the ideal WIP determines both the slope of the Output Rate Operating Curve as well as the required WIP buffer. The fourth basic law of production logistics addresses this.

The variance and mean of the work content determine the logistic potential of the shop.

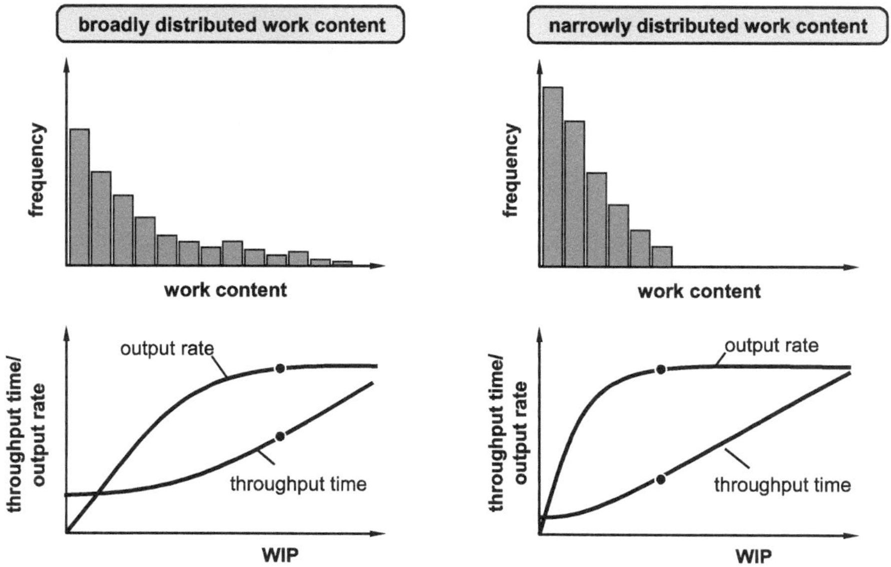

Fig. 5.4 Fourth Basic Law of Production Logistics

Figure 5.4 illustrates the impact of a broadly distributed work content: The greater the standard deviation, the lower the slope of the Output Rate Operation Curve and therefore the greater the required throughput time and WIP level.

Assuming that the transport times are insignificant, the logistic potential of a shop is significantly dependent on the degree to which the orders' work content can be harmonized on an as low as possible level. As can be seen in Eq. 4.5 this harmonizing has to be emphasized especially here, as the variance of the work content is squared when it enters into the ideal minimum WIP. If other logistic potential is developed in order to reduce the throughput time and WIP, then it is often more effective to harmonize the work content by selectively splitting the order lots rather than equally reducing the lots size of all the orders.

5.5 Fifth Basic Law of Production Logistics

The theoretical limits of the key performance figures are clearly described by the ideal minimum WIP and the ideal Logistic Operating Curves based on it. However, on the shop floor there are a number of influential factors which impact how the real Logistic Operating Curves deviate from the ideal ones. If the situation on a workstation is characterized by rigid capacities and simultaneous strongly fluctuating loads, then large WIP buffers are required in order to prevent a loss of utilization. However if the capacity can be flexibly adjusted to the load, or if the load can be shifted time-wise in order to correspond to the system's possible ca-

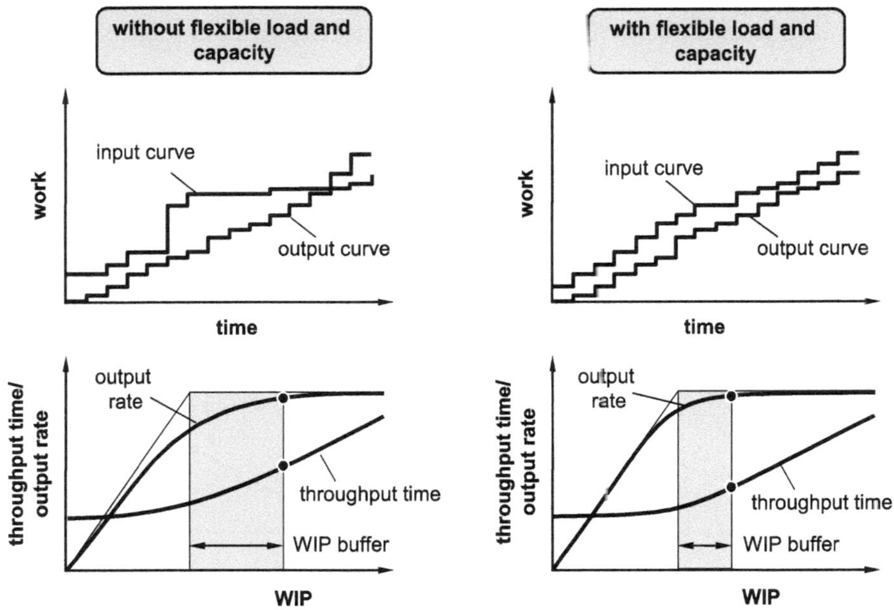

Fig. 5.5 Fifth Basic Law of Production Logistics

pacities, then the required WIP buffer is clearly reduced (Fig. 5.5). These conclusions are summarized by the fifth basic law of production logistics:

The size of the WIP buffer required to ensure the utilization of the workstation is mainly determined by the flexibility of the load and capacity.

In the Logistic Operating Curves Theory, these circumstances can be taken into consideration by changing the stretch factor α1 (see also Sect. 4.7.1, Fig. 4.21).

5.6 Sixth Basic Law of Production Logistics

In general, the Logistic Operating Curves depict the mean values of the key logistic parameters. This is sufficient for many problems. However, differentiated statements are preferred for the individual throughput times, especially for scheduling. Laws six through nine, explained in the following, provide a basis for this.

The sixth basic law of production logistics states:

When the orders are processed according to the FIFO principle, the inter-operation time is independent of the operation's individual work content.

The inter-operation time between two operations is primarily determined by the WIP level and not by the individual work content when the First In – First Out principle is applied (Fig. 5.6). The inter-operation time is therefore a workstation

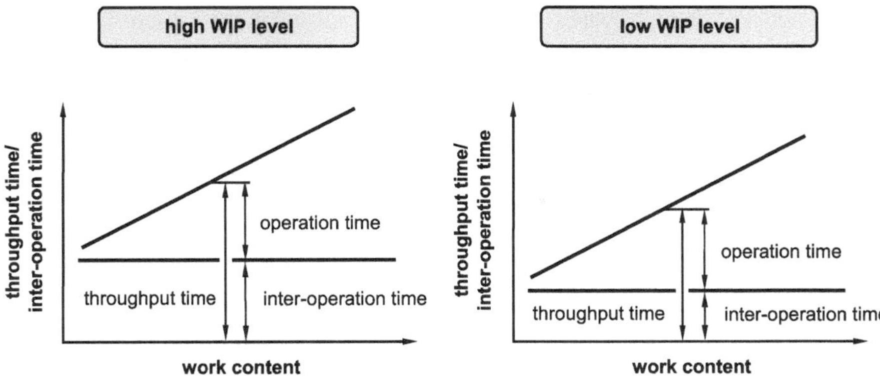

Fig. 5.6 Sixth Basic Law of Production Logistics

specific parameter. Assuming this sequencing rule, the individual throughput time increases linearly with the work content corresponding to the operation time.

If, however, other sequencing rules are applied, then it is not possible to draw conclusions about the individual inter-operation and throughput times, because, with regards to the deciding factors for the respective sequencing rules, both are determined to a great extent by the structure of the actual pending work (e. g., work content or remaining slack).

5.7 Seventh Basic Law of Production Logistics

The mean (unweighted) throughput time can also be influenced by the sequencing rules on a workstation. The level of the existing WIP as well as the variance of the work content are crucial here (Fig. 5.7). The sequencing rules can only impact the

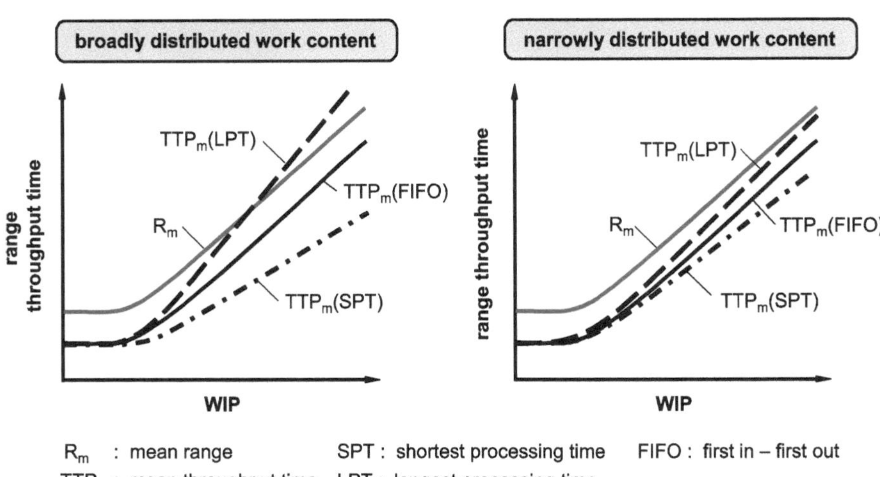

Fig. 5.7 Seventh Basic Law of Production Logistics

throughput time though, when there are enough orders on the system i. e., when the WIP level is high (see also Fig. 4.13). The lower the WIP is the more the natural sequencing rule FIFO has to be followed. Furthermore, the more uniform the work content is the less significant the sequencing rules are for the throughput time (see Eqs. 4.42 to 4.46). These conclusions are summarized in the seventh basic law of production logistics:

The mean throughput time can be influenced significantly by sequencing rules only when there is a high WIP level and a broadly distributed work content.

5.8 Eighth Basic Law of Production Logistics

The variance of the throughput time cannot fall below a certain level even when there is a generally constant WIP level and the orders are processed according to the FIFO principle. This lower bound is essentially based on the distribution of the work content and therefore the operation time. Each of the sequencing rules that deviate from the FIFO principle, increases the variance of the throughput time. The more WIP there is on the workstation, the greater this increase will be (Fig. 5.8; see also [Ludw-94]). Furthermore, throughput time variance can also be caused by the WIP variance. The eighth basic law of production logistics summarizes these correlations:

The throughput time variance is determined by the applied sequencing rule, the WIP level and the distribution of the work content.

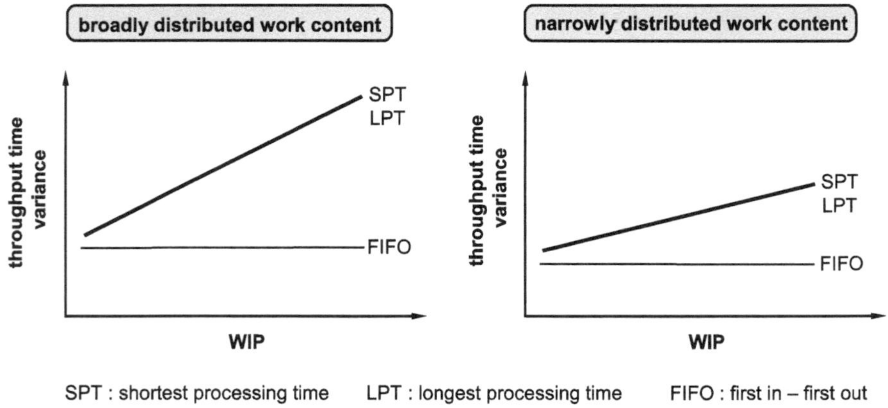

Fig. 5.8 Eighth Basic Law of Production Logistics

5.9 Ninth Basic Law of Production Logistics

The reliability of a production area's logistic processes is significantly determined by whether or not it is possible to achieve low and above all uniform throughput times. Only then can the production process be planned:

The logistic process reliability is determined by the mean value and distribution of the throughput time.

Greater throughput time variance inevitably leads to uncertainties when scheduling the orders (Fig. 5.9). In order to counteract this, the planned throughput time is often increased in order to ensure a larger time buffer. By intervening in the processing sequence, urgent orders should be able to be completed on time. Doing so however, further increases the throughput time variance. The vicious cycle of production planning and control, which was outlined in Chap. 1, thus develops and can only be stabilized at a high level. This vicious circle can be most reliably broken through low and stable throughput times.

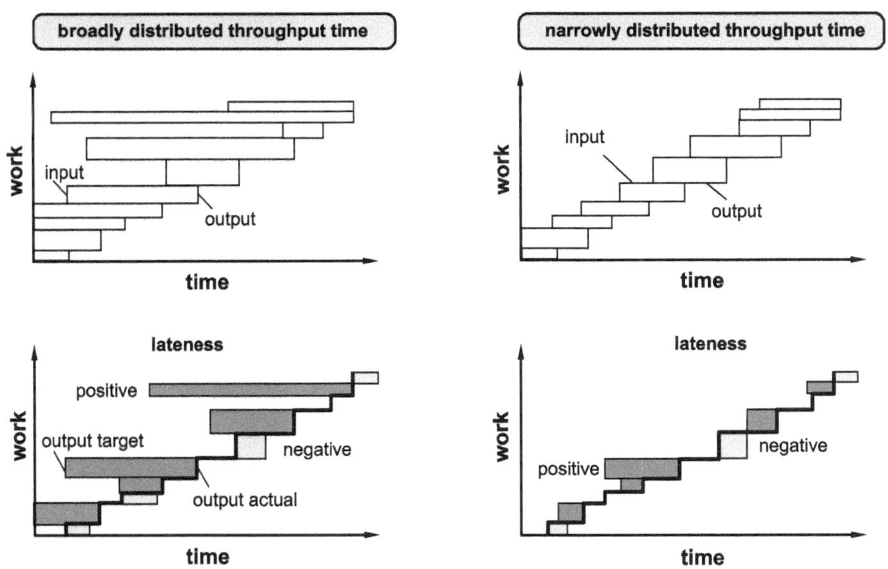

Fig. 5.9 Ninth Basic Law of Production Logistics

Chapter 6
Applications of the Logistic Operating Curves Theory

In order to apply the Logistic Operating Curves Theory it is essential to sufficiently consider the model premises and to ensure that the data entered is error free. When this is not the case it has to be possible to at least evaluate the imprecision of the model's outcomes that results from the data errors. In order to increase the reliability of such evaluations we will now extensively discuss developing and implementing calculated Logistic Operating Curves based on a concrete example (Sects. 6.1 and 6.2). In doing so we will first draw upon simulation data. This not only allows us to ensure that all of the required data is complete and of high quality, but it also makes certain that the application prerequisites are exactly adhered to.

In practice the necessary data is not always complete nor error free. In particular the capacity data frequently does not correspond with the situation on the shop floor. This is also true to some degree for other data such as the work content or transport time. Nevertheless, it is generally possible to develop and utilize Logistic Operating Curves. In individual cases, it then needs to be taken into consideration that – depending on the type of data error – the possibilities for drawing conclusions as well as the accuracy of the model are more or less limited (Sects. 6.3 and 6.4).

When the application prerequisites are met there is an exceedingly wide variety of possible applications for the Logistic Operating Curves. In Sect. 6.5 we will provide an overview of decision making models that are supported by them. Methods based on these models, therefore, also have LOC techniques integrated into them and as such are characterized by the fact that the logistic objectives are relevant parameters within the decision making process.

6.1 Developing and Analyzing Calculated Logistic Operating Curves

In the following section, we will use feedback data from an order sample processed on a workstation with two workplaces as an example for developing and analyzing Logistic Operating Curves. The workstation, a multi-spindle drill is part of the workshop at the automobile supply manufacturer we already introduced in Sects. 3.2.2 and 4.1.

6.1.1 Calculating the Logistic Operating Curves

In order to calculate Logistic Operating Curves the ideal minimum WIP ($WIPI_{min}$), maximum possible output rate ($ROUT_{max}$) and stretch factor α_1 first have to be determined.

The $WIPI_{min}$ is derived according to Eqs. 4.5 and 4.55 from the mean work content (WC_m), the coefficient of variation for the work content (WC_v), the mean transport time (TTR_m) and the number of workstations. In order to determine both the WC_m and the WC_v, the work content of the orders to be processed on the workstations was evaluated. The results are depicted as a distribution graph with the relevant key figures in Fig. 6.1. The transport time between two operations was consistently 6 minutes (0.1 hrs).

Based on this data, the ideal minimum WIP per workplace $WIPI_{min,i}$ according to Eq. 4.5 is:

$$WIPI_{min,i} = WC_m \cdot \left(1 + WC_v^2\right) + TTR_m$$

$$WIPI_{min,i} = 6.68 \text{ hrs} \cdot \left(1 + 0.86^2\right) + 0.1 \text{ hrs} = 11.72 \text{ hrs}$$

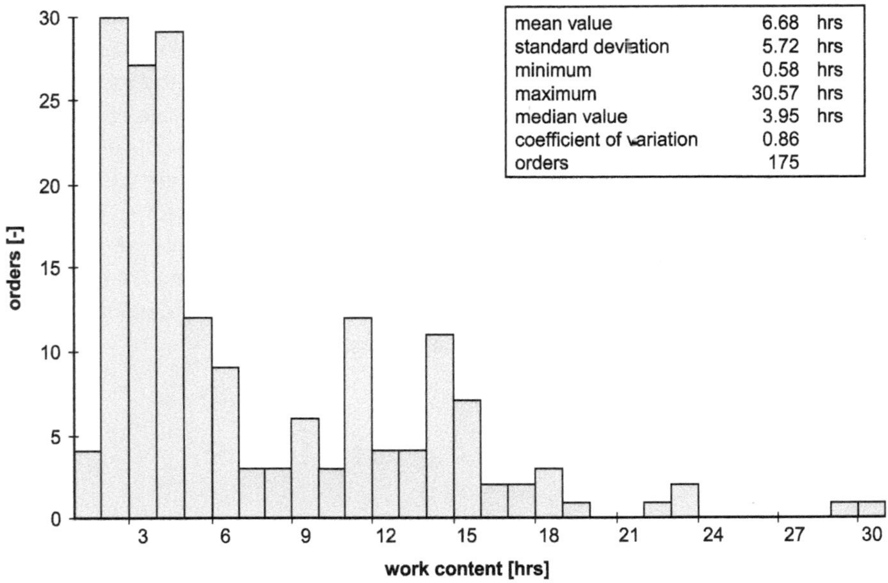

Fig. 6.1 Distribution of the Work Content for the Multi-Spindle Drill Workstation

and since the workstation has two workplaces, in total (see Eq. 4.58):

$$WIPI_{min} = 2 \cdot 11.72 \text{ hrs} = 23.44 \text{ hrs}$$

The capacity for each of the workplaces was 8 hrs/SCD. Capacity disruptions were not considered in the simulation and the rate of efficiency was assumed to be one (100%). The total $ROUT_{max}$ thus corresponds to the capacity 2·8 hrs/SCD = 16 hrs/SCD.

In the simulation, both of the workplaces were supplied by a common WIP buffer. The orders were allocated solely according to the available capacity. With such a constellation, it is to be expected that in comparison to two independent workplaces the WIP buffer can be halved without negatively impacting the reliability of the utilization (see Sect. 4.8.2). This can be reflected when calculating the Output Rate Operating Curve by halving the stretch factor. Instead of the default value of $\alpha_1 = 10$ that was recommended in Chap. 4, the calculation should therefore be made with $\alpha_1 = 5$. Furthermore, the orders were sequenced according to the slack rule, thus when calculating the Throughput Time Operating Curve we can assume that the sequencing rule was independent of the work content.

With this information the Logistic Operating Curves for the output rate, throughput time, and inter-operation time can now be determined. Based on the equations introduced in Chap. 4, a total of 10 variate pairs were calculated for different values of t, the results are displayed in Table 6.1. In accordance with Sect. 4.2.2 (Fig. 4.7) the running variable was chosen with a different increment. The slope of the curve could thus be depicted well in both the proportional and transition zones despite few calculated points.

In Fig. 6.2, the values calculated in Table 6.1 for the output rate, range, throughput time and inter-operation time (see Eqs. 4.30 to 4.41) are plotted over the WIP and are joined to one another in order to form the corresponding Logistic Operating Curves. Furthermore, the ideal Output Rate Operating Curve is depicted using the values that were determined for the ideal minimum WIP and maximal possible output rate. It is also worth noting that the value range for the calculated Logistic Operating Curves is generally limited to $(\alpha_1 + 1) \cdot WIPI_{min}$ (see also Sect. 4.2.2). This is usually sufficient enough for describing the situation on the shop floor, however, it is also not problematic to cover a larger WIP range if a specific case requires

Table 6.1 Calculating the Logistic Operating Curves for the Multi-Spindle Drill Workstation

No.	t	U_m (t)	$ROUT_m$ (t)	WIP_A (t)	WIP_B (t)	WIP_m (t)	R_m (t)	TTP_m (t)	TIO_m (t)
	-	[%]	[hrs/SCD]	[hrs]	[hrs]	[hrs]	[SCD]	[SCD]	[SCD]
1	0.0001	34.4	5.50	8.06	0.01	8.07	1.47	0.85	0.01
2	0.001	54.3	8.69	12.73	0.12	12.85	1.48	0.86	0.03
3	0.01	78.1	12.50	18.32	1.17	19.49	1.56	0.94	0.11
4	0.03	88.4	14.14	20.72	3.52	24.23	1.71	1.10	0.26
5	0.0625	93.8	15.00	21.98	7.33	29.30	1.95	1.34	0.50
6	0.1	96.3	15.41	22.58	19.72	34.30	2.23	1.61	0.77
7	0.2	98.8	15.81	23.16	23.44	46.60	2.95	2.33	1.50
8	0.4	99.8	15.97	23.40	46.88	70.28	4.40	3.78	2.95
9	0.7	100.0	16.00	23.44	82.04	105.48	6.59	5.98	5.14
10	1.0	100.0	16.00	23.44	117.21	140.65	8.49	8.17	7.34

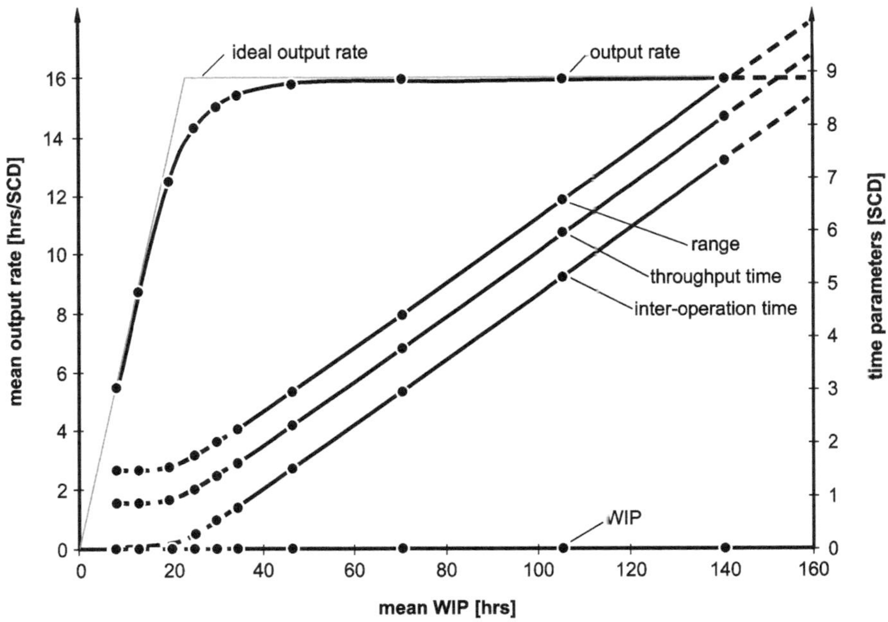

Fig. 6.2 Calculated Logistic Operating Curves for the Multi-Spindle Drill Workstation

it: Since the output rate has already reached its maximum value and is thus no longer dependent on the WIP, the corresponding LOC can be extended parallel to the axis. The Logistic Operating Curves for the various objectives can also be extrapolated by applying the Funnel Formula to calculate the range.

The graph indicates that $R_{min} \approx 1.47$ SCD and that a loss of utilization is first to be expected on the workstation when WIP_{total} goes below 45 hrs and R below 3 SCD. When it is assumed that the sequencing rule applied was independent of the work content, then the mean throughput time, independent of the WIP, is approximately 0.61 SCD less than R (see Table 6.1). The minimum inter-operation time is 0.1 hrs.

6.1.2 Applying Logistic Operating Curves for Analyzing a Simulated Manufacturing Process

In order to verify the above statements, we will now evaluate a chosen simulated manufacturing process using the PROSIM III simulator introduced in Chap. 3. In Fig. 6.3 a Throughput Diagram based on the feedback data from a 60-day evaluation period (three intervals, each 20 SCD) created in the simulation is illustrated. In addition, the throughput elements, which describe the flow of the individual orders on the workstation, are plotted. The most important figures for the intervals and the total evaluation period are summarized in Table 6.2.

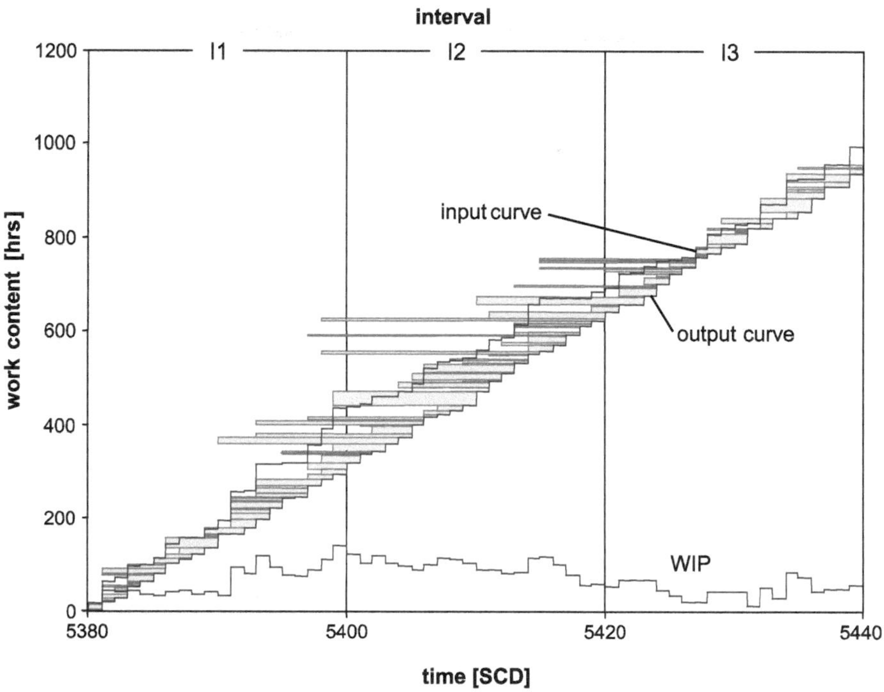

Fig. 6.3 Throughput Diagram for the Multi-Spindle Drill Workstation (Intervals I1 to I3; Simulation Results)

Table 6.2 Key Interval Related Figures for the Multi-Spindle Drill Workstation (Intervals I1 to I3; Simulation Results)

key figures		intervals			
		I1	I2	I3	I1 to I3
output	[hrs]	319.2	322.9	315.6	957.6
mean output rate	[hrs/SCD]	16.0	16.1	15.8	16.0
mean WIP	[hrs]	69.1	91.5	46.0	67.8
mean range	[SCD]	4.3	5.7	2.8	4.3
mean weighted throughput	[SCD]	2.7	6.5	3.5	4.3
mean unweighted	[SCD]	2.2	6.0	3.0	3.8
mean inter-operation time	[SCD]	1.4	5.1	2.1	3.0

As can be seen in the diagram during the evaluation period there was never a steady process state. However, there were also no long phases where there was a WIP dependent loss of output rate. During the first interval (I1) a stockpile developed, but was subsequently reduced before the end of the second interval (I2). Only during the third interval (I3) did the WIP become relatively stable. Since the simulation was run with a constant capacity throughout the evaluation period, the WIP fluctuations can for the most part be traced back to an irregular input. Moreover, the throughput elements especially during I2, although to some degree also in I3, illustrate that frequent order sequence interchanges resulted due to the usage of slack time sequencing.

The WIP in interval I3, as well as in the first half of interval I1, was evidently just high enough that both workplaces could be constantly supplied with work. WIP dependent losses of utilization could have occurred only briefly during days 5381, 5431 and 5433. The throughput elements in the diagram also show that on these days, there were obviously only one order on the entire workstation and therefore one of the workplaces could not be utilized. However, since feedback data was only collected on a daily basis, there is a level of uncertainty connected to these outcomes.

In Fig. 6.4 the values for the output rate, WIP and range of the separate intervals, as well as for the entire evaluation period were transferred to the previously calculated Logistic Operating Curves. The states which are described with regards to the WIP and output rates for the different intervals are also plotted on the LOC. The values for I2 are clearly in the overload zone of the LOC, whereas those during I3 are in the transition zone. Finally, the values in I1 are between those for the other two intervals. The conclusions drawn from the Throughput Diagram are therefore confirmed.

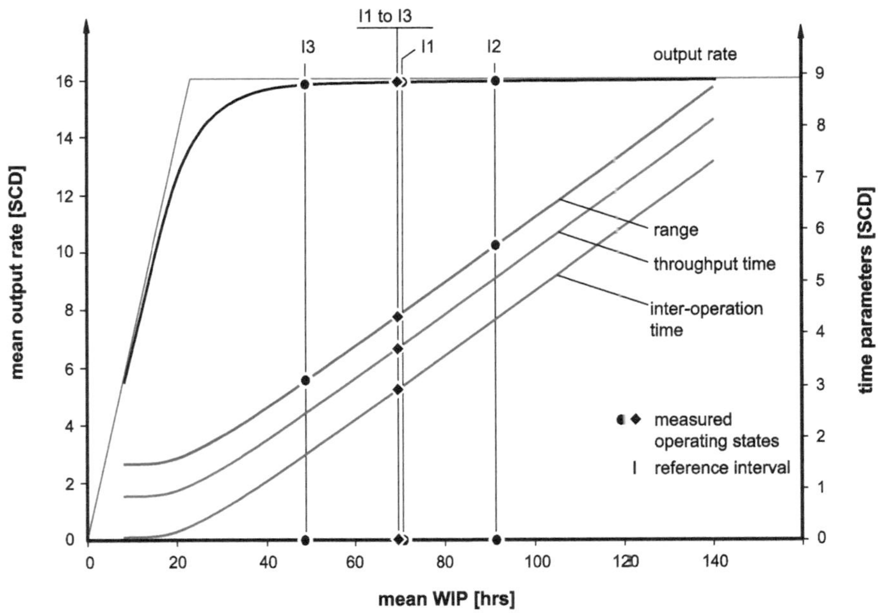

Fig. 6.4 Key Interval Related Figures in the Calculated Logistic Operating Curves (Interval Range I1 to I3)

In calculating the Throughput Time Operating Curve with Eqs. 4.37 and 4.38, it is assumed that during the evaluation period there was no tendency for the WIP to significantly change (see Sect. 4.4). Particularly for short evaluation periods it has to be ensured that the key throughput time figures are not strongly influenced by the order sequencing rules. A simple and at the same time quite reliable method for verifying this prerequisite is to compare the key values for the mean range and mean weighted throughput time: If these values largely agree with one another, the condition is generally met.

Based on the figures in Table 6.2 it can be seen that the throughput time figures strongly deviate from one another during the different intervals. In contrast to I1 and I2, the WIP level in period I3 is relatively constant. However, at the beginning a number of orders that had above average waiting times were processed (see Fig. 6.3) and therefore, the key figures do not agree with another during this interval either. For this reason, we have not plotted the key throughput time figures that were measured for the separate intervals in the Throughput Diagram.

Nevertheless, the mentioned prerequisites have been met for the entire evaluation period (I1 to I3) as is reflected by the strong agreement between the measured points and the calculated outcomes (see Fig. 6.4). Therefore, the equations for calculating the throughput time and inter-operation time are also confirmed.

As the above example demonstrates, given the work content and capacity structures, the level of WIP that is attainable on a workstation without a loss of utilization can be easily shown using the calculated Logistic Operating Curves. The mathematical and simulation models for the output rate strongly agree with each other during both the individual intervals as well as during the entire evaluation period. In comparison, the figures for the throughput time and inter-operation time behave somewhat differently. When the implemented sequencing rule is independent of the work content, the LOC Theory indicates the throughput time that can be expected given the WIP level. As was the case in our example, it is generally possible for the values to temporarily deviate from those calculated and anticipated, especially when the evaluation period is short. The longer the evaluation period is, the more reliable the results calculated for the throughput time and inter-operation time are.

6.2 Evaluating Alternative Methods for Developing Potential for Logistic Improvement

The Logistic Operating Curves calculated for the analyzed workstation show that the practical minimum WIP is approximately 45 hours. Without changing the basic structural conditions, it is not possible to further reduce the WIP and throughput time without accepting losses of utilization.

In this next section we will base our discussion on the following scenario: Due to extreme competitive pressure, the company wants to reduce the WIP to approximately 25 hrs. When no supporting structural interventions are taken, the output rate will decrease from 16 hrs/SCD to approximately 14 hrs/SCD at the demanded WIP (see Fig. 6.2 and Table 6.1). However, the company finds this extent of utilization loss unacceptable; additional measures therefore have to be taken in order to increase the system's logistic potential.

The LOC Theory can also be utilized here in order to *derive and evaluate alternative measures*: A Logistic Operating Curve can always be calculated when the necessary parameters exist. Since a new LOC results every time one of these parameters are changed it seems obvious to use the approximation equation to describe how the different measures influence the logistic potential of a workstation and by doing so to support business decisions.

Based on the chosen example, Fig. 6.5 illustrates the changes in the Logistic Operating Curves that result from intervening in the structural conditions. If the minimum WIP is reduced through smaller and/or more uniform work content or by accelerating the transport (reducing the minimal inter-operation time: Fig. 6.5, upper), the slope of the Output Rate Operating Curve is then steeper and the minimum throughput time is shorter. If the total output rate does not change, comparing the LOC can immediately determine whether or not the respective measures create any throughput time or WIP potential. Furthermore, the extent to which the logistic objectives can be influenced by supplying additional capacities or changing the employed technology (leading to reduced standard hours) can be assessed. If a new manufacturing technology which allows a faster processing rate is employed, then the work content is decreased and consequently the ideal minimum

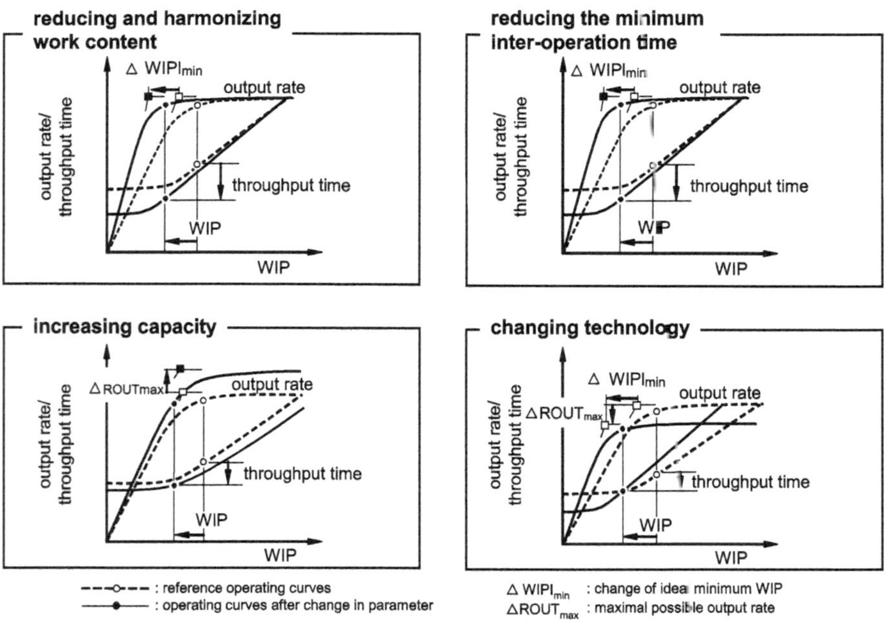

Fig. 6.5 Impact of Measures for Influencing the Logistic Potential on Calculated Logistic Operating Curves

WIP as well. Therefore, as long as the structure of the order is not changed (e. g., through additionally acquired orders) the system's maximal possible output rate will also be reduced, because when the order volume remains unaltered there is less work content to process (Fig. 6.5, lower).

As an example, in the following section we will examine two possible measures qualitatively and quantitatively using the LOC Theory. In particular we will demonstrate how to *evaluate measures using Logistic Operating Curves* and point out a number of the method's special characteristics. Furthermore, we will demonstrate the necessity of verifying if and to what extent the variables $WIPI_{min}$, $ROUT_{max}$ and α_1 can be changed either as a direct or indirect consequence of the discussed measure, for each of the measures being evaluated.

6.2.1 Varying the Work Content Structure

The particular influence that the distribution of the work content's structure has on a manufacturer's logistic potential is explained by the Logistic Operating Curves Theory. From the fourth basic law of production logistics it follows that the attainable throughput time and WIP level can be strongly influenced by specifically manipulating the work content structure.

In addition to numerous small orders, a number of very large orders with a work content of much more than 10 hours had to be processed on the observed workstation (see Fig. 6.6a). These orders were primarily responsible for the large standard deviation and coefficient of variation for the work content. It stands to reason

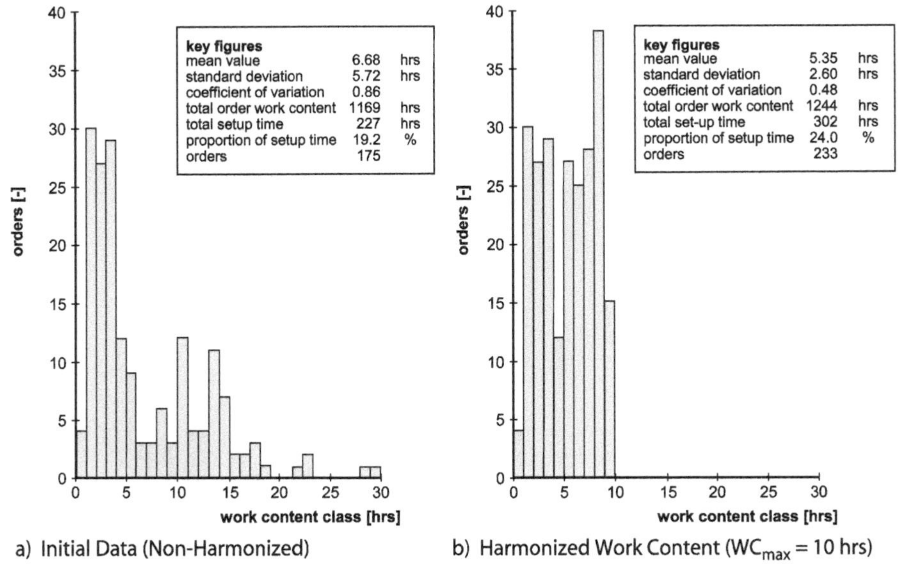

Fig. 6.6 Comparison of Harmonized and Non-Harmonized Work Content Distributions ($WC_{max} = 10$ hrs)

then, that specifically these orders should be divided in order to achieve a lower minimum WIP as well as to reduce the required WIP buffer. This procedure is called *harmonizing the work content.* In Fig. 6.6b, it can be seen how the shape of the distribution for the sample application could be influenced by such a harmonization. All of the orders whose work content was greater than 10 hours were divided into two to four sub-orders. In this example, 51 of 175 orders were affected by these measures. The mean work content WC_m was reduced relatively weakly from 6.68 hours to 5.35 hours (approx. 20%). In contrast, it could be determined that the standard deviation WC_s in particular was strongly influenced: The targeted manipulation of the distribution shape resulted in a reduction from 5.72 to 2.60 hrs (almost 55%). Even so, the coefficient of variation WC_v as a measure of the relative variability was decreased by 44%. Thus, according to Eq. 4.2 the minimum WIP was reduced from 23.4 to 13.4 hours (43%).

However, splitting the lots does not have only positive aspects: Due to frequent order changes more effort was required for the setup of the workstations. Consequently, the setup time portion increased from 19.2% to 24%. Therefore, the total work content increased 6% from 1169 hours to 1244 hours. In order to cover the additional effort each of the workstation's capacity had to be increased by 0.5 hrs/SCD (≈ 6%). The stretch factor however, should not be influenced by harmonizing the work content, because neither the principal course of the load nor the flexibility of the capacity is specifically changed.

The Logistic Operating Curves that result under these premises are presented in comparison to the initial situation in Fig. 6.7. In order to simplify the interpretation, a target operating point was defined for both the states with a relative WIP of 200% and plotted in the diagram. In the graph, it can be seen that by harmonizing the

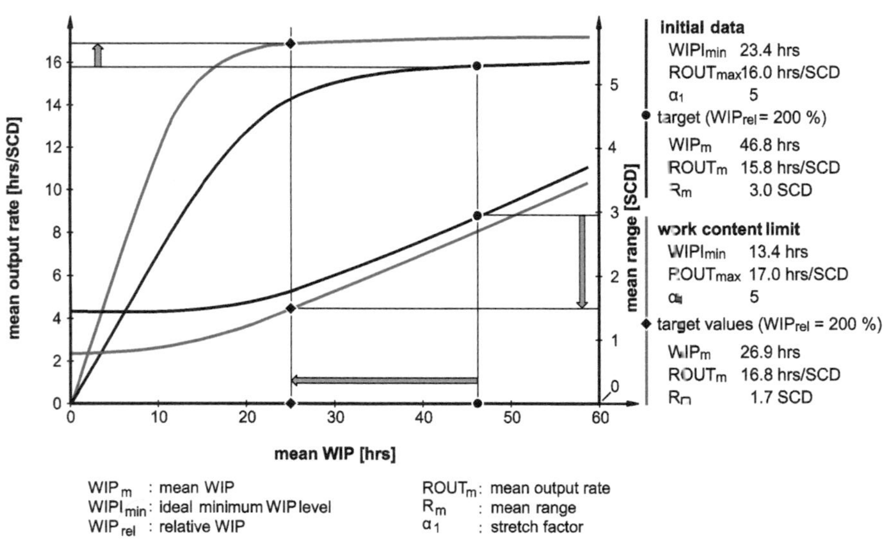

Fig. 6.7 Influencing the Logistic Potential by Harmonizing the Work Content

work content and increasing the capacities accordingly, it is possible to reduce the mean WIP from approximately 47 hours to 27 hours (approx. 45%) and the mean range from 3.0 SCD to 1.7 SCD (approx. 43%). The initially mentioned logistic objectives can therefore clearly be attained by implementing these measures.

6.2.2 Varying the Capacity Structure

It is also possible to support the stated objectives by changing the capacity's structure. Thus, instead of having one shift on two workplaces, one of the workplaces can be shutdown and the other one can be operated in two shifts. This measure is clearly indicated when both the upstream and downstream manufacturing areas also have two shifts. By employing this measure the entire system's capacity as well as the structure of the work content remains unchanged. Nonetheless, it can be expected that the logistic performance of the system will be significantly impacted.

This is mainly due to the fact that the ideal minimum WIP is halved, because it only has to be provided for the workplace that continues to be used. Based on the Funnel Formula it follows that the minimum range R_{min} is also halved when the maximal possible output rate $ROUT_{max}$ and the reduced ideal minimum WIP $WIPI_{min}$ remain the same. This can also be clearly explained: The capacity with which an order is processed doubles by rearranging the capacity structure. Thus, with the initial state (where the capacity is 8 hrs/SCD per workplace) the operation time of an order with a work content of 24 hours is 3 SCD. After the changes are made,

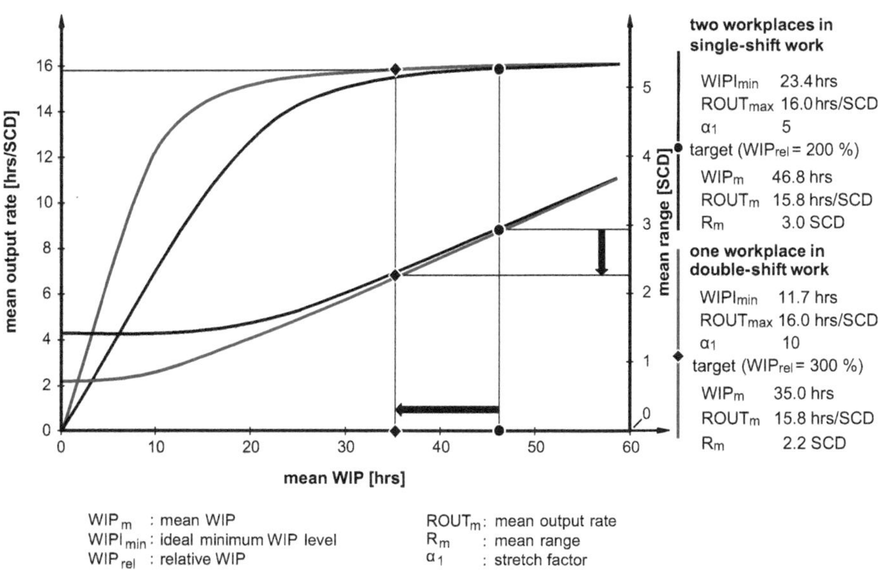

Fig. 6.8 Influencing the Logistic Potential by Varying the Capacity Structure

the operation time is reduced to 1.5 SCD because there is now a total capacity of 16 hrs/SCD available for processing the order.

Nevertheless, the WIP level cannot be decreased to 50% of the initial value without a loss of utilization, because the stretch factor α_1 also has to be adjusted to the changed situation: In the initial state the orders could be flexibly allocated to either of the workplaces and it could therefore be calculated with the reduced stretch factor $\alpha_1 = 5$ (see Sect. 6.1.1). By shutting down the one machine, the reason for reducing the stretch factor is no longer valid, thus the default value $\alpha_1 = 10$ is once again applicable

In Fig. 6.8 (above), the Logistic Operating Curves that result from applying the new parameters are shown. It needs to be noted though, that although the $WIPI_{min}$ is halved, the WIP buffer remains unchanged due to the simultaneous doubling of the α_1 value. The target values for the new structure of the capacities, which were defined for the comparison, were thus determined with a relative WIP of 300% in order to base them on similar output rates. Based on the graph, it also follows that in comparison to the initial state, the WIP and the range can only be reduced by 25% by changing the capacity structure. Therefore, the above stated logistic objectives cannot be fully achieved by applying these measures without supportive interventions.

6.3 Calculating Logistic Operating Curves with Missing or Incorrect Operating Data

Generally, the data required for the Logistic Operating Curves Theory is readily available since it is also required for other applications. The work content and capacity data are for example, also the basis for essential PPC functions. Nonetheless, the quality of the data is often unsatisfactory (see e. g., [Domb-88]).

In the following, we will individually address the relevance of possible data errors for the LOC Theory. On the one hand, we will investigate the question of how the different types of data errors influence the Logistic Operating Curves and on the other hand, we will discuss what influence each of the data errors exert on the production processes' key logistic performance figures (WIP, output rate and throughput time). Based on that we will investigate whether or not data errors jeopardize the application of the LOC Theory in general or if only individual outcomes are affected, and whether or not the LOC can be used to correct the errors with the help of a model.

6.3.1 Incorrect Work Content and Transport Time Data

The ideal minimum WIP has a decisive role within the context of the Logistic Operating Curves Theory. When this value is incorrect, the resulting Logistic Operating Curves is either too steep or too flat in the proportional zone. This indicates that the WIP buffer in the transition and overload zone, is incorrectly dimensioned. Therefore, the outcomes regarding the attainable minimum WIP and throughput time are also incorrect. The degree and the significance of the deviation though is strongly dependent on which parameter caused the error. According to Eq. 4.5, for an individual workplace only the mean work content (WC_m), the coefficient of variation for the work content (WC_v) and the mean transport time (TTR_m) come into consideration as a cause. We will now examine the significance of each of these three potential sources of error.

6.3.1.1 Case 1: WC_m incorrect; WC_v correct; TTR_m correct

First, we will assume that the data for the mean work content is incorrect, but that the shape of the distribution for the work content and thus the coefficient of variation are known with a good approximation. This can be the case for example, when there is a systematic error while determining the planned work content. Secondly, we will assume that the data for the mean transport time is correct. This 'Case 1' is comparatively uncritical for the application of the Logistic Operating Curves Theory, because it effects the parameters for the LOC equation in the same way it influences a few of the manufacturing processes' central measured values. Thus, the mean output rate, which according to Eq. 2.12 results from the sum of

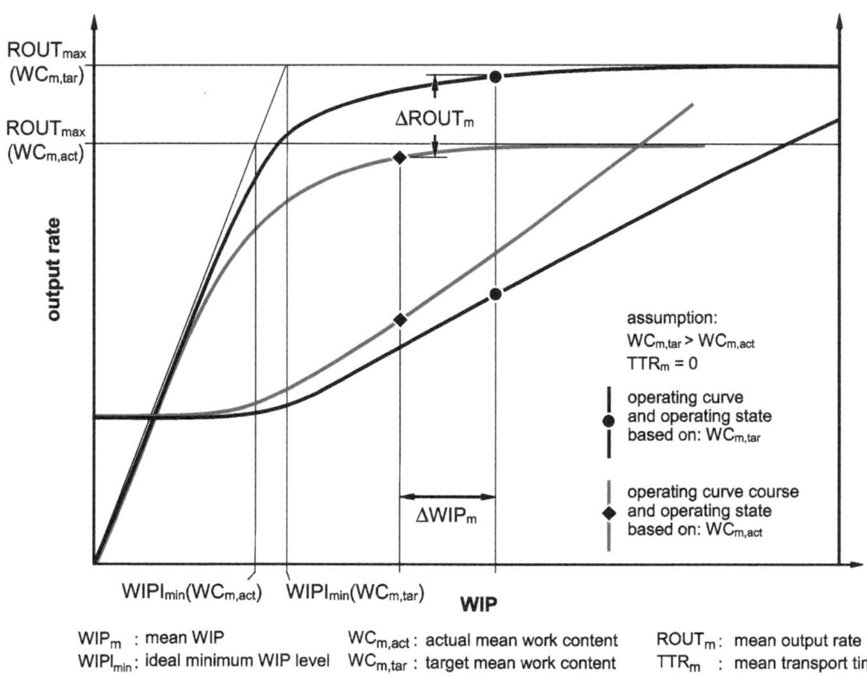

Fig. 6.9 Impact of Incorrect Work Content Data on Calculated Logistic Operating Curves and Operating Points (WC_m incorrect; WC_v correct; $TTR_m = 0$)

the work content, behaves proportionally to the error. The same applies for the output and input and thus the WIP (Eq. 2.14). In the Funnel Formula (Eq. 2.16) the error is reflected in both the denominator and numerator; the range and consequently the weighted throughput time (Eq. 2.18) are therefore not influenced. Finally, the unweighted throughput time, as a measured value, is generally independent of the work content (Eq. 2.10).

When this type of error occurs, the resulting absolute deviation of the $WIPI_{min}$ behaves proportional to the incorrect value. In contrast, the relative deviation is also determined by the transport time component. Ultimately, the $ROUT_{max}$ is also influenced proportional to the error: If the incorrect work content is not recognized and corrected, the capacity data or the value of the maximum possible utilization has to be adjusted at least for the mid-term. When it is not adjusted the measured output rate will permanently deviate from the maximum possible output rate even in the overload zone.

In consideration of all the abovementioned effects, errors in the mean work content are barely visible to the user, especially when the transport time is negligible. In this case, the relative WIP, which describes the position of the planned or attained operating point on the Logistic Operating Curves, is independent of the error. This is also valid for data about existing potential for improving the logistics as well as that which is to be developed through specific measures. The significance of this error with regards to the shape of the LOC and the measured operating point is summarized above in Fig. 6.9.

6.3.1.2 Case 2: WC_m correct; WC_v incorrect; TTR_m correct

For the second type of error we will presume that the mean work content and transport time are correct, but that there was a false assumption regarding the span of the work content's distribution (expressed by the coefficient of variation). This type of error occurs particularly when the distribution of the work content is not determined from past data but based only on representative orders or just estimated.

The measured value for the output rate and the unweighted throughput time is generally not influenced by errors of this nature. If and to what extent the WIP and therefore the range and weighted throughput time are influenced by this type (especially when the evaluation period is brief) cannot be determined exactly though. Nevertheless, since the behavior of the input and output are only temporarily influenced by the error and not their basic shape, significant deviations are also not expected for these key figures.

This is not the case for the ideal minimum WIP. With a given mean value for the work content, the ideal minimum WIP is decisively influenced by the coefficient of variation for the work content. Figure 6.10 shows how the ratio of the minimum WIP to the mean work content behaves when the coefficient of variation is changed. If, as is typical on the shop floor, this value range is between 0.5 and 2, (see also Fig. 2.2), then it follows that the $WIPI_{min}$ is between 1.25 and 5 times greater than the WC_m when the transport time is insignificant. All of the outcomes

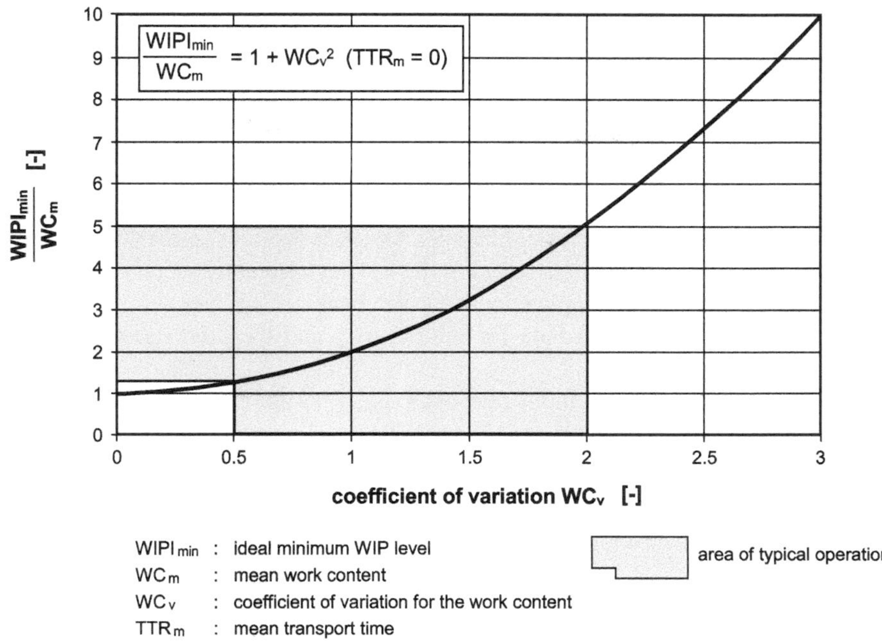

Fig. 6.10 Impact of the Coefficient of Variation for the Work Content on the Ideal Minimum WIP (single workplace, TTR = 0)

stemming back to the ideal minimum WIP, such as the data for the relative WIP and thus when applicable, the potential for logistic improvement, are therefore connected with a corresponding degree of uncertainty. On the one hand then, applying the Logistic Operating Curves Theory to assess and design manufacturing processes has to be critically evaluated when errors of this kind are involved. On the other hand, the outcomes regarding how the WC_v influences the workstations' required WIP and throughput time levels in particular, make it possible to integrate these aspects early into the design and layout of the manufacturing process.

6.3.1.3 Case 3: WC_m correct; WC_v correct; TTR_m incorrect

In the third error case, we will assume that the data regarding the transport time is false, but that the parameters for the work content are known with sufficient accuracy. As can be easily verified with the corresponding equations, the measured values for the key logistic performance figures (WIP, output rate, range, and throughput time) are not affected by such errors. Nevertheless, the resulting deviation of the ideal minimum WIP is as large as the difference between the assumed and actual mean throughput times. Consequently, all of the outcomes based on the incorrect ideal minimum WIP are also false. Nevertheless, this type of error is of secondary importance when the proportion of the ideal minimum WIP caused by the transport is small.

6.3.2 Missing or Incorrect Data for the Maximal Possible Output Rate

In the operational practice the maximum possible output rate parameter is largely unknown. Furthermore, there is frequently no dependable data about existing or available capacity through which a workstations output rate should be described. Despite this parameter's crucial significance for many of the company's planning functions, the quality of the data stored in the PPC system is frequently not maintained. Even in the sample applications that we evaluated in Sect. 4.4.2 there is a strong deviation between the data for the capacity and the workstation's actual yielded output rate.

Although the maximum output rate ($ROUT_{max}$) is one of the three input parameters for the Logistic Operating Curves Theory's key equations, errors of this type are not necessarily problematic. It is however necessary to correct the value assigned for $ROUT_{max}$ in special cases; extensive support for doing so is provided by the LOC Theory.

Next, we will discuss the impact of incorrect information about the $ROUT_{max}$. In Fig. 6.11 two pairs of Logistic Operating Curves are presented in which the $WIPI_{min}$ and the stretch factor α_1 are identical, but which differ with regards to the $ROUT_{max}$. It is assumed that the required value of $ROUT_{max}$ is larger than the actual value. In addition, an operating state is plotted in the graph at which the order input (the system's load) and therefore – when assuming a stationary manu-

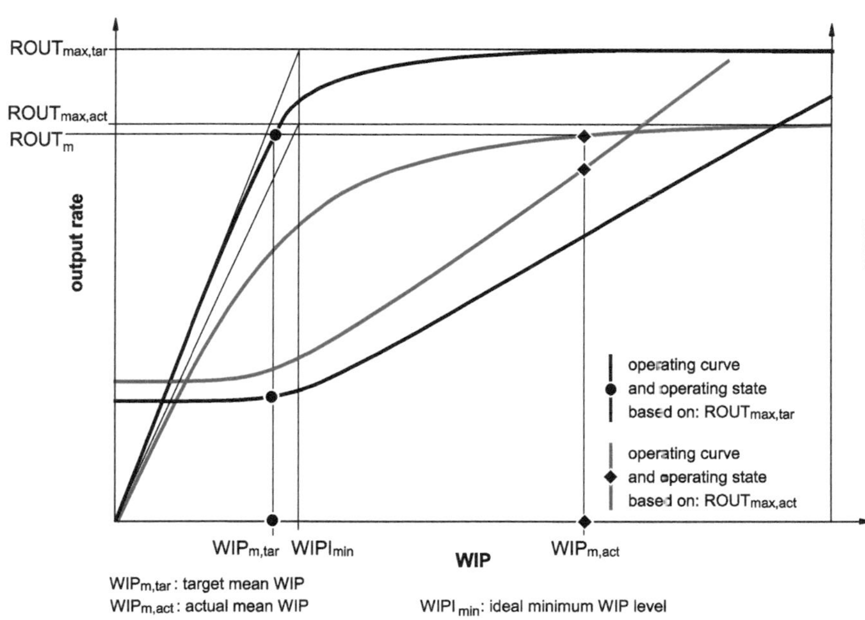

Fig. 6.11 Impact of the Maximum Possible Output Rate on the Logistic Operating Curves and Operating Points

facturing process – the output rate $ROUT_m$, is equally high in both cases. When the $ROUT_{max}$ is equal to the target value, the workstation will obviously be operating in the underload zone. Due to the resulting low utilization, it is possible to attain correspondingly low WIP and throughput time levels. Nevertheless, this is dependent on the operators being continuously available and newly arriving orders being processed without delay.

On the shop floor such operating states are rare, unless the workstation is fully automated. Generally, when there are currently no orders waiting operators are pulled from one station and employed on another. If a new order then arrives and the operator cannot be immediately ordered back (e. g., because the task begun on the one station needs to be finished first), it results in waiting times and a corresponding stockpile. Therefore, in the Logistic Operating Curves only the operator's actual attendance time should be considered when determining the maximum possible output rate (see below with regards to possible methods for determining $ROUT_{max}$). Even when the value of $ROUT_{max}$ is reduced comparatively minimally, if the same output rate as in the first case is to be yielded there will be a disproportionately higher WIP and throughput time level (see Fig. 6.11). The operating state will no longer be positioned in the overload zone, but will instead be in the transition or underload zone.

Thus, if the data for $ROUT_{max}$ is incorrect during the planning or when analyzing the process, the calculated WIP and throughput times will be unrealistic. This can be directly recognized particularly when analyzing the flow, since for example, the measured and calculated WIP value or even the corresponding value for the output rate will not agree with one another. The measured operating state therefore cannot be positioned on the Logistic Operating Curves. If the stretch factor as well as the measured ideal WIP and output rate are correct, then this deviation can only result from incorrect data for $ROUT_{max}$.

Here, the Logistic Operating Curves Theory itself offers the possibility to determine a typical, realistic value for the maximum possible output rate, based on the measured output rate and WIP data. This is quite simple when the workstation is clearly operating continuously in the overload zone, that is, when there is always WIP on the workstation. In such a case there can be no WIP dependent loss of utilization, therefore, $ROUT_{max}$ corresponds to the measured $ROUT_m$. An example of such a situation is presented in Sect. 4.7.2 (Overload Operating Zone, Fig. 4.26).

If on the other hand the observed workstation is operating in the underload or transition zone, then the WIP dependent loss of utilization can be determined with the help of the normalized Logistic Operating Curves (Sect. 4.5). According to the normalized LOC, at a relative WIP of 150% (defined as the ratio of WIP_m to $WIPI_{min}$), the WIP dependent loss of utilization is approximately 7%, and in comparison at a relative WIP of 300%, only approximately 1% (the numerical examples apply for a stretch factor of $\alpha_1 = 10$). The $ROUT_{max}$ can then be inferred from the determined loss of utilization and the measured mean output rate $ROUT_m$.

However, in addition to deriving the loss of utilization from the normalized Logistic Operating Curves (either with graphs or tables), they can also be calculated. Since the Output Rate Operating Curve is based on a constant function with an entirely positive slope, the $ROUT_{max}$ can be determined using the *Newtonian Iteration Method* (see also [Bron-95], [Möll-96]). First, the value of the running variable t for the normalized LOC equation (Eq. 4.48), with which the measured relative WIP is set in the operational practice, has to be found. This is obtained through the following equations:

$$F(t_k) = \left(1 - \left(1 - \sqrt[4]{t_k}\right)^4\right) + \alpha_1 \cdot t_k - \frac{WIP_{rel}}{100} \tag{6.1}$$

$$F'(t_k) = \left(\frac{\left(1 - \sqrt[4]{t_k}\right)^3}{\sqrt[4]{t_k^{\,3}}}\right) + \alpha_1 \tag{6.2}$$

and

$$t_{k+1} = t_k - \frac{F(t_k)}{F'(t_k)} \tag{6.3}$$

where $F(t_k)$ ancillary function ($F(t_k) = WIP_{rel}(t) - WIP_{rel}$; if $F(t_k) = 0$ solve for t),
$F'(t_k)$ first derivation of the ancillary function,
t_k, t_{k+1} value for the running variable t according to k or k + 1 iterations,
WIP_{rel} relative WIP [%],
$WIP_{rel}(t)$ relative WIP calculated for t [%],
α_1 stretch factor [-].

With this application, the function $F(t_k)$ generally converges with zero within 10 or less iterations, whereby it nevertheless needs to be determined that the initial value for t is smaller as the sought value. Thus, when using this method it is practical to start with a very small initial value for t (see Table 6.3). An iteration can be discontinued when:

$$F(t_k) = 0 \tag{6.4}$$

By transforming Eq. 4.30 the maximal possible output rate can be determined with the value found for t_k:

$$ROUT_{max} = \frac{ROUT_m}{\left(1 - \left(1 - \sqrt[4]{t_k}\right)^4\right)} \tag{6.5}$$

where $ROUT_{max}$ calculated maximal possible output rate [hrs/SCD],
$ROUT_m$ mean output rate (measured value) [hrs/SCD],
t_k the value of the running variable t after k iterations.

In order to better understand this method we will now discuss a specific application. The analysis data presented in the example for the overload operating zone in Sect. 4.7.2 will be used as our starting point. During the underlying investiga-

Table 6.3 Implementation of the Newtonian Iteration Method for Calculating the Maximum Output Rate (Example)

Iterations k	t_k	$F(t_k)$	$F'(t_k)$	t_{k+1}
1	0.000000001	-1.87769536	5529087.41	0.000000341
2	0.000000341	-1.80680995	65920.10	0.000027750
3	0.000027750	-1.63950932	2096.35	0.000809829
4	0.000809829	-1.36948012	129.67	0.011371031
5	0.011371031	-0.99198334	18.77	0.064216674
6	0.064216674	-0.31865124	10.96	0.093290593
7	0.093290593	-0.00713866	10.53	0.093968509
8	0.093968509	-0.00000218	10.52	0.093968716
9	0.093968716	0.00000000	10.52	0.093968716
		$ROUT_{max}$ (Eq. 6.5)	10.15	hrs/SCD
initial data:	WIP_m = 20.1 hrs		WIP_{rel} = 190 %	
	$WIPI_{min}$ = 10.6 hrs		$ROUT_m$ = 9.75 hrs/SCD	
			$\alpha 1$ = 10	

tion, it became clear that the capacity data did not correspond to the workstation's actual output rate. The company reported that the capacity for the chosen workstation was 13.5 hrs/SCD, however the feedback data indicated it was only 9.75 hrs/SCD. Despite this apparent under utilization, there was almost constantly WIP on the workstation – the relative WIP was approximately 190% (see Fig. 4.23).

The iteration method was applied using this data and the chosen stretch factor $\alpha_1 = 10$ (Table 6.3). The method was stopped here after 9 iterations. Using the resulting value for the running variable t_k, the maximum possible output rate $ROUT_{max}$ was then calculated according to Eq. 6.5 as 10.15 hrs/SCD.

With this procedure, the measured operating point is inevitably situated on the Output Rate Operating Curve calculated using $ROUT_{max}$. Whether this is realistic or not can be determined by comparing the calculated loss of utilization with the portion of time where there was actually no WIP on the workstation. The example in Sect. 4.7.2 documents this procedure. If the validation process indicates that the calculated results are not plausible, then it can only be traced back to an error in determining the relative WIP (see Sect. 5.3.1) or an error in the chosen stretch factor α_1 (refer to the following section).

6.3.3 An Incorrect Stretch Factor α_1

The stretch factor α_1 can only be calculated empirically, which essentially means that a level of uncertainty is connected to it. General advice on how to select the value of α_1 can be found in Sect. 4.7. If the underlying value of α_1 is incorrect,

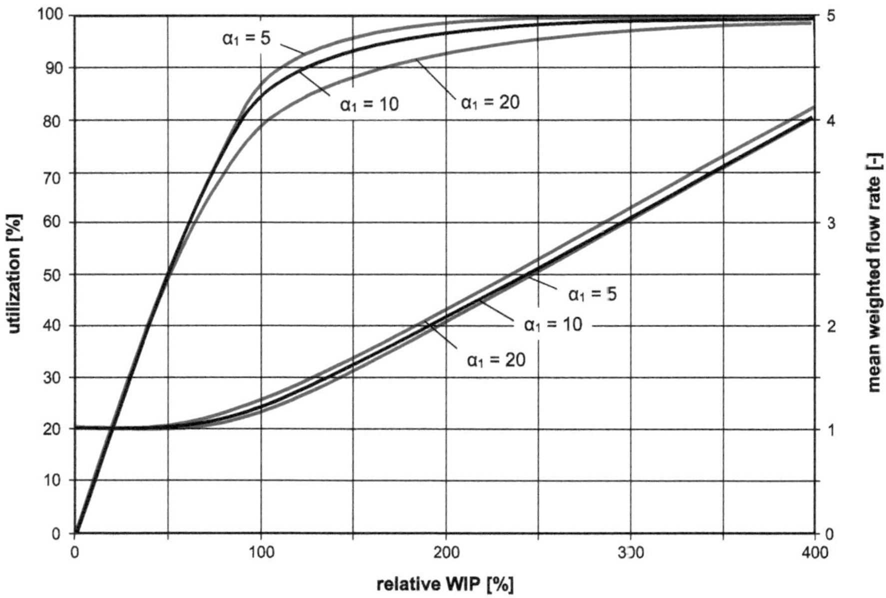

Fig. 6.12 Normalized Logistic Operating Curve for Various Values of the Stretch Factor α_1

then the resulting WIP buffer will either be smaller or larger than that which is required to ensure a specific utilization. The significance of this type of error is largely dependent on the respective problem.

In Fig. 6.12 three pairs of normalized Logistic Operating Curves are plotted for three different values of α_1. The graph shows that the outcomes for the mean output rate in the overload zone are strongly dependent on the chosen stretch factor. Given a relative WIP of 200%, the utilization will be calculated depending on the stretch factor as 98.8% ($\alpha_1 = 5$), 96.5% ($\alpha_1 = 10$) or 92.8% ($\alpha_1 = 20$). If for example, the workstation capacity is to be dimensioned based on the LOC Theory, then it is necessary to know exactly what the correct stretch factor is – though only when the workstation is supposed to be operating in the overload zone. If the Logistic Operating Curves are to be used to describe the relations between the (relative) WIP and the flow rate or as the case may be the throughput time, then it is barely noticeable when the stretch factor is incorrect.

It is however possible for example, to conduct a special simulation trial in order to determine which stretch factor needs to be applied in specific cases so that the system behavior can be described as accurately as possible. By analyzing the results of the simulation, an appropriate value can then be established. This procedure is quite time consuming and should thus be limited to exceptional cases. Furthermore, the value that is determined can only be as good as the emulation of the actual system's behavior in the simulation.

A second possibility for verifying or when necessary adjusting the stretch factor is provided by the measured WIP and output rate data. If the maximum possible output rate and the ideal minimum WIP are known, the value of α_1 can be adjusted until the measured operating state is positioned on the calculated LOC. If the maximum possible output rate is also unknown, then it should be verified whether or not the losses of utilization calculated through the relative WIP are plausible at an initially assumed value of α_1. If this is not the case, then the α_1 value can be changed accordingly (to reduce or increase it). However, this method is only possible, when there are WIP dependent losses of utilization on the observed workstation. If the workstation is operating in the overload zone, the reference points required to adjust the parameter are not available. Especially in such cases though, it is often enough to initially work with an estimated stretch factor (e. g., with the default value 10), because at the then relatively high WIP level, at least moderate WIP reductions that are independent of the stretch factor, only effect the utilization minimally, if at all. As soon as the loss of utilization can be determined, then the stretch factor can be adjusted using the described method.

6.4 Impact of an Unsteady Process State on Developing and Interpreting Logistic Operating Curves

When applying the Logistic Operating Curves Theory it has to be ensured that the underlying evaluation period is long enough that the input data is representative for the workstation. At the same time, it has to be made certain that there is a steady process state i. e., that the factors which influence the objectives do not significantly change during the observed time. Based on two examples, we will now address the importance of these requirements and the consequences of not observing them.

6.4.1 Time Related Changes to the Work Content Structure

The processes on a workstation are shaped in numerous ways by random influences. Without a doubt two of the most important of these is the work content of the orders that are to be processed and the variability of it. In the Logistic Operating Curves Theory these are summarized by the ideal minimum WIP.

Especially because of the variability of the work content, it is important to choose a long enough evaluation period. Furthermore, when establishing the scope of the analysis, the underlying problems have to be considered. If during the analysis, all of the required feedback data for the chosen time period is going to be acquired completely, then both the actual parameters for the analysis (e. g., WIP, throughput time and output rate) as well as the ideal minimum WIP are calculated based on the respective data. The model can then technically be correctly applied also with shorter analysis periods. The outcomes derived from the model though, are then only valid for the evaluated time period. The knowledge gained can quickly lose its relevance due to the stochastic influences on the process.

However, the planning process has to be based on a range of data large enough to target a defined level of planning certainty. We can find a typical value for the amount of required data, by referring to the following statistic: If the standard deviation of a (normal) distribution should be verified with 95% certainty and 20% accuracy (the permissible relative deviation between the expected and found value), then the resulting minimal sample size is n = 49; with a 10% accuracy the number increases to n = 193. The same applies to the mean value of the distribution (e. g., the mean work content), nonetheless, no general conclusions can be made here since the result is also determined by the specific shape of the distribution [Sach-84].

A practical example that was evaluated with regards to these aspects confirmed these results. Figure 6.13 illustrates the chronological behavior of the mean work content, standard deviation and the resulting ideal minimum WIP (plotted as a daily mean) for a workstation, calculated on the basis of the relevant processed orders. The daily means vary quite strongly although on average approx. 24 orders were processed per day. In order to smooth these fluctuations a sliding mean was calculated for each of the three key figures. For each of these values the data from 5 SCD (on average approx. 120 orders) was utilized.

The sliding means are also subject to certain fluctuations, however quite good outcomes are obtainable through the work content structure (limited at first to the time period represented in the graph as interval I). A longer evaluation period or a correspondingly larger number of orders is required in order to obtain a higher accuracy. Additional investigations which are not presented here confirm that by creating sliding means over 10 SCD, the individual means on these workstations

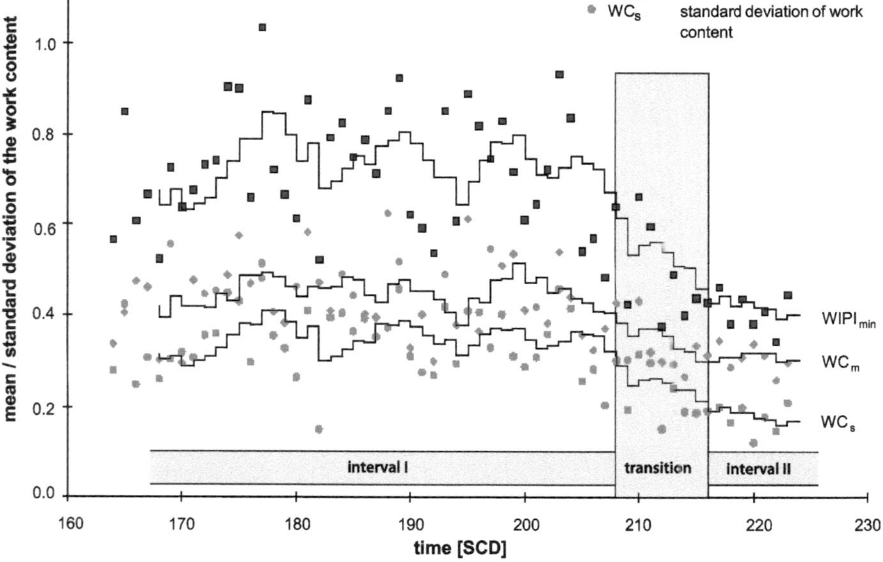

Fig. 6.13 Chronological Behavior of the Work Content Structure Figures (Plated-Through-Hole Workstation)

do not indicate any noteworthy variability up to day 208. The certainty of outcomes based on them is therefore accordingly higher.

In choosing the evaluation period it should be ensured that the parameters for the analysis do not clearly change in one direction during the time. Such a case is illustrated in Fig. 6.13. The manufacturer changed the batch-sizing strategy during the analysis of the data, in order to develop new logistic potential. Around approximately SCD 205, smaller standard lot sizes were introduced. The first of the orders affected by this entered the observed workstation on SCD 208. The mean work content was reduced by 22% due to this measure and the standard deviation by approximately 50%. The ideal minimum WIP was thus reduced by approx. 40%.

The behavior of the key work content structure figures indicate that applying the Logistic Operating Curves Theory over the entire evaluation period is not practical in this example. Whereas the work content structure was constant during intervals I and II, the time span from SCD 208 to 216 exhibits an unsteady transition zone, which needs to be factored out when applying the LOC.

6.4.2 Time Related Changes in the WIP Level

An approximately constant WIP level on the observed workstation is required both for empirically determining the stretch factor α_1 via simulation as well as for determining it by analyzing the processing data. WIP fluctuations are however only critical when they lead to temporary but significant losses of utilizations. States such as this are primarily problematic because the length and position of the interval period strongly determines the analyses of the results to the extent that the basic system behavior can no longer be depicted.

We will clarify this problem based on the simulation example introduced in Sect. 6.1. However, from now on a longer evaluation period will be applied. Figure 6.14 depicts the Throughput Diagram for the multi-spindle drill workstation during four intervals of 20 SCD. Thus, in comparison to our discussion in Sect. 6.12, we are observing an additional interval here. This fourth interval is characterized by a decreased input through which the WIP is strongly reduced, resulting in an output rate that is clearly lower than in the previous intervals (Table 6.4).

The basic system conditions (in particular the work content structure and the workstation capacity) were not changed during the simulation. The Logistic Operating Curves calculated in Sect. 6.1.1 are thus also valid for this extended period and are thus referred to again in Fig. 6.15. In addition, the key figures for the individual intervals and the entire evaluation period are also plotted in the graph. The simulation results for the fourth interval also correspond very well with the calculated Logistic Operating Curves. However, if one considers the entire period of observation (I1 to I4), then greater deviations can be determined between the calculated and simulated results. According to the LOC Theory given a mean WIP of 56 hours, the loss of utilization has to be significantly lower than the 8% that was measured during the simulation.

Table 6.4 Key Interval Related Figures for the Multi-Spindle Drill Workstation (Simulation Results)

key figures		intervals				
		I1	I2	I3	I4	I1 to I4
output	[hrs]	319.2	322.9	315.6	211.5	16.0
mean output rate	[hrs/SCD]	16.0	16.1	15.8	10.6	14.6
mean WIP	[hrs]	69.1	91.5	46.0	17.3	56.0
mean range	[SCD]	4.3	5.7	2.9	1.6	3.8
mean weighted throughput time	[SCD]	2.7	6.5	3.5	2.0	3.8
mean unweighted throughput time	[SCD]	2.2	6.0	3.0	1.4	3.3
mean inter-operation time	[SCD]	1.4	5.1	2.1	0.6	2.5

In principle, it is possible to choose a larger stretch factor and to thus develop a different Output Rate Operating Curve in which the measured operating states would be situated on the calculated Logistic Operating Curves during the entire evaluation period of the calculated curve. Given the fixed capacities and the comparably strong load fluctuations, such a measure at first seems permissible. Nevertheless, it has to be taken into consideration that the determined value is rather random and is only valid for the interval range from I1 to I4. Should any other point in time be chosen, another value for the stretch factor could result. For example, if only intervals I3 and I4 are considered together, the losses of utilization have a much stronger effect in the fourth interval.

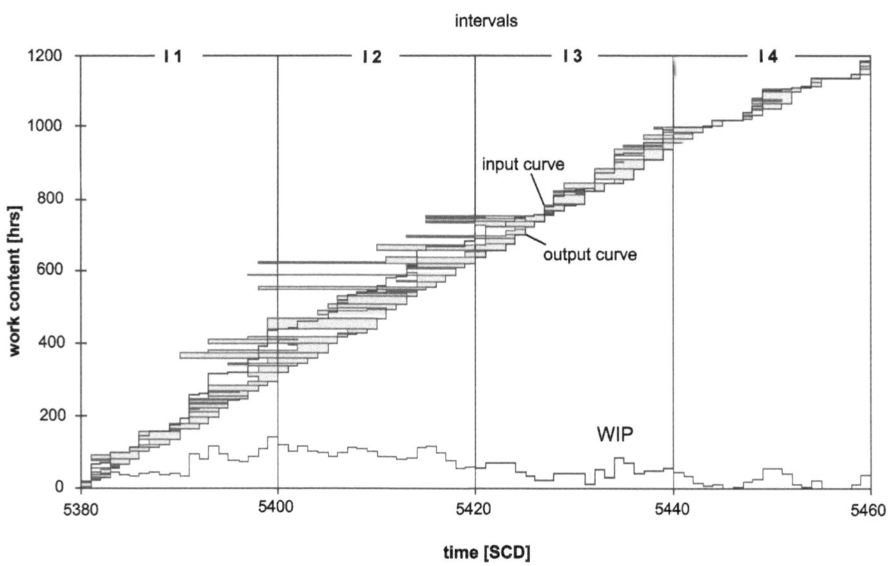

Fig. 6.14 Throughput Diagram for the Multi-Spindle Drill Workstation (Interval Range I1 to I4; Simulation Results)

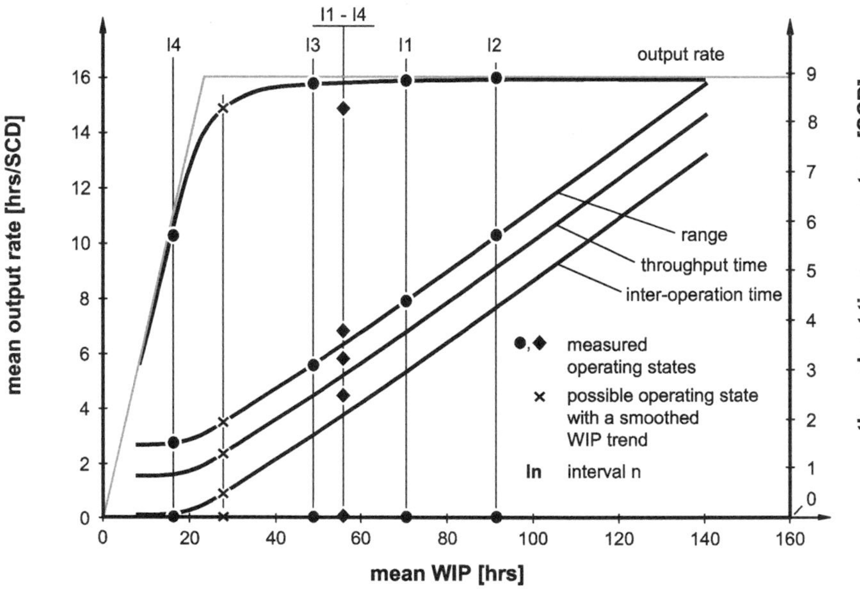

Fig. 6.15 Key Interval Related Figures in the Calculated Operating Curves (Interval Range I1 to I4)

The stretch factors dependency on the position and length of the evaluation period is however neither practical nor plausible. Furthermore, the results from the individual intervals confirm that the basic system behavior can be described very well with the underlying stretch factor. Therefore, the deviations between the simulation results and calculated results can be traced back to the change in the WIP level (in connection with the non-linearity of the Output Rate Operating Curve) and not an incorrect parameterization of the LOC equation.

This information can also be used in order to draw upon the deviations when evaluating the process. Thus, from Fig. 6.15 it can be concluded that: When it is possible to make the input and output conform better with one another by increasing the flexibility of the capacity and/or appropriately controlling the load, then the WIP and therefore the range can be halved without the overall output rate being affected.

The Throughput Diagram confirms this. It indicates that there was generally sufficient capacity in order to process all of the orders. If, the input flow is unchanged and it is possible to prevent the WIP from climbing in the second half of the first interval by adjusting the capacity at the right time (increasing the capacity in I1 and reducing it in I2), then the WIP level is more constant and low during the entire time period. A similar effect arises if the WIP is smoothed by an appropriately timed load shift. In the extended evaluation period the measured throughput time values deviated considerably from the calculated parameters. All of the measured states are located above the respective LOC. These differences are explained in that due to the abovementioned reasons, the simulation's yielded output rate and thus – by the given WIP – also the range, deviates from the calculated results. Consequently, because the range is inserted directly into the calculation for the Throughput Time Operating Curve, (see Eqs. 4.37 and 4.38), deviations to the same order of magnitude also result in the throughput time and inter-operation time.

This sample application makes it clear that when the WIP is strongly fluctuating, the choice of the evaluation period is crucial to the level of certainty for the model's outcomes. Temporary phases of overload can lead to the established operating state not being positioned on the LOC. The deviations can in principle then be interpreted differently. First, they can be seen as an indication that the equation was incorrectly parameterized (in particular the value of α_1). However, especially in simulations they can also be interpreted as indicating that the modeling of the order input and/or of the capacity is incorrect. Which of these is actually the case, has to be examined in the individual cases. Finally, the potential that can be developed by making the capacity or load more flexible, can be evaluated through the deviation between the measured operating state and the calculated LOC. In the latter case however, it has to be assumed that the generally attainable system behavior will be realistically described with the stretch factor's default value 10.

We have consciously gone into great detail in considering errors and the significance of unsteady production processes here, in order to prioritize the goal of being able to identify, evaluate and when necessary correct data or application errors when applying the Logistic Operating Curves Theory. Furthermore, the extent of the discussion makes it clear that the LOC Theory is generally very tolerant of errors. Even when the input data for the mathematical model is incorrect, it is still possible in many cases to practically interpret the results. Furthermore, we have demonstrated that in some cases the model can be used to adjust parameters, for example, when there is incorrect data for the maximal possible output rate or for the stretch factor α_1. Such an adjustment or correction is greatly facilitated when the plausibility of it is evaluated through supportive analyses. Depicting the process flow in the form of a Throughput Diagram can be very helpful for this purpose.

As numerous investigations have shown, the described data and application errors are usually of secondary significance when applying the Logistic Operating Curves Theory in the operational practice. Thus, incorrect work content information is frequently found in companies, but as long as these can be traced back to random errors, the figures for the work content distribution (mean and standard deviation) and thus the ideal WIP as well, will be influenced minimally if at all. It was also evident in the practical investigations, that the process behaviors on the shop floor were generally well mapped with the default stretch factor $\alpha_1 = 10$. Nevertheless, a more exact result can surely be achieved in one or the other application cases by changing the value of α_1. Similarly, it can be determined that even with a (limited) incorrect value, valuable information and help as well as orientation can be gained, which otherwise can only be attained with great effort. It is nonetheless advisable to verify and where necessary adjust the parameters when first applying them as well as when there are strong changes in the load or capacity conditions. A permanent critical examination of the data is only required with regards to the maximal possible output rate $ROUT_{max}$. By employing the

Newtonian Iteration Method introduced in Sect. 6.3.2 the $ROUT_{max}$ is easily verified and when necessary corrected.

6.5 Possibilities for Employing Logistic Operating Curves in Designing and Controlling Production Processes

The Logistic Operating Curves Theory offers a wide range of possibilities for quantifying potential for logistic improvement and for designing and controlling production processes oriented on logistics. In order to ensure a higher efficiency of the measures that are to be introduced, it is crucial to coordinate the individual measures. Permanently reducing the WIP proves to be the key logistic strategy. It is therefore the job of production control to first reduce the level of WIP to an acceptable size (Fig. 6.16). The limits of the achievable throughput time and WIP can subsequently be shifted to lower values using scheduling measures, particularly through a more harmonized work content. Next, further logistical potential can be created by shortening the processing time using advanced manufacturing technology such as new processing methods or with regards to factory planning, through restructuring measures. Re-adjusting the corresponding control parameters ensures that the thus gained room to maneuver can be used to lower the WIP.

Since the Logistic Operating Curves Theory describes the interdependencies between the logistic objectives as well as how they can be influenced, it represents an ideal basis for developing and monitoring a company's processing reliability

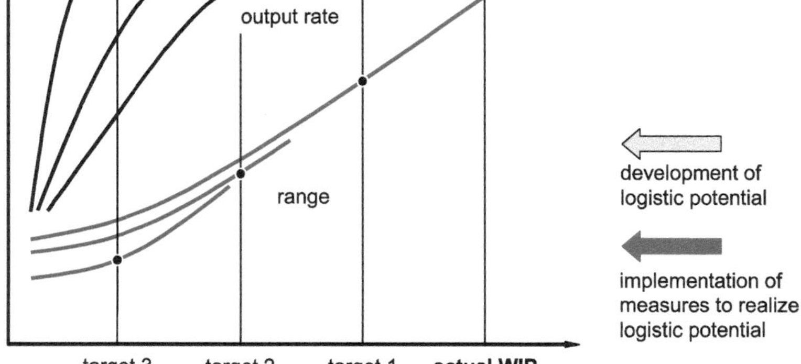

Fig. 6.16 Steps for Decreasing Throughput Time and WIP on the Shop Floor

monitoring and developing process reliability	**monitoring and developing process capability**
• **logistic positioning** - setting adequate operating ranges - verifying consistency of logistic objectives	
• **production monitoring** - evaluating processes - determining potential for logistic improvement • **production planning and control** - determining consistent planning parameters (flow rate, throughput time, inter-operation time, WIP)	• **selecting planning and control strategies** - lot size determination - scheduling - order release • **designing production structures** - planning required capacities - planning layout - choosing technology

Fig. 6.17 Fields of Application for the Logistic Operating Curves in Production Planning and Control

and capability (Fig. 6.17). Logistic Operating Curves can therefore be utilized while monitoring production to evaluate manufacturing processes. They indicate for example, the throughput time and WIP levels which can be achieved with the existing structural conditions, without having to expect a significant disruption in the material flow and with that a loss of output. In order to apply them as part of the production planning and control (PPC) process, the system parameters can be derived and adjusted to conform with the goals. Presenting the logistic objectives in a diagram makes it possible to decide which of the objectives should be assigned the greatest weight according to the actual operating and market situation and depending on the workstations' specific boundary conditions. At the same time it can be shown how changing a parameter influences the logistic key performance indicators. The dilemma of operations planning is thus solved using a novel approach. Instead of seeking an imaginary optimum (usually the minimum cost), a company can start with a primary logistic objective determined for the most part by the market. For example, a desired throughput time. The remaining target values, such as the utilization and WIP inevitably result from it.

If it turns out during the application that the set target values cannot be met without supporting measures, the Logistic Operating Curves can also be utilized to maintain and evaluate the planning activities corresponding to the imagined possibilities (Fig. 6.17). Alternative strategies which can be implemented for planning and control can thus be evaluated and chosen according to logistic criteria. The LOC Theory can also be directly integrated into methods for determining lot sizes, scheduling or releasing orders. Aligning the PPC with the logistic objectives is therefore, continuously and methodically supported. When planning a factory, the Logistic Operating Curves help evaluate alternative manufacturing principles or new logistic concepts. Furthermore, it is possible to evaluate investment decisions

(e. g., installing a new transport system, introducing new manufacturing technologies) and to expand the business process model. The Logistic Positioning, which determines the targets and thus also represents the link between all the individual functions, is fundamental for all of the mentioned applications.

In the next section, we will introduce a variety of applications based on the LOC Theory. However, because these applications are for the most part already described in detail in other professional resources, we will only outline them here briefly. Further explanations can be obtained in the corresponding recommended literature.

6.5.1 Logistic Positioning

On the shop floor, the seemingly unsolvable contradiction between short throughput times and low WIP levels on one hand, and a high utilization level on the other hand, is a constant topic. This dilemma of operations planning is commonly known but can generally not be quantified. A target oriented positioning between these competing objectives is thus at best possible based on empirical values.

In this case, the Logistic Operating Curves are an effective tool, because they describe the interdependencies between the logistic objectives qualitatively and quantitatively. Based on this information the objectives can be weighted depending on the actual operating situation – which may also differ locally. In turn, it can be derived from this (for example in the following described decision making model) how the parameters should be adjusted. Thus, depending on the situation, an active adjustment can be made between the company's internal demands and those of the market.

The desired operating points (WIP, output rate, throughput time) can be determined as a function of the existing work content and capacity structures as well as the required delivery time, capacities and cost structures (Fig. 6.18). We will refer to this from now on as a *Logistic Positioning* (or positioning) in place of the otherwise frequently used term 'optimizing'. In the process, depending on the type of problem or application, absolute or relative values can be set for the logistic objectives output rate, WIP and throughput time. Furthermore, the given target values and when necessary the permissible tolerance range can also be checked for consistency. Therefore, if the utilization of a workstation is provided, the corresponding key values for the various throughput time parameters as well as for the absolute or relative WIP value can be directly derived from it.

By conducting a Logistic Positioning such as this, it can be shown whether or not the desired goals are at all achievable with the existing conditions. When this is not the case, the target values are not situated on the calculated Logistic Operating Curves. Measures can then be taken in order to develop new logistic potential. During the prototypical LOC application described in Sect. 6.2, it was demonstrated for example how to recognize and evaluate points at which the throughput time and WIP could be reduced in such cases. Further examples of this can be found in Chap. 7.

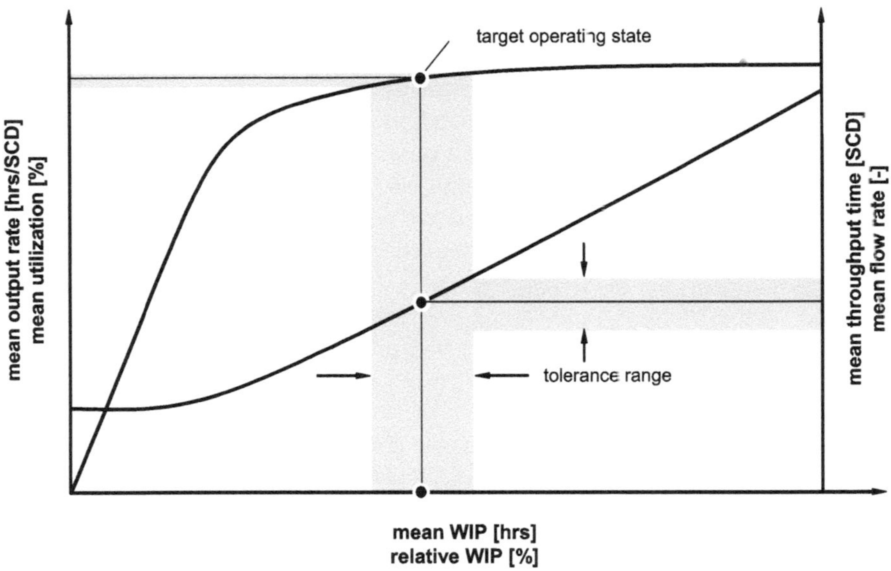

Fig. 6.18 Logistic Positioning with Logistic Operating Curves

In Fig. 1.8, the term 'costs' was named as the reference processes' fifth essential logistic objective. This eludes to the fact that the costs which can be influenced by the logistics should also be minimized. From the perspective of the Logistic Operating Curves Theory, it is necessary to represent the costs as a function of the WIP, since only then is it possible to directly compare it with the other WIP dependent objectives output rate, throughput time and schedule reliability.

Kerner [Kern-02] who further developed the work of Weber [Webe-87] and Großklaus [Groß-96] suggested this approach and thus established a pragmatic solution based on calculating activity based costs. According to this approach, the sub-processes of the reference processes first have to be defined and the various types of costs caused while executing them have to be ascribed to them. Subsequently, for every sub-process the cost driver and rates are identified. Finally, the latter are converted into equations as a function of the WIP objective. Once this is accomplished it is then possible to calculate and graphically depict a Cost Operating Curve (COC). We will now clarify the individual steps that are involved in doing so.

The main production sub-processes are the setup and manufacturing operations, whose duration can be drawn from the operational sheets. The costs related to the setup portion include manufacturing wages, material costs, and capacity costs for the setup components, which the machine's use requires.

Finally, in addition to the production processes' setup, manufacturing and WIP dependent costs the production planning and control related costs still need to be considered. Here, both the number of orders that are to be manufactured and the number of operations required are the cost drivers. The required sub-processes which were identified by Kerner include: planning the production program, plan-

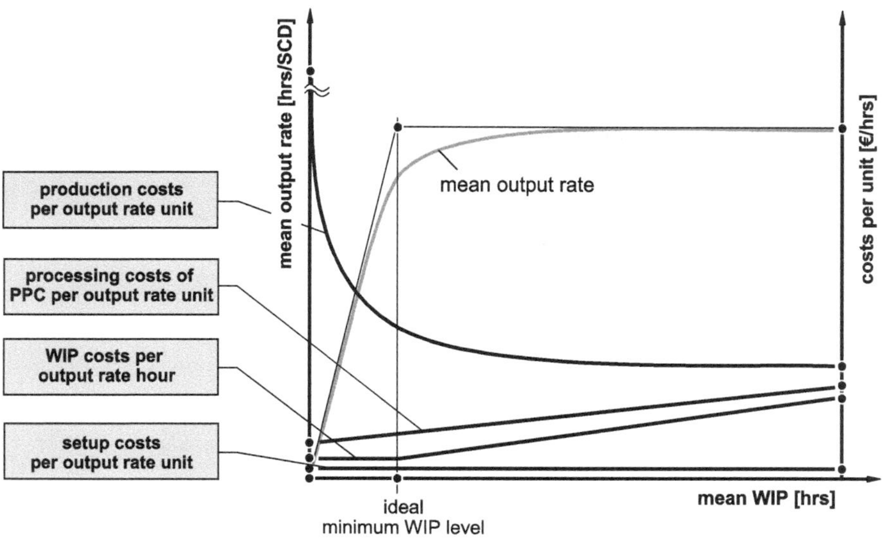

Fig. 6.19 Trends of the WIP Dependent Costs for the Production Processes and WIP

ning the lot sizes, planning the production sequence, developing the shop documents, releasing the orders, planning the machine load, controlling the manufacturing progress, and reporting back the finished orders. The corresponding costs for operators and space, as well as the depreciation and interest costs for the PPC system can then be determined for each order.

Figure 6.19 depicts the principle trends of the costs resulting from the four types of processing expenditure for the production process as functions of the WIP. For the purpose of comparison the ideal and theoretic Output Rate Operating Curves are also plotted.

The *Production Cost Operating Curve (PCOC)* lays between two values: When the maximal output rate is achieved there are no idle costs, the production costs per output rate unit (i. e., a standard hour) reach a minimum and correspond to the planned hourly rates. Should the output rate decrease, the fixed costs have to be distributed among continually fewer hours, the limit of the processing costs soars almost infinitely. In contrast, the setup cost's behavior is independent of the workstation's output and WIP, because the setup processes is always bound to the respective orders being processed. The WIP costs are assumed to be constant up to the point of achieving the ideal minimum WIP and from there increase linearly with the WIP. Finally, assuming a WIP independent minimum value, the costs which are caused by the planning and control are linearly dependent on the WIP.

The sum value of these four types of costs results in the total Cost Operating Curve. It along with the Output Rate Operating Curve, the Throughput Time Operating Curve and the Scheduling Reliability Operating Curve are shown in Fig. 6.20.

One can see that each Logistic Operating Curve has either a minimal or maximal value, and that each is allocated to a certain WIP value. In principle, the WIP

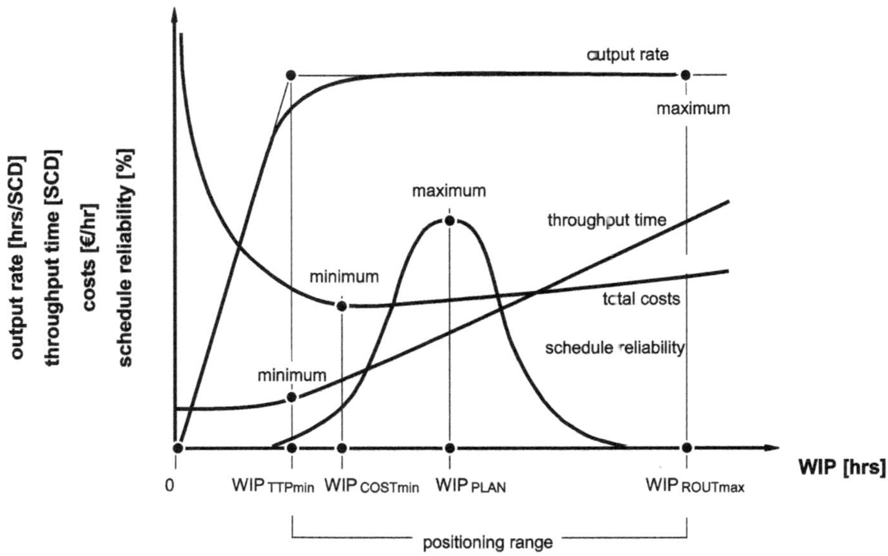

Fig. 6.20 Principle Trends of the WIP Dependent Output Rate, Throughput Time and Costs

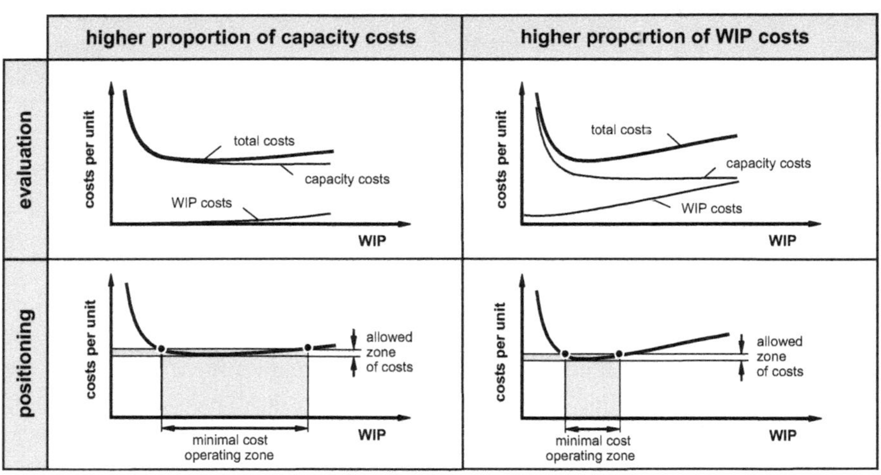

Fig. 6.21 Impact of the Cost's Structure on the Minimal Costs Operating Range (according to Jainczyk)

values for the minimal throughput ($WIP_{TTP,min}$), the minimal costs ($WIP_{COSTmin}$), the maximal output rate ($WIP_{ROUTmax}$) and the maximal schedule reliability (WIP_{RelS}) cannot agree with one another. Thus, a possible range of positioning for the WIP results. Which value will ultimately be chosen for the WIP, is then dependent on the primary logistic objective being pursued.

As can be seen in Fig. 6.21, due to its WIP components, the shape of the total Cost Operating Curve is essentially defined by the ratio of capacity costs to WIP costs. The more the capacity costs dominate over the WIP costs, the more flat the zone for the total minimum cost is. If a permissible level of deviation for the costs from the minimal value is then applied, for example 5%, then an operating zone for the WIP results which becomes narrower the more the WIP costs dominate. In this case it is better to orient on the minimum costs. If however, the capacity costs are predominant, then the minimal zone for the cost with the same cost tolerance is clearly larger, it is thus better to strive for a lower throughput time.

Existing methods for calculating the Cost Operating Curve have already been applied in the operational practice. In addition to the positioning ([Fast-94], [Groß-96], [Wien-96b]) the most common application is for extending the traditional calculation of capital costs by including aspects of economically evaluating the production logistics [Jain-94].

What should be noted here is that the evaluation of the throughput time and schedule deviations can still not be described according to costs since they are more related to revenue. Further work is thus required here. Independent of a quantitative description, it can be determined however that short throughput times and minimal schedule deviation generally result in a shifting of the optimal cost zone in the direction of lower WIP levels.

6.5.2 Implementing Logistic Operating Curves in Production Control

One instrument that company's employ for orienting their business activities on success is a *control process*. This is typically conducted in six coordinated individual steps (Fig. 6.22):

- *Target Setting*: The different objectives have to be quantified within the target system in consideration of the interdependencies (target consistency).
- *Deriving Target Values*: Target values serve as reference variables. It is the job of the control to help determine them through a goal oriented derivation of control parameters.
- *Recording Actual Values*: The processes' behavior can be determined based on actual feedback data. Setting the measuring point, figures and methods has to be oriented on the parameters used for planning.
- *Comparing the Target and Actual Values*: By comparing the target and actual values, excessive deviations in the actual process caused by unforeseeable process disruptions or planning errors can be determined.
- *Analyzing Deviations*: Should excessive deviations surface, the root cause of the deviations has to be analyzed in order to correct the process.
- *Deriving Measures*: Controls should offer support in deriving appropriate correction measures. Prioritizing the measures according to the preferred cost/benefit ratio is essential in this context. Finally, the results of the activities have to be reported to management.

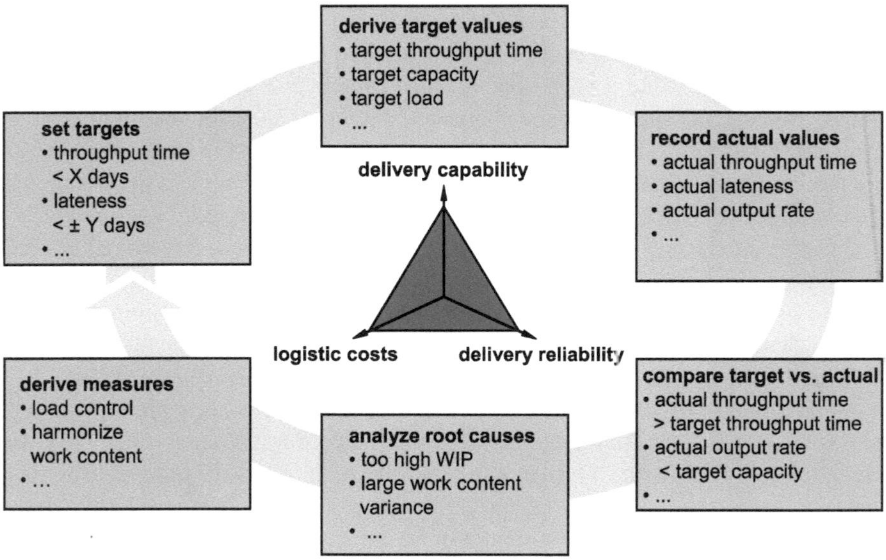

Fig. 6.22 Control Loop (according to Hautz, Siemens)

The example presented in Sect. 6.1 and 6.2 demonstrates that the individual steps of a production control can be comprehensively supported by the Logistic Operating Curves Theory. Once the logistic objectives have been set and the target values have been determined through a Logistic Positioning (see above), the LOC together with the Throughput Diagram can be applied in conjunction with the acquired and evaluated data in order to compare the target and actual values as well as analyze the deviations and derive measures (see also [Ullm-94], [Wyss-95]).

An application, such as the one presented in Sects. 6.1 and 6.2, is basically oriented on resources. In order to evaluate the logistic performance of the entire production, it is also necessary to describe how the individual workstations are embedded in the production flow. This can be accomplished by analyzing the relationships of the material flow which result from the various operation sequences. By combining a number of suitable analysis methods it is possible to transition to a *Bottleneck Oriented Logistic Analysis* (BOLA) (Fig. 6.23) ([Wien-93a], [Wien-95b], [Wien-98] and [Wind-01]).

This method of analysis makes it possible to transparently present the production flow and to demonstrate the logistic bottlenecks in the material flow. Both *capacitive bottlenecks* (yield restrictions) as well as *throughput time/delivery time determining workstations* can in this way be localized. Furthermore, the significance of the individual workstations for the entire order throughput can also be quantified. By applying the Logistic Operating Curves technique it can be shown which type of possible measures for reducing the throughput time and WIP can be practically implemented on which workstations. It can thus be examined for example, where in the analyzed production area, the throughput time can be reduced through a targeted WIP control and on which workstation supporting measures for the capacity struc-

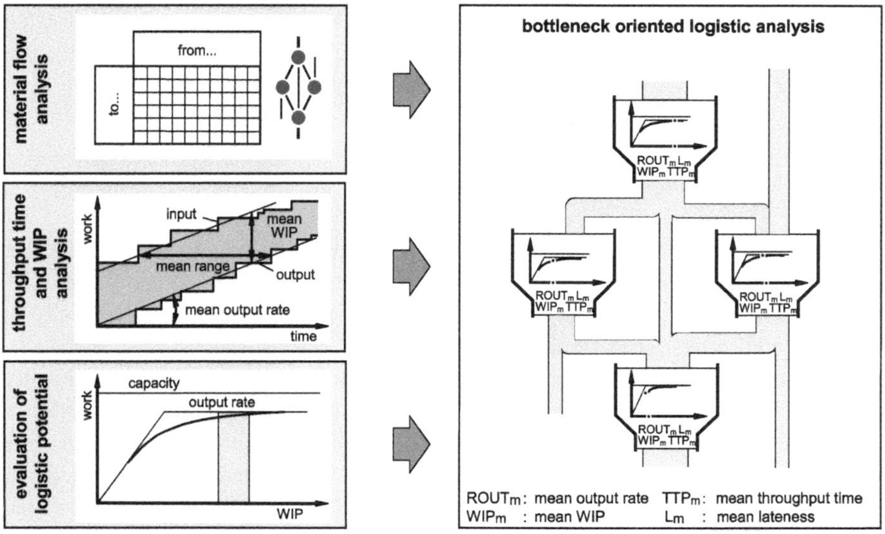

Fig. 6.23 Principle of a Bottleneck Oriented Logistic Analysis (BOLA)

ture, work content structure or structural integration of individual systems are required. It can also be shown how reducing the throughput time on one single workstation influences the entire order throughput. Due to the particular significance that can be attached to the BOLA in logistic oriented production design, we will discuss it extensively in Chap. 7 on the basis of two sample applications.

6.5.3 Logistic Oriented Design and Parameterization of Planning and Control Strategies

Traditional production planning and control is performed in increasing levels of detail whereby the results are fed back to the next higher level. The steps program planning, material planning, schedule and capacity planning as well as the release and control of orders through the production are thus based on continually actualized data. The quality of the planning developed from it is however generally unsatisfactory. External signs of existing problems can be seen for example in fluctuating schedule compliance (see [Ludw-95] and the references cited there), high WIP levels and highly variable throughput times. In order to meet these problems head-on, there is increasing demand for modern PPC systems to place production logistic goals at the forefront of all planning steps ([Pete-95], [Wien-97]). In order to facilitate realistic and coordinated guidelines across all the planning stages and to be able to track them, the different planning functions should be based on a common model. The Logistic Operating Curves Theory can be integrated into the corresponding PPC procedures and can serve as a link between them. In the next section, we will thus introduce various planning functions that are based on LOC theory and can therefore assist in aligning both the production and production flow with the logistic objectives.

6.5.3.1 Throughput Oriented Lot Sizing

The main task of material planning for in-house manufacturing is to determine the requirements for all items of the bill of material and convert these into manufacturing and assembly orders. In this context, lot sizing has a key role especially in small and mid-sized series production.

In most enterprises, lot sizes are determined based mainly on economical criteria. The majority of methods for determining the size of the batches consider in more or less detail the contradicting costs for storage on the one hand and the costs to change orders (in essence the setup costs) on the other hand. However, methods such as these are increasingly met with criticism. In particular, it has been stated that these traditional methods for determining the lot size fail to consider the influence of the batch size on the throughput time and the inventory tie-up in the production ([DeBo-83*], [Knol-90]).

The Logistic Operating Curves Theory proves that this critique is justified. It shows that the lowest possible throughput time is decisively determined by the work content's mean and variance and thus through the lot size of the production orders. Therefore, both the operation times as well as the waiting times during the

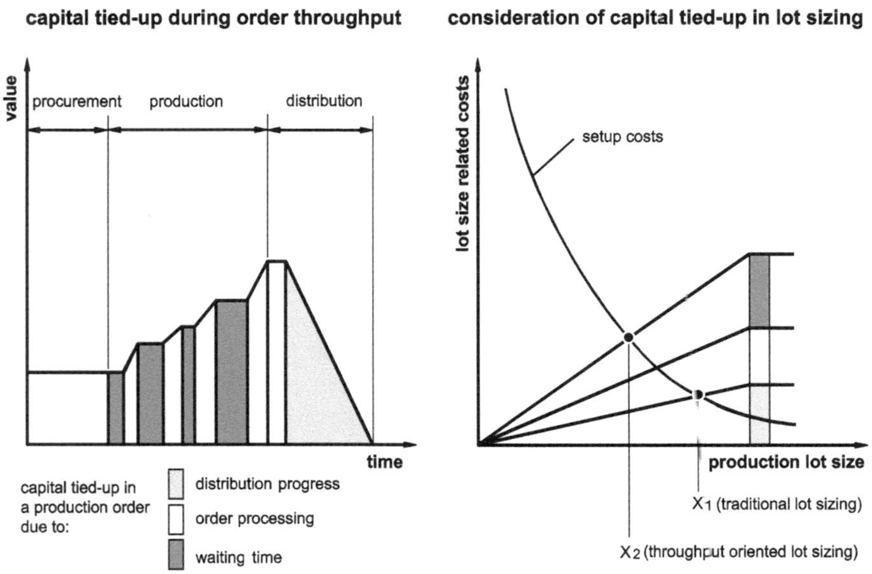

Fig. 6.24 Principle of Throughput Oriented Lot Sizing

production (inter-operation time) where the capital is tied-up due to the lot size, have to be included in calculating economical lot sizes.

This is the point at which *Throughput Oriented Lot Sizing (TOLS)* [Nyhu-91] is applicable. Through it, the information derived from the LOC Theory regarding the correlation of the work content and thus the lot size on the one hand, and the logistic objectives on the other hand, are taken into consideration. The method distinguishes itself from others in that the capital costs tied-up during the order's entire throughput are considered (Fig. 6.24). The duration which a lot is bound in the production is calculated based on the flow rate set during the Logistic Positioning. The main effect is that, compared to traditional approaches, smaller economically optimal lot sizes are calculated. In doing so, lots which have a relatively large work content are more strongly reduced than those with small work content. The Throughput Oriented Lot Sizing method therefore, serves to reduce and harmonize the work content (see [Nyhu-91] and [Gläs-91]). According to the fourth basic law of production logistics (Sect. 5.4) the method thus contributes twofold to improving the manufacturing's logistic potential.

6.5.3.2 Flow Rate Oriented Scheduling

The main goal of throughput scheduling is to determine the planned start date of an order given a planned completion date. In addition, completion dates should often be established for the different operations in order to be able to chronologically allocate the workstations' load through the individual operations.

On the shop floor, different methods of scheduling are employed. In the simplest case, the order throughput times are estimated or established based on past data. When necessary, they are classified according to product groups or other criteria. If this is insufficiently accurate, or if completion dates on the operational level are required, the throughput times are then generally determined through the workstation's mean throughput times, or through the sum of the operation times and the workstation related inter-operation times.

Numerous studies have shown however, that the quality of the planned throughput time in the operational practice is often very unsatisfactory, even with the more extensive scheduling methods. The main reason for this is that the PPC system does not offer any appropriate support for maintaining or determining the planning basis for the throughput time or inter-operation time respectively. Considering how significant the planned throughput times are for a company's entire scheduling system, methods which can close this gap are required.

At this point, we can once again draw upon the results of the Logistic Positioning. If the positioning was conducted for each of the workstations there are then throughput values available, which have been directly derived from the goals, that are realistic and can be utilized for scheduling.

Particularly in cases where the order's individual operation times constitute a significant part of the throughput time due to a low WIP level, it can be necessary to investigate the planned throughput time in detail.

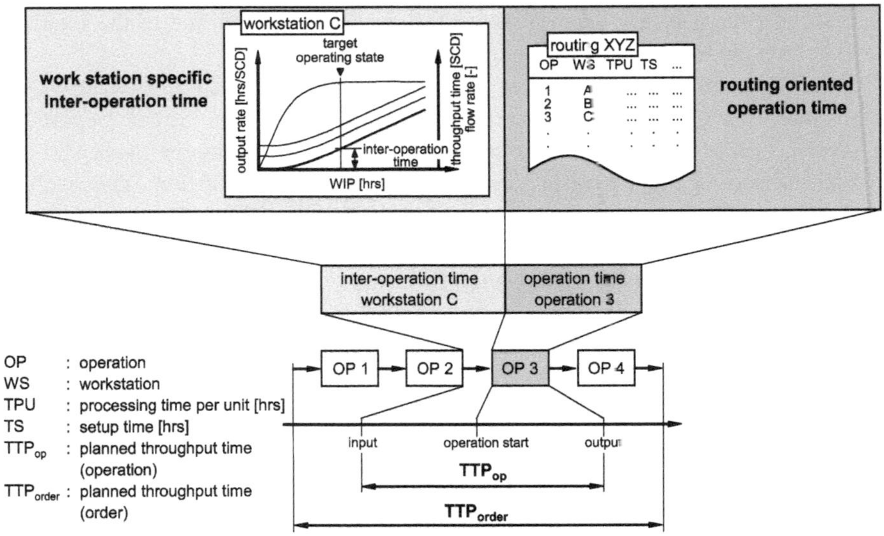

Fig. 6.25 Principle of Flow Rate Oriented Scheduling. (Ludwig/Nyhuis)

This can be accomplished using *Flow Rate Oriented Scheduling* (FROS) (Fig. 6.25) in which not the mean throughput times, but the workstation related mean inter-operation times are first determined by positioning and setting a planned flow rate. In order to do so, it is assumed that the inter-operation time between two operations is essentially determined by the WIP level required on the workstation and not by the order's individual operation times (see also Sect. 5.6: sixth basic law of production logistics). This applies as long as work content independent sequencing rules (FIFO or slack) can be assumed. Generally this assumption can be made for scheduling, because the planning process is usually based on work content independent sequencing rules. The planned throughput time of an order which is to be scheduled is thus derived from the sum of the inter-operation times and the time calculated from the work plan for the operations.

This procedure results in planned throughput time distributions for the individual workstations, which essentially correspond to the operation time distributions, but which are oriented around the inter-operation times. In simulations, it has been proven that a higher planning accuracy can be achieved with Flow Rate Oriented Scheduling ([Ludw-92], [Ludw-93]). Flow Rate Oriented Scheduling helps to derive workstation specific throughput times in consideration of the work content and capacity structures as well as the respective targeted WIP situation. Each change to the logistic objectives or the workstation specific conditions e. g., the work content structure is directly reflected in the determined throughput times. It is thus ensured that they are constantly adjusted to the actual situation.

6.5.3.3 Integrating the Logistic Operating Curves Theory in Load Oriented Order Release

During the 'order release' process it is determined whether or not the scheduled manufacturing orders can be completed on time and if a punctual delivery is generally realistic. Furthermore, it is also verified whether or not the materials, resources (tools, devices and testing equipment) and necessary operator and equipment capacities are available.

One method of order releasing, which is centered on the fundamentals of production logistics described in Chap. 2, is the *Load Oriented Order Release* (LOOR). The ideas behind this method are presented in the first and second basic laws of production logistics: Only as much work should be released for each workstation within a planning period as that which can be processed, based on the expected output rate within that period. The mean WIP on the workstation and therefore indirectly the mean throughput time are thus controlled through it ([Bech-84], [Wien-95a*], [Wien-87], [Wien-97]).

The control system analogy of the Load Oriented Order Release is summarized in Fig. 6.26. The method is based on two key sub-steps. In the first, all of the orders are chosen which, due to the internally conducted scheduling, need to be considered urgent. Afterwards, the release based on the actual WIP situation is investigated. At this point it is established under which circumstances new orders can be released, without unwanted WIP developing. It is assumed in this method that as a result of an existing capacity calculation, the required capacities are available.

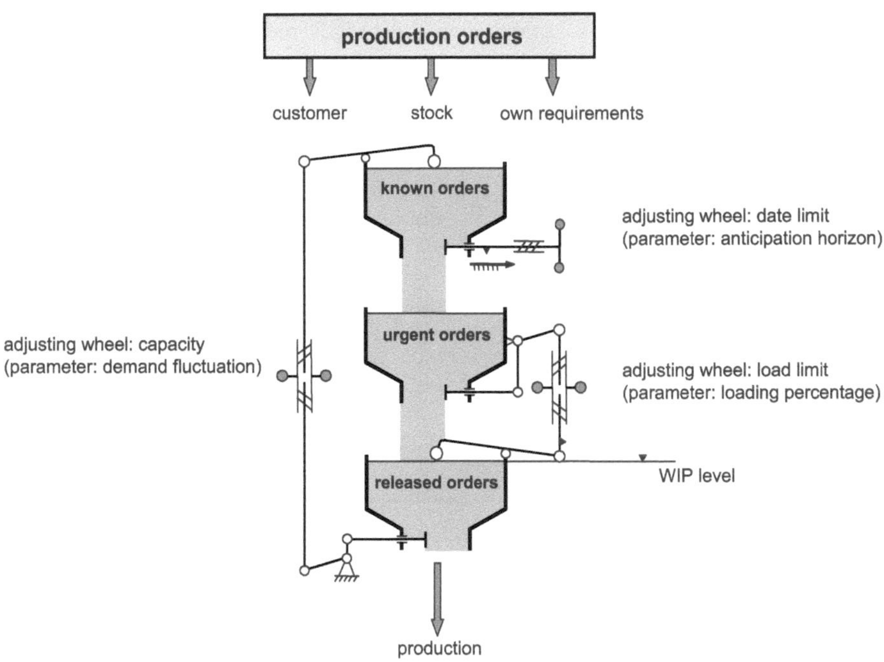

Fig. 6.26 Control System Analogy for the Load Oriented Order Release

In the operational practice however, finding appropriate values for the adjustable parameters and in particular for the loading percentage (LP) is problematic. The LP is used to set the target WIP level for the individual workstations. The methods recommended for establishing this parameter in [Wien-87] – especially various methods of estimation as well as simulation – were not able to establish themselves in the field due to either related uncertainties or the required effort. Thus, up to now when applying the Load Oriented Order Release method, the parameters have been set exclusively by trying to constantly lower the loading percentage step-by-step until the utilization limit is reached. Since changing the parameter too drastically can lead to unplanned losses in utilization, it needs to be conducted very cautiously [Wien-87].

It would appear that an active and situation driven parameterization of the LOOR can only be partially realized using the described trial and error method and thus, it cannot be employed to its fullest extent. In order to compensate for this shortfall the Logistic Operating Curves Theory can be integrated into the technique for Load Oriented Order Release (see also [Spen-96]). The basic idea behind this combination is presented in Fig. 6.27.

Once the Logistic Positioning has been conducted, the planned range can be established using the Logistic Operating Curves Theory. This can in turn be directly converted into the loading percentage (see Fig. 6.27, right). The throughput time as well as the inter-operation time can also be utilized from the Logistic Operating Curves and employed in the preceding scheduling steps. Finally, it is possible to extend this technique by also integrating short to mid-term capacity planning into it. If the available capacity is established based on the actual backlogs of the planned input and target utilization, then the LOOR method is expanded into a combined backlog and WIP control (Fig. 6.27, left).

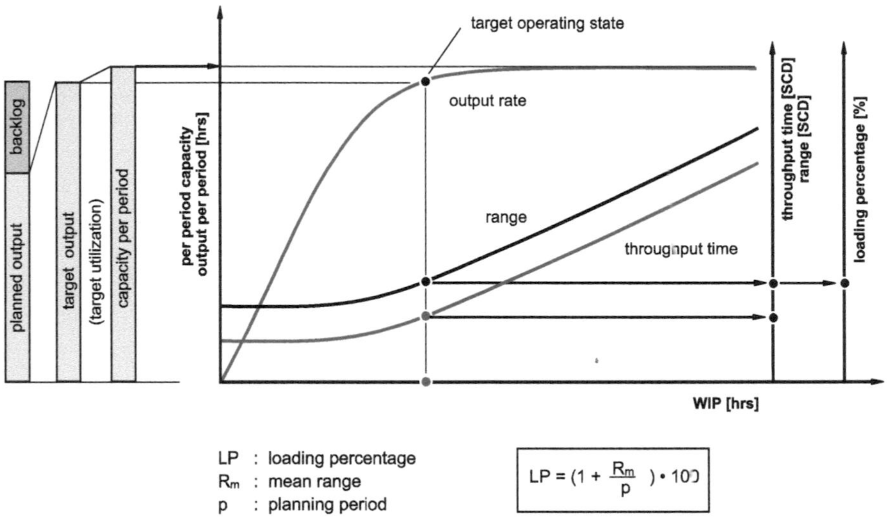

Fig. 6.27 Configuration of a Combined Backlog and WIP Control Based on Calculated Logistic Operating Curves

6.5.4 Logistic Oriented Production Design

In order to timely influence a company's logistic abilities, it is necessary to support the designing of alternative production processes and structures which meet the requirements already during the planning phase. In the following we will introduce two techniques based on Logistic Operating Curves and conceived with this situation in mind.

6.5.4.1 Employing the Logistic Operating Curves in Factory Planning

Factory planning has to solve complex problems within ever shorter time periods. In addition to the traditional tasks of optimizing resources with regards to the capacity and utilization, as well as designing the layout with regards to material flow, it increasingly has to consider the logistic performance of the to be planned production area. Logistic Operating Curves can also be employed here and make it possible to continuously evaluate the planning results across all levels ([Möll-95], [Möll-96]).

The first step of planning a factory (Fig. 6.28) is the operational analysis based on given targets. This serves to record the actual state, document critical areas and derive further tasks. As far as the logistic processes are to be analyzed within the production, the Bottleneck Oriented Logistic Analysis can be implemented in order to identify existing potential for logistic improvement as well as to indicate where improvement measures can be employed. In this way, concrete areas of action for the planning can be derived.

In the second step, the structure planning, the basic production concept and logistic strategies for pursuing it as well as the resources and capacities which will

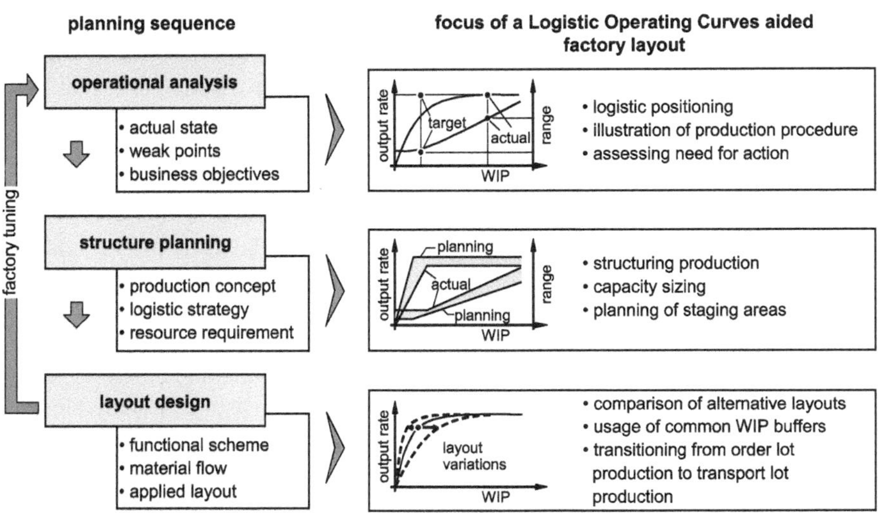

Fig. 6.28 Possibilities for Implementing Logistic Operating Curves in Factory Planning

be required by the production in the future are determined. The Logistic Operating Curves can be applied here in order to evaluate the logistic potential of alternative structures with minimal effort. When employing alternative measures for dimensioning the production capacity it is possible that significant differences in the logistic performance will result. These can be evaluated using the Logistic Operating Curves techniques.

It is often beneficial at this time to explore the logistic impact of alternative layouts. Thus for example, additional logistic potential can be developed by integrating a number of workstations with the same processing tasks not only organizationally but also spatially, so that all the systems can access a common and thus smaller WIP buffer. Furthermore, the LOC can also illustrate the logistic advantages of overlapped manufacturing (e. g., in a manufacturing cell), for which a strong process oriented layout with very short transport routes is essential [Scho-96].

6.5.4.2 Logistic Oriented Evaluation of Supply Chains

With the aim of achieving greater customer orientation and to accordingly align the required tasks along the value-adding chain, numerous models for illustrating business processes have been developed in the last years. In a comparative study [Goeb-96] the diverse models were extensively discussed and compared on the basis of various criteria (model development, model content and evaluation of objectives). According to this investigation, Kuhn's process chain model was the only method that was developed especially for illustrating and designing logistic processes.

The starting point for this model is the general process chain element, which from the perspective of transformation logic represents a system ([Kuhn-95], [Beck-96]). A process chain element is initially described by processes that help to make goal oriented changes to an object (Fig. 6.29a). From a logistic point of view, the typical processes are production and testing, transportation, and storage and supply. In order to accomplish these tasks a performance potential is required, which is supplied through the resources. The mode of operation and system design is documented in the layout. Each element of the process chain ultimately represents an autonomous control area with its own control functions, which among other aspects is responsible for implementing the logistic objectives. The system behavior of a process chain element is on one hand, influenced by the internal structures and processes, and on the other hand by the exchange with the environment i. e., the source (input) and sink (output).

A process chain element can be constructed for various levels of detail. Thus an enterprise can be conceptualized as a whole via a process chain element, or it can be broken down and particularized across a number of increasingly detailed levels. In the latter case, each element of the process chain is always setup the same, independent of the level of particularization.

Visualizing complex procedures involving a number of process chain elements results in a processing chain plan (Fig. 6.29b). In addition to the key logistic per-

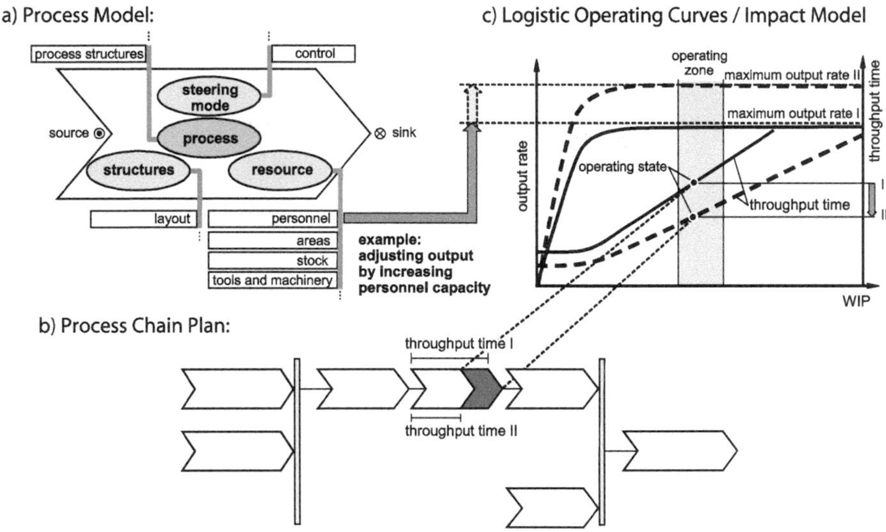

Fig. 6.29 Concept of an Integrated Process and Impact Model (according to Kuhn and Fastabend/Helms)

formance figures (throughput time, output rate, WIP and schedule adherence) the procedures and structures visualized in this way, offer the basis for evaluating the processes and for deriving design measures. These refer primarily to the structural design. Operations, which do not contribute to the customers value, should be investigated with regards to their necessity and where applicable removed from the process. Furthermore, individual processing steps can be outsourced or chronologically shifted and thus, where applicable, also run parallel and/or combined. Finally, the impact of changing the properties of individual process chain elements on the entire chain (e. g., the duration of an operation) can be evaluated.

In process chain models, the throughput of products or orders is considered first and foremost. A resource oriented perspective is not feasible. Therefore it is not possible to recognize and evaluate a system's behavior when there are competing objects, for example when different orders are contending for the limited available capacities [Goeb-96]: The effects of resource oriented measures can not be directly described.

This gap can, however, be closed by combining the Logistic Operating Curves Theory and the supply chain model into an *integrated process and impact model* ([Wien-96a], [Fast-97]). Using the Logistic Operating Curves it is possible to quantify the effects of various interventions on the supply chain elements. In the example illustrated in Fig. 6.29c, the number of operators is increased in comparison to the initial state based on an increasing demand for capacity. This leads simultaneously to a higher output level and a reduced throughput time. The newly resulting key figures can then be transferred over to the corresponding process chain plan. In this manner, alternative methods for improving the throughput time of individual products or product groups can be evaluated.

Chapter 7
Practical Applications of Bottleneck Oriented Logistic Analyses

The Bottleneck Oriented Logistic Analysis (BOLA) (see Sect. 6.5.2) is a control method designed especially for logistically evaluating and improving existing production processes. This type of analysis makes it possible to describe the processes of a production area from a logistic perspective both qualitatively and quantitatively. The specific causes of the problems can thus be localized and represented as interdependencies. Furthermore, existing potential for logistic improvements as well as possible measures for developing them can be illustrated and evaluated.

7.1 Conducting a Bottleneck Oriented Logistic Analysis

The Bottleneck Oriented Logistic Analysis is based upon the systematic analysis of manufacturing process data. By analyzing the orders' processing behavior and describing the available potential for logistic improvement using the Logistic Operating Curves technique, problem specific measures can be derived and evaluated for designing, planning and controlling the production process.

A well-structured, but complex and interconnected problem lies at the core of a logistic analysis. In this context, well-structured means that extensive data is available (in this case essentially the production feedback data) in order to comprehensively document the problem. The complexity and interconnectivity results from the influence the individual workstations have on one another's behavior within the production. Furthermore, a number of logistic objectives, which are also connected with each other and to some extent contradict one another, need to be simultaneously observed. Solving problems of this sort generally requires the following procedural steps:

- Compiling all important information and key figures.
- Reducing the complexity (limitation to the essential).
- Developing a structured and visual representation of the problem situation [Sell-90].

A traditional problem solving process can then be constructed on this basis [Daen-92].

7.1.1 Determining Key Figures

The first step in a Bottleneck Oriented Logistic Analysis is to calculate chosen key figures that may shed light on potential causes of problems as well as possible measures. Generally speaking, when defining or choosing the key figures it should be kept in mind that these "have a functional, practical relation to one another, supplement one another or explain one another and are generally directed at a common goal" [Reic-90]. At the same time, all of the required information should be acquirable from as few key figures as possible, in order to minimize the effort required for the analysis and in particular of the interpretation.

An overview of the most important figures that should be drawn upon in a BOLA in order to evaluate a workstation is presented in Fig. 7.1. These figures fundamentally contribute to the following described tasks:

7.1.1.1 Key Work Content Figures

The key work content figures are the mean work content WC_m, the coefficient of variation for the work content WC_v and the minimal inter-operation time TIO_{min}. They are required in order to calculate the Logistic Operating Curves and thus for evaluating the process. If the analysis shows that a workstation is already operating in the transition zone of the LOC, then these figures also indicate which measure is best suited for increasing the logistic potential. For example, if WC_v is very large, then the possibility of harmonizing the work content should be examined (see Sects. 5.4 and 6.2.1). However, if particularly in comparison to the mean work content, the minimum inter-operation times are large, then improvement measures should be prioritized (depending on the cause e. g., through new technology a modified layout or improving the transport organization).

work content	
• WC_m	: mean work content
• WC_v	: coefficient of variation of work content
• TIO_{min}	: minimum inter-operation time

throughput time	
• R_m	: mean range
• TTP_{mw}	: mean weighted throughput time
• TTP_m	: mean throughput time
• TTP_v	: coefficient of variation of throughput time
• $TTP_{m,calc}$	: calculated mean throughput time

output rate	
• $ROUT_m$	: mean output rate
• n	: number of orders (operations)

WIP	
• WIP_m	: mean work in process (number of hours)
• $WIPO_m$	: mean WIP (number of orders)
• WIP_{rel}	: relative work in process

lateness	
• L_{rel}	: relative lateness
• $L_{rel,s}$	: standard deviation of relative lateness

Fig. 7.1 Key Figures in a Bottleneck Orientated Logistic Analysis

7.1.1.2 Key Throughput Figures

Comparing the key figures for the mean Range (R_m) and mean weighted throughput time (TTP_{mw}) make it possible to conclude whether or not the analysis' results are based on a steady processing state. Only when the two values are approximately equal, can the figures which were derived from the Logistic Operating Curves Theory be drawn upon directly for evaluating the process. The two figures therefore basically function as auxiliary figures in the Bottleneck Oriented Logistic Analysis. Whereas, the mean throughput time TTP_m states how long on average an order dwells on a workstation, the throughput time's coefficient of variation TTP_v provides information about the variability of the throughput times. Finally, the mean throughput time $TTP_{m,calc}$ calculated according to the LOC Theory (see Fig. 6.25) indicates which throughput time is possible when orders are processed according to the FIFO principle and is thus based solely on the workstation's WIP level (Eq. 4.40). When TTP_m and $TTP_{m,calc}$ significantly deviate (and if $R_m \cong TTP_{mw}$), it can be assumed that there is at least a tendency for orders to be processed according to the SPT principle ($TTP_{m,calc} > TTP_m$) or the LPT principle ($TTP_{m,calc} < TTP_m$) (see Sect. 4.4).

7.1.1.3 Key Output Rate Figures

The mean output rate $ROUT_m$ describes the resources required for processing the orders. The number of orders to be processed per workstation during the evaluation period describes the significance of a workstation in the material flow. The more orders processed, the more relevant the workstation is particularly with regards to the mean order throughput time as well as the order's mean and relative lateness.

7.1.1.4 Key Work in Process Figures

The mean work in process WIP_m and relative work in process WIP_{rel} are primarily required for evaluating the measured operating states when applying the Logistic Operating Curves Theory and for quantifying the existing logistic potential by comparing them with the given target values. The mean WIP level in number of orders $WIPO_m$ can be used to comparatively evaluate the WIP across the entire system.

7.1.1.5 Key Lateness Figures

Provided that there are also planned throughput times on a company's operation level, the relative lateness and its distribution should also be evaluated. These two figures, in conjunction with those previously mentioned can be used to verify whether or not the planned dates are realistic for the scheduling and which workstations primarily contribute to any already existing lateness.

7.1.2 Determining Logistically Relevant Workstations

The second step in a Bottleneck Oriented Logistic Analysis is to choose relevant workstations. In order to clearly reduce the complexity of the task and to concentrate the measures that are to be introduced on the essential, the number of workstations are decreased according to how significant they are for the underlying goals. In many cases, improvements attained on individual workstations will require that other workstations' constraints are changed, whereby both positive and negative consequences are possible. Due to the frequently huge scope of the entire task and the objects that need to be considered, it does not seem practical to strive to optimize the process all at once. Instead, it is much more advisable to understand the BOLA as a permanent control's task and to thus construct it as a continuous process for improving logistics.

The choice of the relevant workstations has to be oriented first and foremost on the concrete aims of the analysis, whereby it is assumed that there is always one main goal that is prioritized. The dependencies between this main goal and the other objectives are taken into consideration when evaluating the possible measures (see also [Wind-01]).

The most important criteria for investigations of this sort are summarized in Fig. 7.2 and are explained in detail in the following.

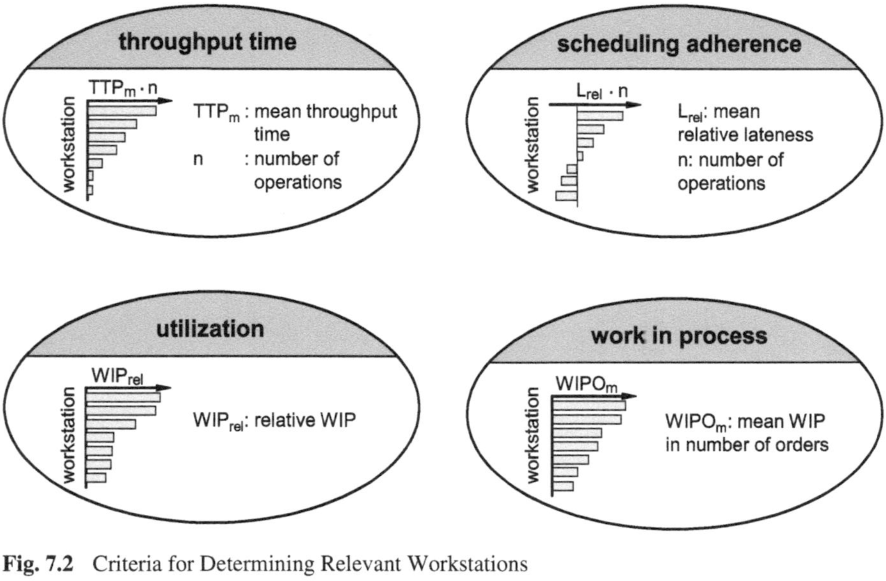

Fig. 7.2 Criteria for Determining Relevant Workstations

7.1.2.1 Goal: Reducing the Order's Mean Throughput Time

Assuming that there is a steady operating state, the sum of all the order throughput times corresponds to both the sum of all the operation throughput times as well as the sum of the product of the mean workstation throughput time and the number of orders processed per workstation:

$$\sum_{k=1}^{m} TTP_{order,k} = \sum_{j=1}^{w}\sum_{i=1}^{n} TTP_i = \sum_{j=1}^{w} \left(TTP_m \cdot n\right)_j \tag{7.1}$$

where $TTP_{order,k}$ order throughput time for the order 'k' [SCD],
TTP_i throughput time for the operation 'i' [SCD],
TTP_m mean throughput time (per workstation),
m number of orders [-],
n number of operations (per workstation) [-],
w number of workstations [-].

The product of TTP_m and n thus directly describes the degree to which the individual workstations contribute to the order throughput time. When this calculation is completed for each of the workstations they can be ranked according to which ones measures for reducing the throughput time should primarily be implemented on.

7.1.2.2 Goal: Increasing Scheduling Adherence

Measuring and evaluating the lateness parameter is based primarily on the completion of an order. However, quite frequently valuable information about the causes of the lateness, as well as possible measures can also be found at the workstation level. As a prerequisite, there has to be target throughput times not only at the order level, but also for the individual operations.

When the scheduling situation is not satisfactory, *the quality of the scheduling data* should be investigated. The starting point for such an analysis is to compare the actual and target throughput times, described for example by the mean relative lateness (Sect. 2.1.3). This figure expresses whether the target value compared to the actual value is on average set too long or too short. It should be emphasized, that the actual throughput time is not inevitably the cause of the lateness. Similarly, it is possible that the target throughput time are not realistic. In individual cases, statements about the existing cause of the deviation is only useful when combined with previously defined goals – such as those set through a Logistical Positioning.

As was the case with the throughput time not all of the workstations are equally significant for achieving the overall goals. It thus seems clear that the figures for each of the workstations should be weighted by the number of orders that are processed on the corresponding station. When developing a list of the relevant workstations it should however be considered that both positive and negative values unfavorably impact the planning's reliability. Therefore, countermeasures need to be adopted in both cases.

7.1.2.3 Goal: Reducing Loss of Utilization

Often when it seems that the market is making stronger demands on a company to reduce its delivery times and thus generally also its throughput times, it is emphasized that utilization, as an objective has fallen into the background. However, particularly in a high-wage country such as Germany the significance of the economical use of available capacities (especially operator capacity) cannot be forgotten.

The utilization of a workstation is generally determined by comparing the capacity and the output rate which has been reported back. The definition of capacity is critical for evaluating the key utilization figure. If for example the definition concerns a type of bottleneck (e. g., what is the maximum possible output rate, given a peak demand), then it needs to be kept in mind that in the medium term, the output rate should not be greater than what is needed to meet the existing demand (first basic law of production logistics; Sect. 5.1). When the load is less than the allocated capacity, then the output rate also has to be correspondingly lower. A greater utilization of the capacities could only be attained through an over production. A key utilization figure based on a capacity limit thus generally indicates whether the system is already working at the limit of its capacities or not and whether it determines the output rate when the load is increased. The thus defined key figure should therefore be primarily used to fix the resource capacities. In contrast, the yielded output rate based on the planned output is better suited to process controlling.

Whereas the capacity of resources in general can only be limitedly adjusted even over the medium term, operator capacities can be adapted comparably quickly to changes in the load situation, especially when flex-time models are considered. In particular, when the throughput time and WIP level are low breaks in the material flow can result due to the randomness of the production process. This in turn translates into temporary shortages of work for operators. The scope of the resulting loss of utilization can be described by the relative work in process WIP_{rel}. From the normalized Logistic Operating Curves it can be inferred that with a $WIP_{rel} > 250\%$ there is no significant loss of utilization (see Sect. 4.5). However, if the WIP_{rel} is clearly lower, the workstation is operating in the underload zone. Nonetheless, this should only be considered critical when as a result the assigned operator capacity is not used.

7.1.2.4 Goal: Reducing the WIP

From a logistics perspective, the WIP is rarely an independent objective. In general, decreasing the WIP serves only as a means for reducing the throughput time. However, if the goal is to minimize tied-up capital or release available floor space, then the WIP can be analyzed separately from the throughput time.

In general, a number of different measurement parameters which have to be oriented on both the prioritized goals as well as on the existing data, are available as criteria for investigating relevant workstations. Thus, in order to describe the tied-up capital, it is necessary to compile the order value of each operation and to include it in the calculation. On the other hand, if issues with floor space are the main focus, then supplementary information about the actual space required for each specific operation is needed (e. g., in number of containers, units or square meters).

Should the corresponding data not be available, equivalent parameters can also be drawn upon to some degree in order to determine the relevant workstations. For the WIP objective it at first seems obvious to use the WIP_m as a criteria for the decision making. This parameter is however, generally inappropriate because the level of the WIP_m (in standard hours) is predominantly determined by the amount of work on the individual workstations. Due to the underlying definition of the WIP, workstations with long processing times also distinguish themselves through a high WIP value when they have only a few orders. On the other hand, if the work in process is based on the number of orders $WIPO_m$, then the workstations can generally be compared directly with one another. Nevertheless, it needs to be verified whether or not the reference parameter changes during the processing (e. g., different transport containers, increased value of the order due to value-adding) and if so how this influences the underlying problem.

7.1.3 Determining Measures

After choosing the workstations relevant to the existing targets, it is crucial to the success of the manufacturing process' logistic oriented design to coordinate the measures to be introduced with one another and in consideration of the goals (see Sect. 6.5).

The starting point for deriving the specific measures is the *Logistic Positioning*. The target values for the logistic objectives schedule adherence, throughput time, WIP and utilization need to be set with regards to the actual market demands (see Sect. 6.5.1). By using the Logistic Operating Curves Theory, it becomes clear whether or not the target values are consistent and achievable with the given work content and capacities, or if it is necessary to develop additional logistic potential. Through the positioning, the existing interactions between the logistic objective are taken into account. Even when the relevant workstations are only examined on the basis of one main objective, the positioning ensures that the other logistic criteria are appropriately considered when developing and implementing measures.

Comparing the measured relative work in process WIP_{rel} with the corresponding target value $WIP_{rel,tar}$ is a fundamental indicator when determining and evaluating possible measures. When $WIP_{rel} > WIP_{rel,tar}$ it can generally be said that even when the WIP is reduced to the target value, the loss of utilization will not exceed the limit given within the context of the Logistic Positioning. In order to exploit the available potential only two basic measures are possible (Fig. 7.3): Increasing the capacity temporarily or shifting the load time or location wise.

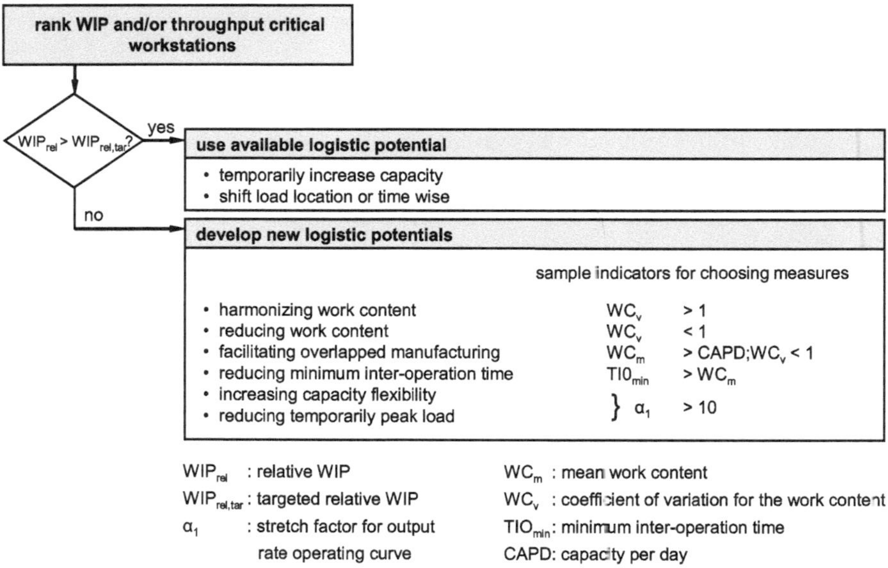

Fig. 7.3 Measures for Determining and Exploiting the Potential for Logistic Improvement (Goal: Reducing the Throughput Time and WIP Level)

Temporarily increasing the capacities is preferable when orders that are in progress have scheduling delays. However, boosting the capacities on a workstation results in the load increasing on the downstream workstations during the period in which it is implemented. It should therefore be investigated whether or not the capacity on these workstations should also be increased.

If it is not possible to boost the capacity because the capacity limit has already been achieved on the observed workstation or one of the successive workstations, then the throughput time and WIP can be reduced by allocating a part of the WIP (an amount equal to the difference between the actual and target WIP) to another technologically similar workstation or an external firm. Obviously, the first measure is only sensible when the alternative workstation has available capacity at the desired time, otherwise not only the WIP, but also the problem is shifted.

When there is no significant backlog, increasing the capacities results in the order being completed too early and thus inventory at a high value level. In situations such as this it is preferable to defer the orders which have not yet entered into production, as long as this seems permissible with regards to the schedule. Such measures also impact the upstream workstations and also only begin to have an impact after the point in time when, according to the original plan, the orders involved should have entered the workstation.

If $WIP_{rel} \leq WIP_{rel,tar}$ a new logistic potential has to be developed before the work in process can be reduced. There are numerous starting points for this, therefore it is helpful to have a preliminary selection of possible measures. In order to assist with pre-determining appropriate measures a number of key figures along with their threshold values are listed above in Fig. 7.3. The threshold values mentioned are only illustrative and should be verified in special application cases and adjusted as required.

One of the preferred measures for reducing the ideal WIP and thus for developing new logistic potential is *harmonizing the work content*, preferably on workstations where a larger coefficient of variation was determined for the work content's distribution WC_v (criteria: $WC_v > 1$). Based on the examples introduced in Sect. 6.2.1 it could be shown that with a comparably low effort a large impact could be attained by a targeted restriction of the work content. The mathematical correlations between the ideal WIP, the mean work content and the coefficient of variation described in Fig. 6.10 demonstrate the significance of this criteria.

However, if there is already a relatively uniform work content on one of the workstations critical to the throughput time or WIP (criteria: $WC_v < 1$) then the attainable effects of harmonizing the work content are limited or linked to a significant increase in the setup effort. In addition, it should then be investigated whether or not the work content can be decreased by implementing *new manufacturing technologies* or measures for *reducing the setup time*.

When the work content is large but consistent (criterion: $WC_m > CAPD$; $WC_v < 1$) only a relatively few orders are being processed on a workstation. Especially when the material flow is directed, *overlapped manufacturing* – when necessary supported by arranging workstations accordingly (keywords: manufacturing cells/segments) – should be considered as an alternative to reducing the work content.

If the minimal inter-operation time TIO_{min} is large in comparison to the work content, then there are also points here from which the logistic improvement potential can be developed. Which of the measures will be useful is determined by the cause of the minimal inter-operation time. If these consist mainly of 'pure' transport times, then *transport system changes* or in certain situations a new *material flow based layout* can be implemented. If on the other hand, TIO_{min} is due more to technology or to the process, then measures that are adjusted to the corresponding main causes should be investigated with respect to their feasibility, the time and effort required for them and the attainable benefits.

The WIP buffers that are required to avoid breaks in the material flow are expressed in the Logistic Operating Curves Theory by the stretch factor α_1. According to the results of previous research, the real system's behavior is generally described well with the value $\alpha_1 = 10$ even in the transition zone of the Logistic Operating Curves. If, when evaluating a model it turns out that a workstation has to be set at a higher α_1 value, then it indicates very pronounced peak loads, which are not absorbed by adjusting the capacities. As long as these peak loads result directly from customer behavior then it is sensible to aim for a more *flexible capacity* as a measure. When this is not possible or the peak load results from particular characteristics in the planning and control e. g., from bundling a number of individual demands into an optimal lot size, the goal should be to try and provide a smoothed input through *active load control* prior to the actual production.

The procedure introduced here for deriving measures is oriented on the primary goal of reducing the throughput time or WIP. However, in principle it can also be applied when the prioritized goal is to reduce the relative lateness L_{rel} and thus increase the scheduling reliability. In all cases in which L_{rel} is positive (when the actual $TTP_{act} > TTP_{targ}$), the throughput time also has to be reduced so that the already introduced method can be directly implemented. In doing so, realistic target throughput times, which for example are determined by a previous Logistic Positioning, are assumed. However, if L_{rel} is negative, the target throughput times are usually set too high. If this is confirmed during the positioning, then it should be sufficient enough to adjust the target value in order to increase the planning reliability.

We will now introduce two practical applications for the Bottleneck Oriented Logistic Analysis. In the first example both the procedure for the analysis as well as choosing and evaluating measures are described in-depth. In the second example, which has already been extensively documented elsewhere, the discussion is limited to presenting experiences when applying it.

7.2 Bottleneck Oriented Logistic Analysis in a Circuit Board Manufacturer

The company investigated here produces printed circuit boards of varying complexity and has approximately 300 employees. Circuit boards of all different kinds are found in almost every electronic device. The company's product spectrum stretches from non-through-hole plated to double-sided through-hole plated up to so-called multi-layers (multi-ply, through-hole plated circuit boards). The plant was newly built a few years before the study and thus was equipped with state-of-the art equipment and technology.

7.2.1 Analysis' Objectives

At the time of the study, the company's goal was to maintain a better logistic performance than their competition and to thus secure and expand their position in the market. Their primary goal was to improve their delivery capabilities and reliability. The manufacturing is based on customer specifications and there are frequent changes to the product design. It was therefore not possible to achieve the goal by using stores. Instead, the order throughput times clearly needed to be shortened and stabilized over the long term. The mean targeted throughput time of the orders was 14.6 SCD (with an average of 18.1 operations per order). The actual throughput time was on average 2.5 SCD longer. By analyzing the order throughput and conducting a Bottleneck Oriented Logistic Analysis it was to be shown if and with which measures the order delivery time and the manufacturing area's scheduling situation could be clearly improved.

7.2.2 *Data Compilation*

During an evaluation period of five months, a total of 4,270 orders with approximately 77,000 operations were compiled and evaluated. In addition to a shop calendar, the company provided the following data for the analysis:

Workstation Data

- workstation designation
- workstation number
- number of individual workplaces
- capacity of each workplace

Order Data

- order number
- target completion date
- target throughput time
- actual order start date
- priority flags

Operation Data

- order number
- operation sequence number
- workstation number
- end of operation
- work content (standard hours)
- technology dependent waiting time (standard hours)

All of the following analyses are based solely on the operating data mentioned here. One of the peculiarities of this study was the *technology dependent waiting times*. On the one hand, they involved times for additional operations, which did not encumber the affected workstation's capacities. Examples of this are the stacking/packeting of boards in front of the drill (a number of boards are simultaneously drilled) or the fixing of boards on the conveyor belt before the galvanic coating. On the other hand, they also included times in which the product was not allowed to be processed or transported due to technical reasons e. g., cooling times after they were tempered or required storage times before being exposed to light. Within the context of the Logistic Operating Curves Theory, these times were understood as minimum inter-operation times, because they could not be reduced even with a very low WIP level on the workstation.

7.2.3 *Order Throughput Analysis*

From the customer's perspective one of the outstanding characteristics of a manufacturer's logistic performance is the scheduling reliability. On the one hand, this is determined by whether or not and to what extent the production area is in the position to adhere to the planned target throughput time. On the other hand, the scheduling reliability with regards to the order's completion is also determined by the lateness of the order's input. If the orders are already late entering into the production, delays in completing the order are often unavoidable. At least prioritizing and order sequence interchanges are required in order to be able to partially compensate for impending lateness. In order to comprehensively evaluate the scheduling situation therefore, both the order input lateness and the relative lateness have to be analyzed in addition to the output lateness.

The results of these analyses for the investigated circuit board manufacturer are presented in Fig. 7.4. On average the orders entered the manufacturing process 1.3 SCD later than the target. The target start dates were not known, instead they were calculated via the target completion date and the target throughput time of the individual orders. What is quite noticeable, is the degree of the lateness variance. Only 55% of the orders are situated within the permitted tolerance zone of ±5 SCD (considering the target throughput time of approx. 14.6 SCD – this tolerance is quite generous). Approximately every fourth order (26%) entered production more than one week too late. In many of the cases orders were accepted with unrealistic delivery dates and therefore inevitably entered into production behind schedule. In addition, there were problems with supplying materials as well as errors made in planning the throughput schedule of the multi-layered boards. The remaining 19% of the orders entered into production more than a week too early. Too early entries such as these can usually be traced back to balancing the load or optimizing the machine load with the goal of reducing setup times and thus are due to personnel actively intervening.

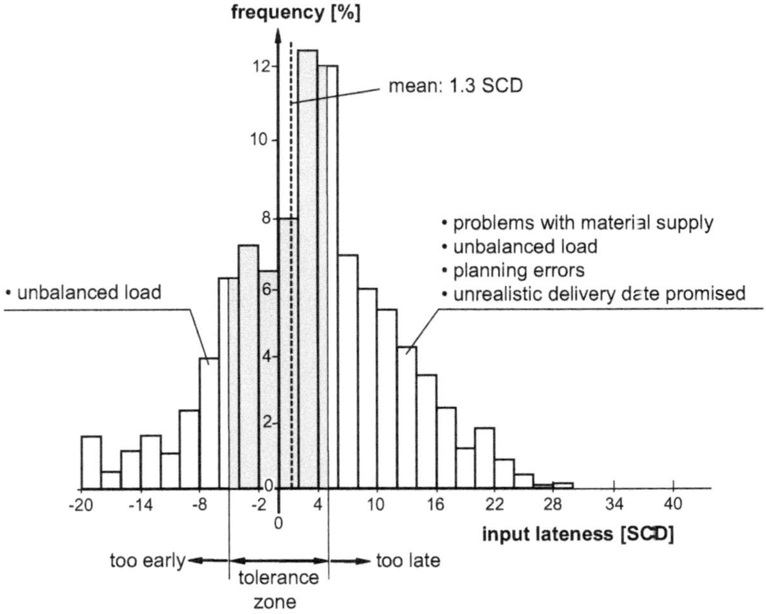

Fig. 7.4 Input Lateness and Causes (Circuit Board Manufacturer)

The relative lateness (Fig. 7.5) corresponds to the difference between the actual and target throughput times. Whereas a negative value indicates that an order was more quickly manufactured than planned, a positive value denotes that the actual throughput time is greater than the target value. The mean relative lateness was 2.5 SCD for all of the orders and once again the particularly broad distribution is noticeable. A few orders were processed more than three weeks quicker than planned. However, many of the orders clearly exceeded the planned throughput time.

The lateness in completing the orders ultimately results from deviations in the input and throughput time (Fig. 7.6). On average the orders were completed 3.8 SCD too late. This value can also be derived by adding both of the mean values for the previously mentioned distributions. It is worth noting here that the distribution of the output lateness is clearly narrower than the input lateness. Almost 75% of the orders were completed within the accepted tolerance of ±5 SCD. Furthermore, orders were rarely completed too early and the delays which were still present at the order input, could at least be partially reduced.

When these results are compiled together, it can be determined that the planning and process reliability was unsatisfactory. In particular the strong variability of the relative lateness reveals the processing plan was obviously not reliable. That the output lateness was problematic but still indicated a more favorable behavior as was to be expected based on the scheduling situation at the order input can essentially be traced back to interventions that were made in the form of targeted, schedule oriented, order sequence interchanges. These can be clearly illustrated in the form of a correlation analysis.

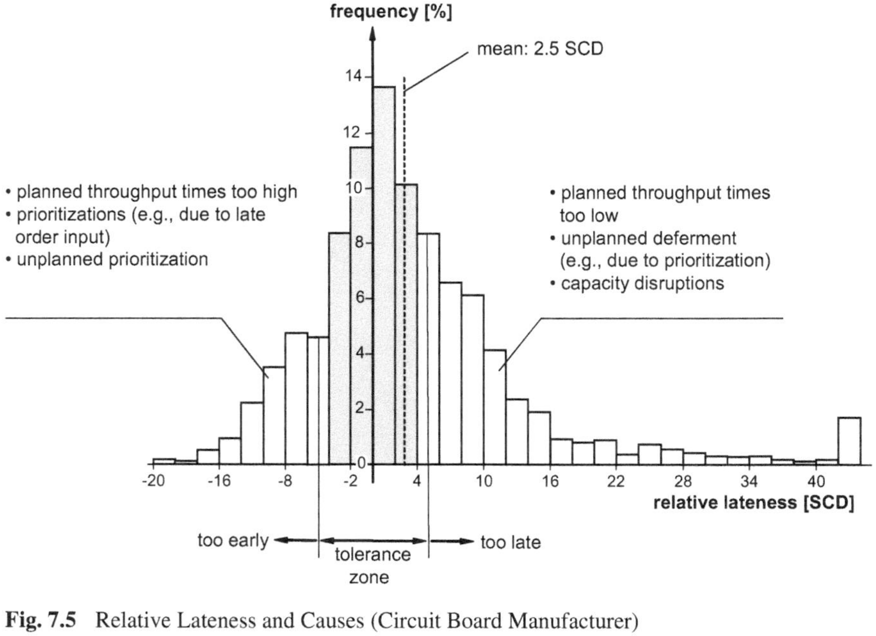

Fig. 7.5 Relative Lateness and Causes (Circuit Board Manufacturer)

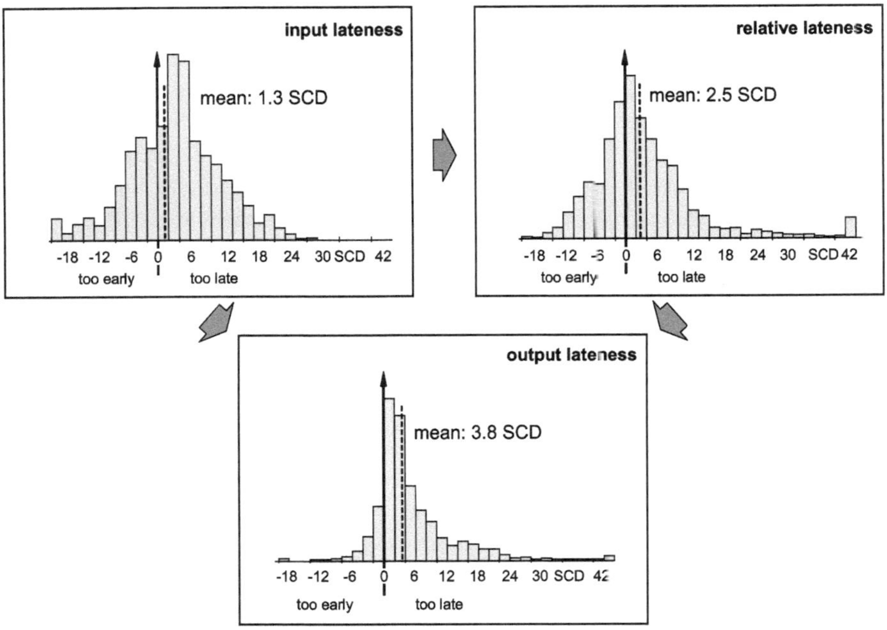

Fig. 7.6 Analysis of the Order Lateness

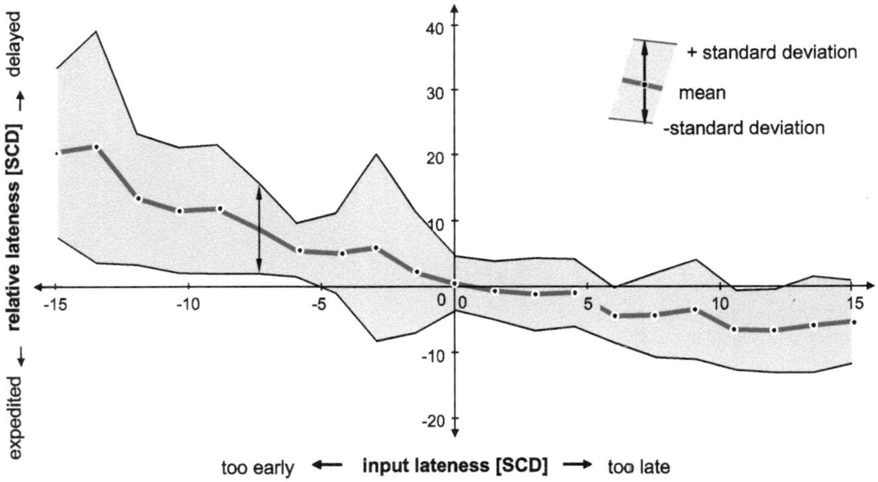

Fig. 7.7 Correlation between the Relative Lateness and Order Input Lateness

The outcomes of such an analysis are reproduced in Fig. 7.7. The relative lateness (as the resulting parameter) was plotted over the input lateness (as an influential factor). Each of the 21 measured points represents 5% of the evaluated orders, the distance between the categories on the x-axis can therefore vary. The evaluation confirmed that the orders which entered too early were strongly delayed during manufacturing. When there is a negative input lateness, the function resulting from the mean values indicates a slope of approx. –1. This means that a too early order input (on average) is fully compensated. This then explains why there were still so few orders that were clearly completed too early (see Fig. 7.4).

However, this was not the case for orders that enter late. These orders could be accelerated, nonetheless the input lateness could not be fully compensated for during the order throughput. The 'curve trace' of the mean values stabilized with a relative lateness of approx. –7 SCD. This limit can also be easily explained: In comparison to the plan, an order can be accelerated at the most by the length of the previously planned available time buffer, therefore generally by the sum of the planned inter-operation time. This value still has to be reduced by the technology dependent waiting times, which also cannot be omitted when prioritizing an order. Furthermore, even with high prioritized orders there are still waiting times, since at the point of the order's input on an individual workstation, the station is usually already occupied with at least one other order.

The results of the analysis proved that the company strongly prioritized orders according to schedule. Nevertheless, it also has to be noted that the depicted correlation only applies to the parameters' means. The standard deviations of the relative lateness, which in each case corresponds to the means, are also plotted in the graph. They indicate that there were other factors that influenced the throughput behavior of the individual orders.

A particularly important factor here was the different priorities, which from the company's side are assigned as soon as the order is placed. In addition to the normal and small batch orders, various special deadlines and rush orders were prioritized. Figure 7.8 illustrates the frequency of the different prioritizations as well as a comparison of the analyzed mean target and actual throughput times.

It can be seen that even though clearly reduced target throughput times are already assumed when planning special and rush orders, the actual throughput times turn out to be even shorter. The success of prioritizing these individual orders occurs however at the cost of the small batch and normal orders, which are inevitably set-back again and again. As a consequence, the throughput times of these are noticeably increased.

Generally, it was confirmed that the company attached outstanding significance to the delivery reliability. Through schedule oriented prioritizing, the lateness which was present at the orders' input was often at least partially compensated for during production. This reactive processing behavior was implemented in conjunction with allocating diverse priority flags as soon as the orders were placed.

In addition to schedule reliability, the enterprise also strove for a high utilization of the installed machinery. This is particularly evident in the order release: A distinctly too early anticipation of orders, whose planned start date has not yet been reached usually indicates an effort to use the available capacities as best as possible.

The information also allows one other conclusion: The intensively used possibilities of prioritizing individual orders indicates that there is a very long throughput time and a high WIP level. If the WIP is decreased and therefore the through-

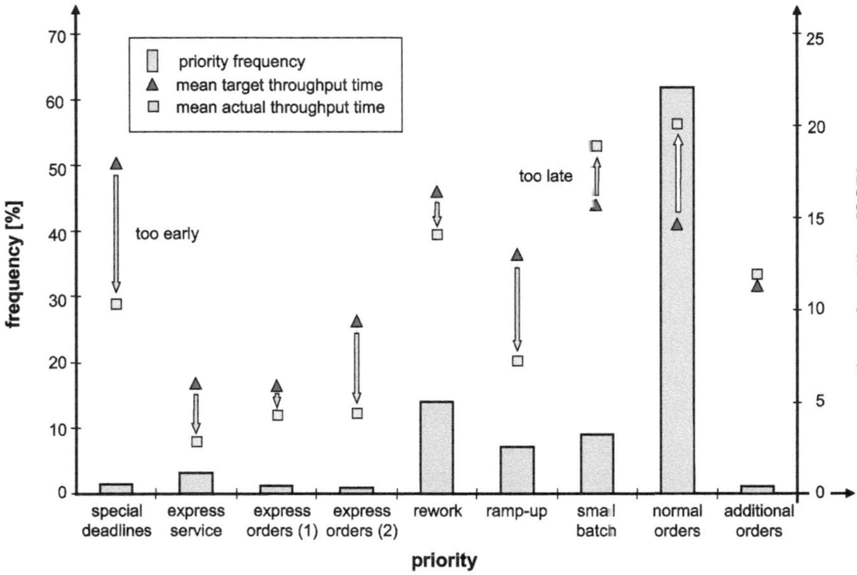

Fig. 7.8 Throughput Time Deviations According to Different Order Prioritizations

put time drastically reduced, then the company's delivery capability and delivery reliability will increase as a result.

7.2.4 Workstation Analysis

In addition to analyzing the order throughput, a Bottleneck Oriented Logistic Analysis was also conducted in order to both locate existing or new points at which the throughput time could be reduced and describe their possible impact. Based on the key figures chosen for the analysis, it was determined which workstations predominantly contributed to the order throughput time and therefore on which workstations the measures preferred for shortening the throughput time should be implemented.

7.2.4.1 Analysis of Key Performance Figures

In order to determine the workstations critical to the order throughput time, the relevant analysis figures, corresponding to the arrangement in Fig. 7.1, were established. However, it was not possible to evaluate the key lateness figures, because the company scheduled the order throughput based on the target throughput times for the whole order. The operation-based target dates required for determining the key lateness figures were thus not available. The assessment period for the workstation related evaluation was limited to three months (62 shop calendar days).

Table 7.1 Overview of Key Figures on a Department Level (Circuit Board Manufacturer; Evaluation Period: 3 Months)

key figures ⇨	throughput time					work content			output rate		WIP		
department ⇩	R_m	TTP_{mw}	TTP_m	TTP_v	$TTP_{m,calc}$	WC_m	WC_v	TW_m	$ROUT_m$	n	WIP_m	WIP_{rel}	$WIPO_m$
	SCD	SCD	SCD	-	SCD	hrs	-	hrs	hrs/SCD	-	hrs	%	-
alkaline etching	0.9	0.9	0.8	1.7	0.9	0.53	0.96	0.97	16.5	1965	15.0	377	24.5
AOI inner layers	1.5	2.0	1.6	3.2	1.5	2.25	0.59	0.17	34.1	929	51.8	1615	18.8
coating/printing	0.5	0.5	0.7	2.5	0.4	0.86	0.86	1.75	20.0	1439	9.8	101	14.4
coating/masking laquer	1.6	1.7	1.4	1.5	1.6	1.41	0.72	0.18	38.9	1704	61.6	1328	36.5
drilling	0.7	0.6	1.0	1.8	0.4	5.64	1.18	4.15	251.4	2793	163.2	103	45.5
brown oxide copper passivation	0.7	0.5	0.6	1.6	0.4	0.69	1.53	0.01	4.6	417	3.2	136	4.5
through-hole milling	1.7	0.8	0.9	3.8	1.4	2.09	1.02	0.62	8.1	262	13.6	280	7.1
through-hole line	0.7	0.8	0.6	3.2	0.7	0.38	0.81	0.48	10.9	1748	7.5	672	17.4
electrical testing	1.6	1.6	1.7	1.9	1.6	1.56	0.65	1.90	38.0	1479	60.7	491	41.1
final testing	0.7	0.8	0.6	3.2	0.4	0.46	2.04	0.51	16.9	2314	11.5	134	20,2
deburring	1.5	1.4	1.5	1.4	1.4	0.22	0.65	0.51	6.2	1835	9.1	1110	43.1
AOI inner layers etching	1.2	1.6	1.4	3.2	1.0	0.96	1.44	0.81	10.5	679	12.4	329	12.4
manual galvanizing	3.7	6.1	3.5	1.9	3.2	6.39	1.07	0.61	47.4	454	173.7	399	17.7
hot air leveling	0.9	1.0	0.8	1.9	0.9	0.70	0.53	5.23	16.1	1431	15.0	244	18.3
conductor pattern	2.0	2.0	2.3	1.4	2.0	0.43	0.83	0.77	11.9	1733	23.7	1593	65.2
mechanical/ potchmilling	0.9	0.9	0.9	1.6	0.9	1.49	0.55	1.01	13.5	564	12.1	412	8.5
mechanical/ contour milling	0.5	0.6	0.5	2.4	0.3	2.90	1.13	0.82	85.2	1807	45.7	124	13.9
mechanical/ cutting	2.7	3.2	3.1	1.4	1.9	1.44	1.15	0.79	2.1	89	5.7	138	4.2
mechanical/ pressing	1.1	1.0	1.0	1.8	0.4	1.50	1.97	0.83	5.4	219	5.9	72	4.4
pressing	1.9	2.0	1.7	1.3	1.8	1.50	0.68	0.17	10.6	435	19.7	831	11.5
resist coating	2.0	2.2	1.9	1.8	2.0	0.32	0.67	1.37	13.4	2603	26.5	1442	75.0
resist structuring	0.9	o.9	0.9	1.3	0.8	0.60	0.67	0.42	27.1	2558	23.4	846	34.5
drying/ tempering	0.2	0.2	0.2	3.4	0.2	1.37	0.41	0.39	21.6	968	4.0	67	3.1
cutting	1.3	1.3	1.2	1.2	1.3	0.10	0.67	0.00	3.4	1988	4.4	2886	37.8
interim testing	0.8	0.8	0.6	2.5	0.8	1.35	0.82	0.32	49.1	2224	39.5	769	23.8

R_m : mean range
TTP_{mw} : mean weighted throughput time
TTP_m : mean throughput time
TTP_v : coefficient of variation for the throughput time
$TTP_{m,calc}$: calculated mean throughput time
WC_m : mean work content
WC_v : coefficient of variation of work content
TW_m : mean waiting time (technically caused)
$ROUT_m$: mean output rate
n : number of orders (operations)
WIP_m : mean WIP in number of hours
$WIPO_m$: mean WIP in number of orders
WIP_{rel} : relative WIP

The key figures for 25 departments (each consisting of up to 9 individual workplaces) are reproduced in Table 7.1. For four other departments (shipping, quality special testing, multi-layered pre-treatment and customer part data) some of the figures could not be calculated because there was no available work content data.

As can be seen in the table, the mean range R_m and mean weighted throughput time TTP_{mw} for most of the workstations correspond well with each other. Therefore, there was generally a steady operating state during the investigation. The key figures differed substantially only for the hand galvanizing and plated-through-hole milling. This indicates that during the evaluation period the WIP on the hand galvanizing workstation noticeably decreased and on the plated-through-hole milling clearly increased.

The mean throughput time range is between 0.2 and 3.5 SCD and it is quite noticeable that the coefficient of variation TTP_v is very high for all of the workstations indicating that there were order sequence interchanges on all of the workstations. The mean measured throughput time TTP_m and the value calculated with the Logistic Operating Curves Theory ($TTP_{m,calc}$ *(FIFO)) either predominantly agree with another or deviate only marginally from one another. Consequently, the order sequence interchanges were obviously mostly conducted independent of the orders' work content. This conclusion corresponds with the results of the order analysis, which indicates that orders were prioritized based primarily on the schedule. Nevertheless, there were exceptions: The large difference between the key figures on the drilling workstation indicates that the sequencing was based on the work content in this department. Since the measured throughput time value is larger than the calculated, it can be assumed that the orders with the larger work content were preferentially processed (LPT dispatching rule; see also Sect. 4.4). This is also true for a number of other workstations, however not to this degree. A detailed examination of this is found in Sect. 7.2.4.3.

The key work content figures indicate that the work content distribution is very inhomogeneous. This is particularly true for the mechanical/pressing and final testing workstations where the standard deviation of the work content is approximately double that of the corresponding mean ($WC_v \approx 2$).

The significance of the technology dependent waiting time TW_m is very different in the material flow. This time component is in most cases noticeably small in comparison to the work content. In the resist coating and hot air leveling departments however, it is approximately 5 and 7½ times as long as the WC_m and therefore is crucial for determining the ideal WIP on these workstations.

The consumption of resources differs greatly between the individual departments. Alone in drilling (9 individual workplaces) approximately 1/3 (251.4 hrs/SCD) of the total output (762.7 hrs/SCD) was produced during the evaluation period. The measured output rate with approx. 28 hrs/SCD per workplace is clearly higher than the theoretical limit of 24 hrs/SCD. This can primarily be traced back to a greater than planned packet density when drilling which resulted in a shorter than planned throughput time. By evaluating the system output according to the number of processed orders, it can then be seen that a greater number of departments need to be assigned a similar significance as drilling.

The key WIP figures and here in particular the data for the relative WIP, ultimately confirms that many of the workstations are operating in the overload zone. In 6 of the 25 departments the WIP was more than 10 times greater than the ideal WIP ($WIP_{rel} > 1000\%$).

Generally speaking then, there were numerous different problem areas in the production. The process was characterized by frequent order sequence interchanges on all of the workstations, inhomogeneous work content and to some degree long technology dependent waiting times. The actual cause of the long throughput times and thus also the central point for reducing the throughput time was however clearly the very high relative WIP level on a large number of the workstations.

7.2.4.2 Identifying Throughput Time Determining Workstations

The basis for choosing and evaluating appropriate measures is generally identifying the relevant workstations using the previously analyzed key figures. Since the goal to reduce the order throughput time stood in the foreground of the investigation presented here, a list of the workstations relevant for determining the throughput time was first developed (Fig. 7.9). The product of the mean throughput time and the number of orders processed on the respective stations was calculated for all of the departments (see Sect. 7.1.2). By sorting these according to the largest

department	$TTP_{m,calc}$	n	$TTP_m \cdot n$	relative proportion of order throughput time [%]
	SCD	-	SCD	%
resist coating	1.97	2603	5128	14.4
conductor pattern	1.97	1733	3414	9.6
deburring	1.44	1835	2642	7.4
coating/masking laquer	1.55	1704	2641	7.4
cutting	1.28	1988	2545	7.2
electrical testing	1.55	1479	2292	6.4
resist structuring	0.84	2558	2149	6.0
interim testing	0.77	2224	1712	4.8
alkaline etching	0.85	1965	1670	4,7
manual galvanizing	3.20	454	1453	4.1
AOI inner layers	1.50	929	1394	3.9
hot air leveling	0.92	1431	1317	3.7
through-hole lines	0.66	1748	1154	3.2
drilling	0.41	2793	1145	3.2
final testing	0.37	2314	856	2.4
pressing	1.80	435	783	2.2
inner layer etching	0.99	679	672	1.9
mechanical processing/ contour milling	0.34	1807	614	1.7
print coating/masking laquer	0.41	1439	590	1.7
mechanical processing/ potchmilling	0.87	564	491	1.4
through-hole milling	1.43	262	375	1.1
mechanical processing/ cutting	1.85	89	165	0.5
drying/ tempering	0.16	968	155	0.4
brown oxide/ copper passivation	0.36	417	150	0.4
mechanical processing/pressing	0.35	219	77	0.2
		total	35583	100.0

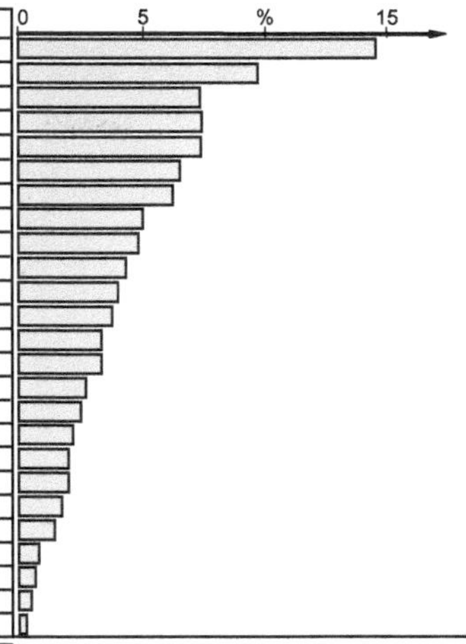

Fig. 7.9 Ranking of the Throughput Time Determining Workstations

value and normalizing these to 100% it was shown which percentage of the order throughput time each of the workstations were responsible for. In order to eliminate the influence of the sequencing rules and to facilitate estimating the potential for logistic improvement later, the calculated value $TTP_{m,calc}$ was used for the calculation and not the measured value TTP_m (see Eq. 4.40).

7.2.4.3 Detailed Analysis of Chosen Work Stations

The list of the workstations that are relevant for determining the throughput time (found in Fig. 7.9) shows that the workstation for resist coating alone, generated on average over 14% of the order throughput time and that in total 46% came from the first five named departments. By comparing these with the key figures presented in Table 7.1, it can be seen that there was a high relative WIP especially on the resist coating station. Therefore, measures for reducing the throughput time should preferably be implemented on this workstation. Before we discuss possible measures and their significance for the total order throughput time though, the detailed results from these and two other chosen workstations should be introduced. These workstations were chosen due to the very different characteristics that they exhibited from a logistic perspective.

7.2.4.4 The Resist Coating Workstation

During the resist coating process a photosensitive foil is applied, exposed and developed on the following workstation (resist structuring). The special significance of this workstation from a logistical perspective became clear once the

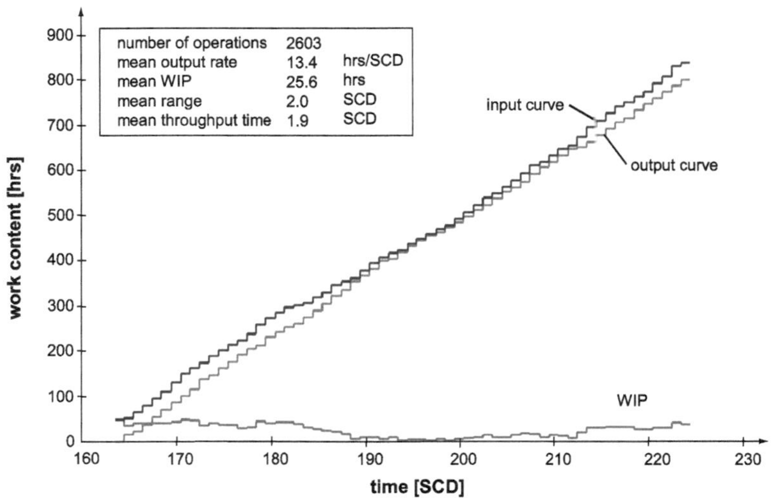

Fig. 7.10 Throughput Diagram for the Resist Coating Workstation

throughput time relevant workstations were identified. The measured mean throughput time on the workstation was 1.90 SCD, the throughput time that was calculated through the Logistic Operating Curves Theory and independent of sequencing rules was 1.97 SCD. The influence that the workstation had on the order throughput time (average 17.1 SCD) was disproportionately high. This was the result not only of the relatively long throughput time, but also because the resist coating was applied to all of the boards and to the inner layers of the multi-layered boards. The number of orders that had to be processed was therefore particularly high in comparison to the majority of other workstations.

In Fig. 7.10 the Throughput Diagram is presented with the input, output and WIP over the time. From the diagram it can be inferred that during the evaluation period there was an almost constant output rate. In contrast, the workstation's load, described by the slope of the input, does not behave as consistently: Whereas in the first third of the evaluation period, the input and output ran parallel, the load continually decreased between SCD 183 and SCD 190. Subsequently, the input and output again corresponded well with one another for approximately 10 SCD; however after SCD 200, the input once again clearly increased.

The results of the load variance on the one hand, and the constant output rate on the other hand, is reflected in the WIP trend (Fig. 7.11): Until SCD 190 the WIP is reduced from approx. 50 hrs to approximately 5 hrs. Afterwards, the WIP remains at this level for over approx. 10 SCD and then after SCD 200 it increases continuously. It is especially noteworthy that the low WIP level in the middle of the evaluation period was obviously still sufficient for supplying the workstation continuously with work.

In addition to the WIP trend, both the mean WIP and the calculated ideal minimum WIP have been plotted ($WIPI_{min}$ = 1.8 hrs) in Fig. 7.11. As can be seen in the graph, the WIP was always greater than the ideal minimum WIP – even in the

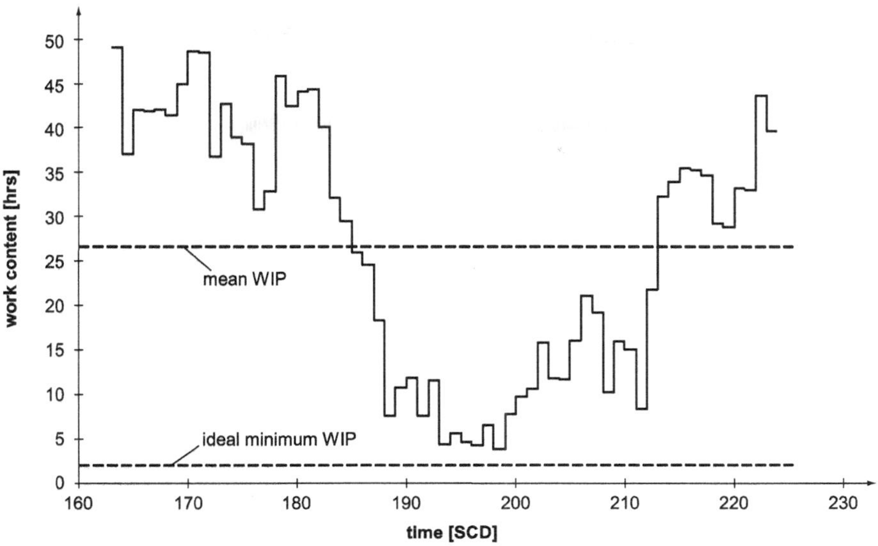

Fig. 7.11 The WIP Trend on the Resist Coating Workstation

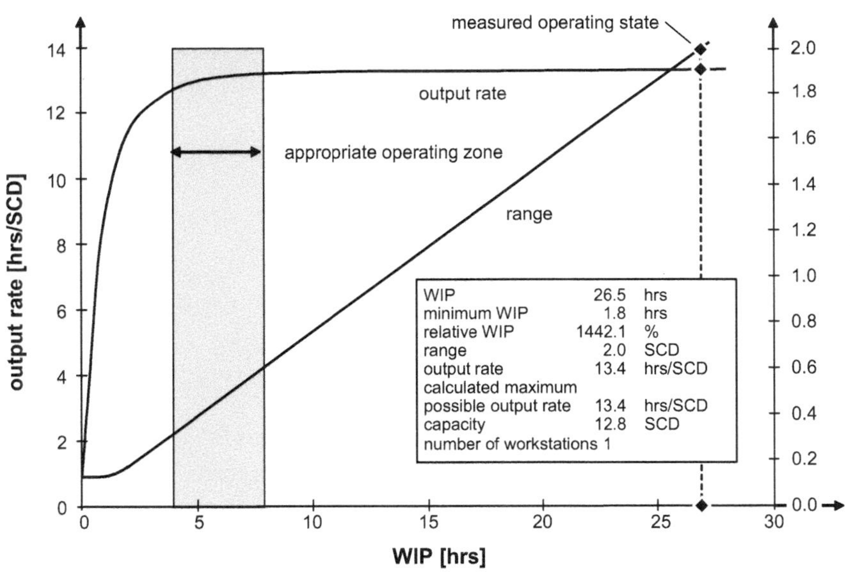

Fig. 7.12 Logistic Operating Curves for the Resist Coating Workstation

middle of the evaluation period. The results of the analysis indicate that generally the existing mean WIP level of approx. 26.5 hrs was noticeably higher than required to maintain the maximum output rate. The Logistic Operating Curves can be applied in order to quantify these conclusions and in particular to derive a target throughput time.

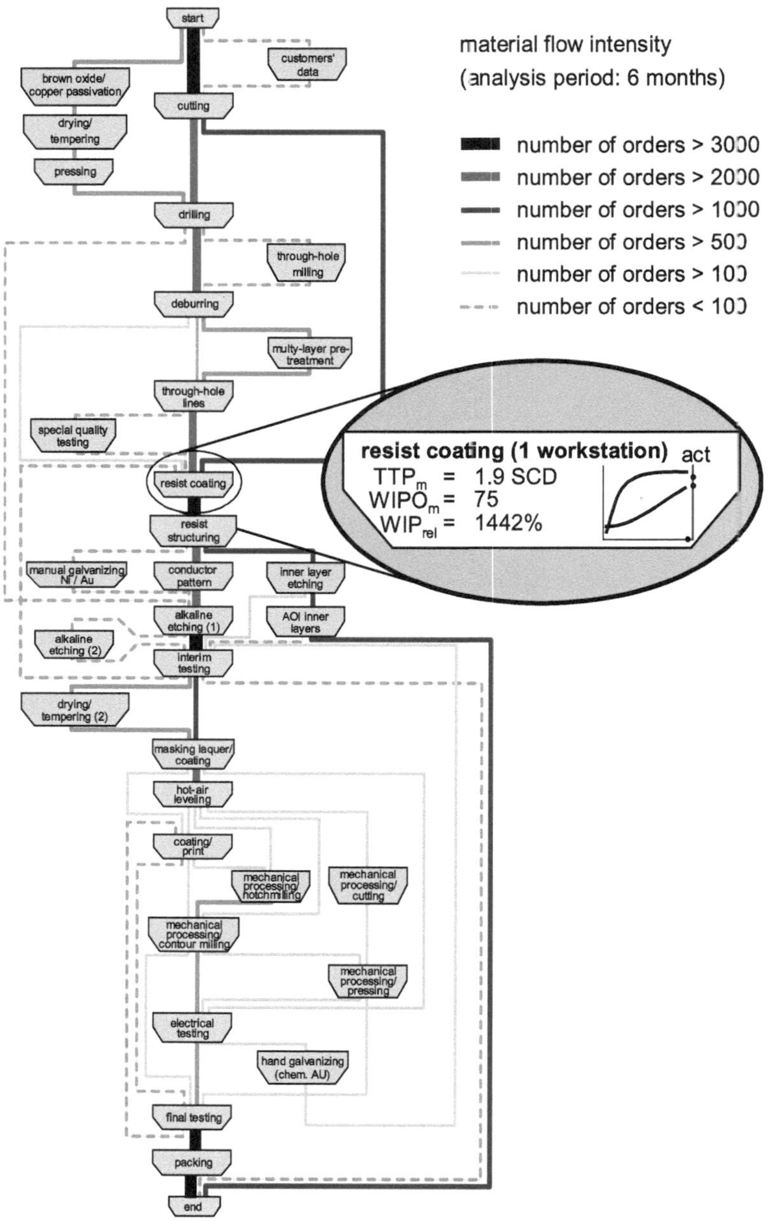

Fig. 7.13 Material Flow Diagram for the Examined Circuit Board Manufacturer

The calculated LOC for the output rate and range as well as the key values that were determined for the output rate, range and WIP are presented in Fig. 7.12. The maximal possible output rate $ROUT_{max}$ which is required in addition to the ideal minimum WIP for calculating the Logistic Operating Curves is determined using the Newtonian approximation method described in Sect. 6.3.2. An *appropriate operating zone* is also plotted for these key figures: In consultation with the company, the limits for this zone were set at double or as the case may be triple the ideal minimum WIP.

The Logistic Operating Curves indicate that even when the WIP level was at 5 hours there would still have been enough WIP for this workstation to extensively avoid WIP dependent losses of utilization. As shown above, the real process confirmed this. If the company was able to maintain this WIP level over the mid and long-term, the mean range and the mean throughput time (the two values differ insignificantly on this workstation) could be reduced by ca 1.6 SCD. Since, as previously mentioned, all of the boards are processed on this workstation, the mean order throughput time could also be reduced by this value (from 17.1 SCD to 15.5 SCD). Therefore, through the targeted reduction of the WIP on only one workstation the order throughout time could be reduced by approx. 10%.

In order to be able to exploit this available logistic potential for improvement, only a temporary capacity adjustment measure or a temporary shifting of the load come into consideration for this workstation (see Sect. 7.1.3). The first of these is best suited here as the schedule delays in the order processing justify a short-term increase in the capacity. Since unlike many of the other workstations during the evaluation period this one was not operated in all three shifts, the possibility to increase the capacity was generally there. However, before implementing such a measure, it needs to be determined whether or not it will have the desired effect and what the consequences for the other workstations in the material flow are. For example, if one of the workstations downstream present a capacitive bottleneck, then the WIP would only be shifted within the production. Improving the manufacturing behavior locally would thus not improve the overall situation.

In such cases it is useful to employ a material flow diagram for evaluating the possible consequences of a measure and when necessary also for deriving a follow-up measure. Figure 7.13 (above) shows the position of the resist coating workstation in the order throughput. It becomes clear that temporarily increasing the capacity would directly influence the input pattern of the following workstation (here, the resist structuring). Thus, in order to have the desired effect it is necessary to temporarily increase the capacity and subsequently to adjust the flexibility of capacities according to the order entry not only for this workstation but also within bounds, to other workstations. If this is not possible, then the WIP can only be reduced through temporarily restricting the order release.

7.2.4.5 The Hot Air Leveling Workstation

In order to protect the copper surfaces (basically the conductor paths and boreholes) as well as to improve the solderability when mounting components a lead-tin alloy is applied. The steps involved in this process are:

- Cleaning boards.
- Applying fluxing agent.
- Fixing boards on carrier.

- Submerging in lead-tin bath.
- Blowing-off excessive lead-tin alloy with hot air, blowing-free the drilled holes.
- Cleaning boards.

Since the "blow-off" sub-process represents the capacitive bottleneck within this chain, it is important that only this operation is given a processing time on the operation sheet. In order to still be able to consider the remaining steps in the procedure the necessary processing times are dealt with as technology dependent waiting times.

The workstation's relative WIP was 244% (see Table 7.1). Accordingly, the workstation must have been operating in the transitional zone of the Logistic Operating Curves. The detailed analysis confirmed this. In the Throughput Diagram (Fig. 7.14) it can be seen that the output predominantly echoed the light fluctuations of the input, so that – apart from a few brief exceptions – there were no large WIP build-ups.

The WIP trend illustrated in Fig. 7.15a confirms the conclusions about the level of the WIP. Intermittently the WIP falls below the ideal minimum WIP. Therefore, the machines would have sporadically been idle due to the WIP. Interestingly, this conclusion cannot be reached if the WIP is measured and provided in number of orders instead of standard hours (Fig. 7.15b). Even during the underload phases, a number of orders are always found on the system. When interpreting these circumstances, it needs to be remembered that a processing time was given for only one of the six operations in the work plan. Although the times required for the procedure's remaining five operations, which were represented by the technology dependent waiting times, are insignificant with respect to the capacities, from the perspective of the minimum throughput time and ideal minimum WIP they are much more significant than the work content ($TW_{TT} \approx 7.5\, WC_m$). Orders, which are for example being cleaned at a specific point in time, are registered as WIP for the

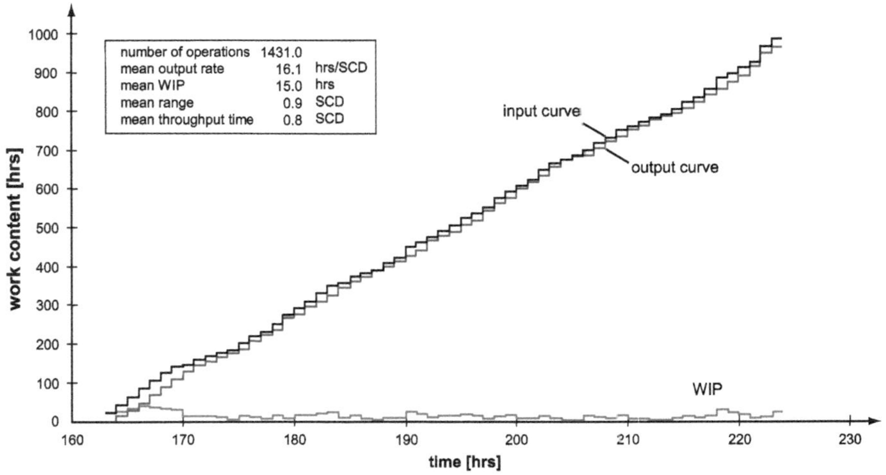

Fig. 7.14 Throughput Diagram for the Hot Air Leveling Workstation

entire workstation. However, at the moment they are not yet (or in case of the second cleaning operation, no longer) available for processing on the "blow-off" station. In order to ensure the utilization it is thus necessary, that there are always several orders on the workstation.

The conclusions reached above, were emphasized by the calculated Logistic Operating Curves (Fig. 7.16). The measured operating states were situated in the ap-

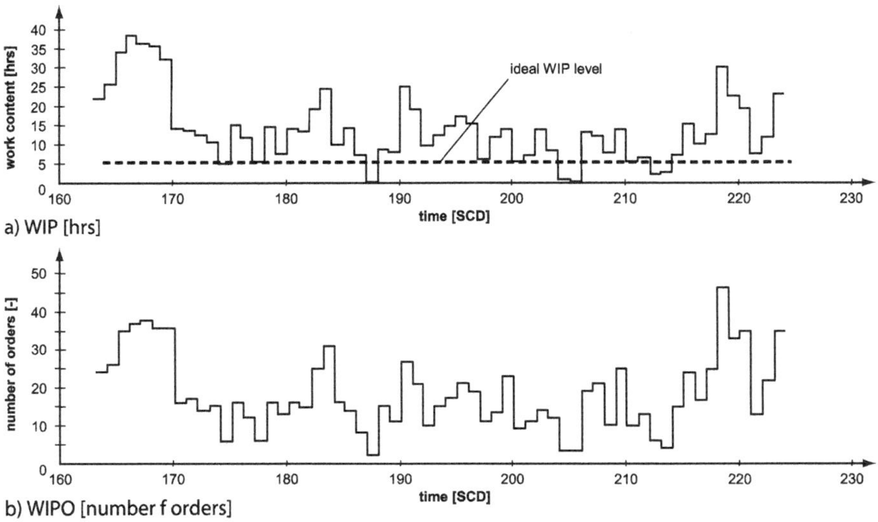

Fig. 7.15 WIP Trend on the Hot Air Leveling Workstation

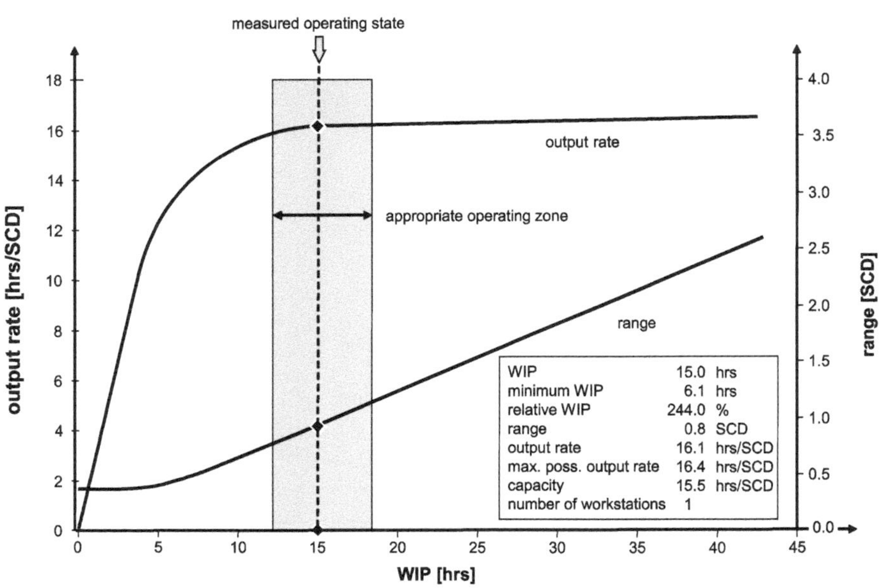

Fig. 7.16 Logistic Operating Curves for the Hot Air Leveling Workstation

propriate operating zone, which is limited to two or three-fold the ideal minimum WIP. In order to not harm the utilization rate, the throughput times and WIP can obviously only be reduced by significantly decreasing the technology dependent waiting times. Considering the multitude of workstations on which the relative WIP is considerably higher than on this station and the relative small proportion (3.7%, Fig. 7.9) it contributes to the order throughput time, it nevertheless does not seem practical to prioritize pursuing such measures. Instead, the existing potential for logistic improvement on the workstations that are operating in the overload zone should first be exhausted by implementing measures aimed at controlling the WIP.

7.2.4.6 Drilling Workstation

The drilling workstation, on which the circuit board's hole pattern is produced, is also of secondary importance for the order throughput time. The system is operated in the underload zone and its relative WIP is only 103%. However, the throughput time and in particular its influence should be more closely examined through the processing behavior.

In the discussion about the workstation key figures (Table 7.1) the huge difference between the measured throughput time and that calculated with the Logistic Operating Curves Theory ($TTP_m = 1.0$ SCD, $TTP_{m,calc} = 0.4$ SCD) was emphasized. Differences in this order of magnitude can generally only be explained through processing sequencing rules that are decisively influenced by the work content. The results of a correlation analysis (for the work content and measured throughput time) illustrated in Fig. 7.17 underscores this conclusion. Every measuring

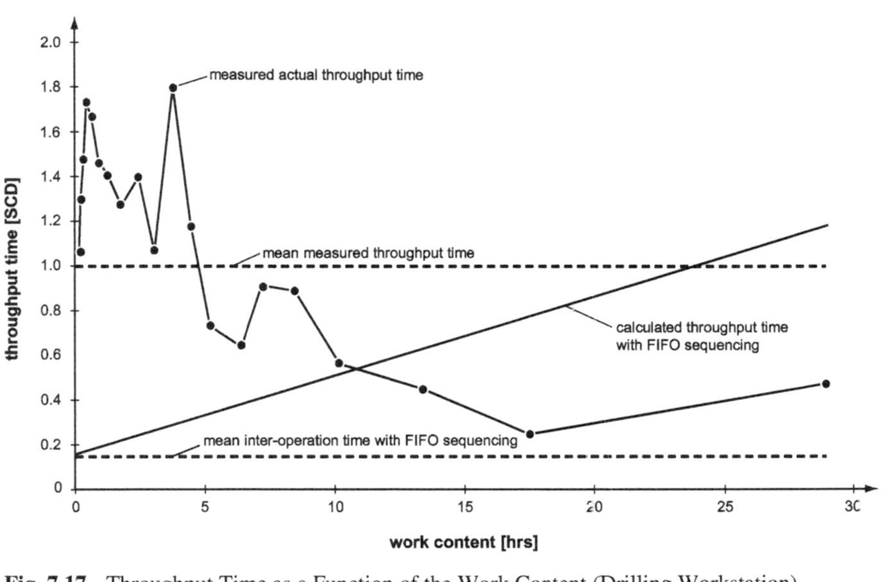

Fig. 7.17 Throughput Time as a Function of the Work Content (Drilling Workstation)

point represented approx. 5% of all the orders processed on the system. Furthermore, the mean inter-operation time TIO_m, which according to the LOC Theory exists at the measured operating state provided the orders are processed according to the FIFO principle, is plotted. By adding the operation time TOP to this inter-operation time, we obtain the trend of the throughput time with FIFO processing as a function of the work content.

In the graph it can be seen that the orders with the largest work content are obviously processed directly after entering the system and thus exhibit a very low throughput time. Furthermore, since the feedback accuracy is only one day, the measured throughput time for many of these orders is 0 SCD. The calculated inter-operation time (TIO = TTP – TOP) in these cases is thus negative. As a consequence of this behavior, the throughput time in 15% of all the orders (represented by the three last measuring points) is clearly reduced in comparison to the natural FIFO processing sequence. In contrast, the processing of the orders with the smaller work content is inevitably deferred. The mean throughput time for the orders with a work content of up to 4 hours is considerably larger than the total mean value of 1 SCD. The difference in the throughput time is even more noticeable with FIFO processing. The behavior of the measured values represented in the graph is thus an expression of a sequencing rule, which to a large extent corresponds with the LPT principle. It should also be noted that the drilling workstation has 9 for the most part structurally similar workplaces. Large orders thus, do not interfere with one another's processing. This therefore reinforces the very distinct processing behavior.

A further conclusion can be made based on the graph and the key figures derived from the Logistic Operating Curves Theory: If a work content independent sequencing rule is ensured on this workstation, then through this measure alone the mean throughput time can be reduced by approx. 0.6 SCD. Since the majority of the product's inner layers for all of the orders have to pass through this workstation, it can also be expected to impact the order throughput time to a similar degree.

7.2.5 Quantifying the Potential for Logistic Improvement

Frequently a company can achieve its goals with a number of alternative measures. The discussion above about the detailed results for example, illustrated the different points at each of the chosen workstations where the throughput time could be reduced. A selection of appropriate measures can then be determined through a cost/benefit analysis. We will now introduce a method for evaluating the attainable improvements.

Measures for reducing the throughput time should preferably be applied to workstations on which there is a high relative WIP. In the following we will assume that it is possible to achieve a specific WIP value (provided by a Logistic Positioning) exclusively through targeted load or capacity control measures. By comparing the measured key values with the target values derived through the positioning it can be shown to what degree the mean order throughput time can be reduced without having to further intervene in the basic conditions for the technical, capacitive or material planning.

In order to calculate the mean target throughput time $TTP_{m,tar}$ per workstation, the following equations derived from the Logistic Operating Curves Theory will be applied. After adapting Eq. 4.40 to the existing problem and taking into account Eqs. 2.5 and 2.6 it can be said that:

$$TTP_{m,tar} = R_{m,tar} - \frac{WC_m \cdot WC_v^2 \cdot w}{ROUT_{max}} \tag{7.2}$$

where $TTP_{m,tar}$ mean target throughput time [SCD],
$R_{m,tar}$ mean target range [SCD],
WC_m mean work content [hrs],
WC_v coefficient of variation for the work content [-],
$ROUT_{max}$ maximum possible output rate [hrs/SCD],
w number of workstations [-].

When converting the key throughput time figures into key work content figures, it was taken into consideration that a workstation can consist of a number of individual workplaces. Since the maximum possible output rate $ROUT_{max}$ is generally given for the entire workstation, it is necessary to convert this data to a single workplace. Only then is it possible to practically convert the work content into the operation time.

The *target range* can also be expressed by the funnel formula:

$$R_{m,tar} = \frac{WIP_{m,tar}}{ROUT_{m,tar}} \tag{7.3}$$

where $R_{m,tar}$ mean target range [SCD],
$WIP_{m,tar}$ mean target WIP [hrs],
$ROUT_{m,tar}$ mean target output rate [hrs/SCD].

Taking into consideration the normalized relative WIP parameter WIP_{rel} and mean utilization U_m it follows that for $R_{m,tar}$:

$$R_{m,tar} = \frac{WIP_{rel,tar}}{U_{m,tar}} \cdot \frac{WIPI_{min}}{ROUT_{max}} \tag{7.4}$$

where $R_{m,tar}$ mean target range [SCD],
$WIP_{rel,tar}$ relative target WIP [%],
$WIPI_{min}$ ideal minimum WIP [hrs],
$U_{m,tar}$ mean target utilization [%],
$ROUT_{max}$ maximum possible output rate [hrs/SCD].

Therefore, it follows for the *targeted throughput time* that:

$$TTP_{m,tar} = \frac{WIP_{rel,tar}}{U_{m,tar}} \cdot \frac{WIPI_{min}}{ROUT_{max}} - \frac{WC_m \cdot WC_v^2 \cdot w}{ROUT_{max}} \tag{7.5}$$

where $TTP_{m,tar}$ mean target throughput time [SCD],
$WIP_{rel,tar}$ relative target WIP [%],
$WIPI_{min}$ ideal minimum WIP [hrs],
$U_{m,tar}$ mean target utilization [%],
WC_m mean work content [hrs],
WC_v coefficient of variation for the work content [-],
$ROUT_{max}$ maximum possible output rate [hrs/SCD],
w number of workstations [-].

The key figures required for calculating the throughput time are for the most part known as a result of the previously conducted Bottleneck Oriented Logistic Analysis. The additionally required target values for $WIP_{rel,tar}$ and $U_{m,tar}$ can be determined for the specific workstations through the Logistic Positioning.

In order to quantify the existing potential for logistic improvement all of the workstations are given a relative work in process of $WIP_{rel} \leq 250\%$. As can be taken from the normalized Logistic Operating Curves, the mean utilization U_m is approx. 98% (for $\alpha_1 = 10$) at this WIP level. Based on the thus defined objectives as well as the measured key processing figures, the target throughput times for all of the workstations can now be determined. The results of the conducted calculations are summarized in Fig. 7.18. For all of the workstations, at which the relative WIP was greater than the given limit during the evaluation period, a targeted

key figures ⇨	current values				target values	
department	WIP_{rel}	TTP_m	n	$TTP_m \cdot n$	$TTP_{m,tar}$	$TTP_{m,tar} \cdot n$
	%	SCD	-	SCD	SCD	SCD
resist coating	1442	1.97	2603	5128	0.34	882
conductor pattern	1593	1.97	1733	3414	0.29	509
deburring	1110	1.,44	1835	2642	0.32	585
coating/masking laquer	1328	1.55	1704	2641	0.27	452
cutting	2886	1.28	1988	2545	0.10	198
electrical testing	491	1.55	1479	2292	0.77	1141
resist structuring	846	0.84	2558	2149	0.24	610
interim testing	769	0.77	2224	1712	0.23	508
alkaline etching	377	0.85	1965	1670	0.55	1978
manual galvanizing	399	3.20	454	1453	1,86	844
AOI inner layers	1615	1.50	929	1394	0.22	201
hot air leveling	244	0.92	1431	1317	0.92	1317
through-hole line	672	0.66	1748	1154	0.24	416
drilling	103	0.41	2793	1145	0.41	1145
final testing	134	0.37	2314	856	0.37	856
pressing	831	1.80	435	783	0.50	219
inner layer etching	329	0.99	679	672	0.71	484
mech. proc./contour milling	124	0.34	1807	614	0.34	614
print coating	101	0.41	1439	590	0.41	590
mech. proc./ potchmilling	412	0.87	564	491	0.52	293
through-hole milling	280	1.43	262	375	1.25	327
mech. proc./cutting	138	1.85	89	165	1.85	165
drying/tempering	67	0.16	968	155	0.16	155
brown oxide/copper passivation	136	0.36	417	150	0.36	150
mech. proc./pressing	72	0.35	219	77	0.35	77
			sum	35583		13814
				100%		39%

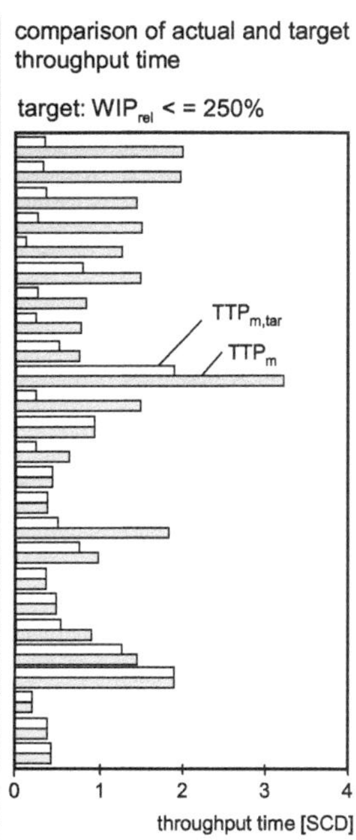

WIP_{rel} : relative WIP
WIP_m : mean WIP
$WIPI_{min}$: ideal WIP

n : number of orders
TTP_m : mean throughput time
$TTP_{m,tar}$: mean target throughput time

SCD : shop calendar day

Fig. 7.18 Estimation of the Existing Potential for Logistic Improvement throughput time was calculated with the help of Eq. 7.5. For the other workstations $TTP_{m,tar}$ was equated to the calculated actual state ($TTP_{m,calc}$). This procedure ensures that the influence of the process sequencing on the throughput time remains ignored for both the actual state as well as the given target values. Furthermore, the product of the mean throughput time and the number of orders processed on the corresponding workstations was also calculated. According to Eq. 7.1 the potential sum of the order throughput time is also described via the sum of this column. Here, the restricting 'potential' is a result of the conditions underlying the equation (i. e., a steady processing state and processing according to FIFO).

Comparing the two sums shows that by otherwise unchanged conditions, the order throughput time in the company can be reduced by approximately 61% (from 17.1 SCD to 6.6 SCD) if the respective targeted throughput times are achieved. The possible potential, which can be developed by changing the processing behavior, harmonizing the work content or reducing the technology dependent waiting time (see previous section) are not yet considered.

In order to develop these potentials, it is necessary to permanently monitor the manufacturing. Only then is it possible to continuously track the target and actual values and to derive system specific measures from the existing deviations as well as the available knowledge about conditions. In this context, the Logistic Positioning takes on a particular significance. Here, the required delivery times and capacities, as well as the structure of the costs for the workstations' target operating state are determined as a function of the work content's and capacity's structures. The resulting values, which are both realistic and conform to the goals, are then available for PPC applications. Finally, the employees need to be educated with regards to the basics of production logistics and PPC, so that they too are always able to align the production process according to the objectives, even in unpredictable situations. Only if the existing potentials are for the most part exhausted, should the barriers impeding a more extensive WIP or throughput time reduction be dismantled. In cases such as these, reducing the technology dependent waiting time and harmonizing the work content are preferred.

7.2.6 Experiences in Applying Bottleneck Oriented Logistic Analyses

Directly after the evaluation, the company implemented a number of the recommended measures. In the first step, the required throughput time values in the PPC system were corrected. Furthermore, the orders were entered into production according to the actual planned start date (not according to utilization criteria) and employees received the necessary training. Through this package of measures alone, the scheduling reliability was clearly increased: With a permissible tolerance range of ± 2 days it improved from the original 28% to 70% and exhibited a tendency to increase further. This success was supported by drastically stabilizing the throughput time (Fig. 7.19c).

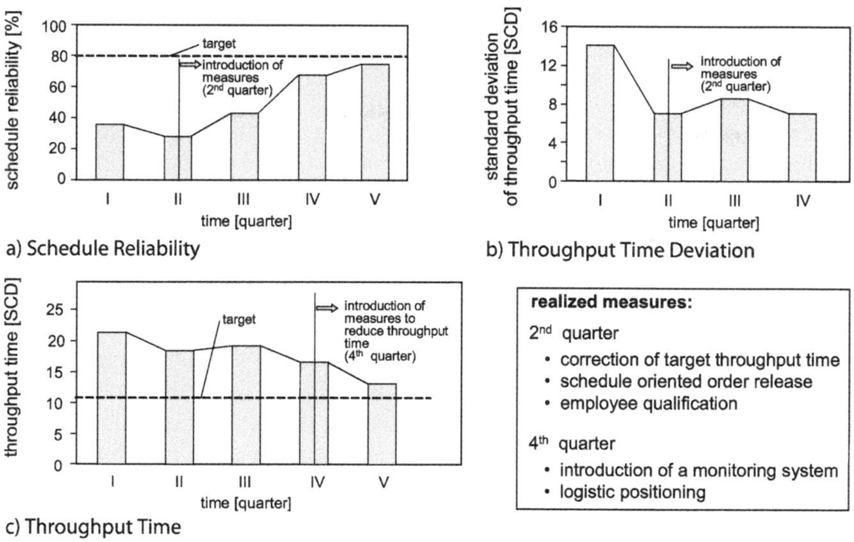

Fig. 7.19 Implemented Measures and Attained Outcomes (Through-Hole Products)

In a second step six months later, a monitoring system for evaluating the processes and conducting the Logistic Positioning (determining the required throughput time and WIP) was introduced. The goal of this measure was primarily to reduce the order throughput time and indirectly to continue increasing the schedule reliability. The throughput time which was reduced by 15% (from 19.5 days to 13) during the fifth quarter for our sample product group (through-hole plating) does not yet conform with the target. However, the attained interim results indicate that the company is on the right path to sustainably improving their delivery reliability and capability.

7.3 Applying the Bottleneck Oriented Logistic Analysis in a Circuit Board Insertion Department

The second application example for the Bottleneck Oriented Logistic Analysis stems from the electronic branch as well. In it, the circuit board insertion department of a company that produces measuring instruments was examined. The printed circuit boards with up to 1000 mounted electronic components differ greatly from one another both with regards to the complexity of the product as well as with regards to the type and number of required operations. As a result of these basic conditions, the production area with 34 cost centers was organized as a traditional job shop production. Here too, the goal of the analysis was to determine the most important access points for halving the order throughput time [Ewal-93].

7.3.1 Determining Throughput Time Relevant Workstations

The starting point of the analysis in this enterprise was also acquiring data and subsequently calculating the most important key logistic figures. Approximately 7000 operations were evaluated for a time period of 6 weeks with 29 shop calendar days. Based on that, the workstations that decisively contributed to the mean work content of on average 38 SCD were determined.

In Fig. 7.20 an excerpt from the list of workstations that were relevant to the throughput time is reproduced. It is noticeable that more than 25% of the entire order throughput time was caused by the manual insertion workstation. From a logistic perspective, the particular significance of the workstation was underlined by the stations central positioning in the material flow (Fig. 7.21). Additional systems, which we will present in detail in the following, are the automatic SMD pick and place system (LS) and one of the central testing workstations (HF testing center).

In the material flow diagram the Logistic Operating Curves that correspond to the three workstations were inserted together with the chosen key figures. It can be seen that despite the higher throughput times and WIP level, the manual insertion workstation was already operating in the transitional zone of the LOC. Therefore, in order to shorten the throughput times, interventions in the structural conditions are required in addition to control measures. However, since changing the work content's or capacities' structures is generally linked to increased costs, the extent to which the order throughput time could be reduced by decreasing the WIP only on the workstations operating in the overload zone, was first investigated before discussing and evaluating other possible measures

database:
production line of circuit board insertion
34 cost centres
6 week evaluation period
7039 orders

department	$TTP_{m,calc}$	n	$TTP_{m,calc} \cdot n$	relative proportion of order throughput time [%]
	SCD	-	SCD	%
manual insertion	7.01	619	4339	25.8
conductor preparation	2.64	846	2233	13.3
functional check	11.94	101	1206	7.2
radial insertion	1.95	561	1094	6.5
HF testing centre	9.08	99	899	5.3
in circuit tester	6.37	135	860	5.1
SMD insertion (LS)	3.17	235	802	4.8
cover assembly	3.11	252	784	4.7
programmer	6.81	111	756	4.5
manual test	6.45	105	677	4.0
manual SMD insertion	1.50	44	66	0.4
		total	16810	100.0

n: number of orders $TTP_{m,calc}$: mean throughput time

Fig. 7.20 Ranking of the Throughput Time Determining Workstations for a Circuit Board Insertion Department (Excerpt)

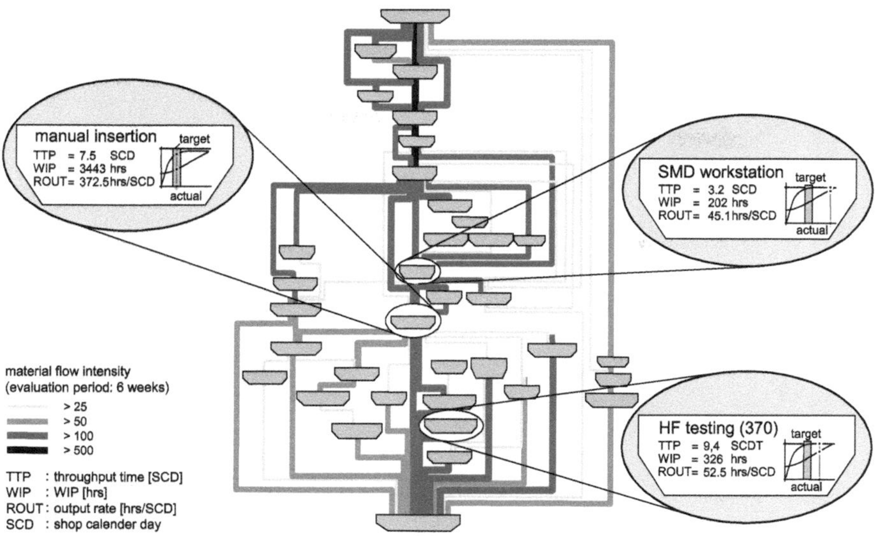

Fig. 7.21 Material Flow Diagram for a Circuit Board Insertion Department

7.3.2 Estimating Existing Potential for Logistic Improvement

The potential for the logistic improvement which was exploitable through measures for controlling the WIP was estimated according to the procedure described in Sect. 7.2.5. A target operating state was defined for all of the workstations and the target throughput times resulting from those were determined (Eq. 7.5). Subsequently, in connection with the number of operations per workstation these throughput time values were projected to a mean order throughput time. According to that, if the work in process was restricted to $WIP_{rel} = 250\%$ at the workstations operating in the overload range, the order throughput time could have been reduced at most by 36%.

With further calculations based on Eq. 7.5 and by applying the normalized Logistic Operating Curves, the correlation between the relative target WIP, the calculated obtainable potential for reducing the order throughput time and the system utilization was examined. From Fig. 7.22 it can be inferred that the order throughput time could only have been halved when the $WIP_{rel} \leq 180\%$ on all of the workstations. However, at this WIP level, pronounced losses of utilization are to be expected. Therefore, the throughput determining workstations were further investigated in order to establish if and when necessary with which measures, additional potential could be developed for reducing the throughput time.

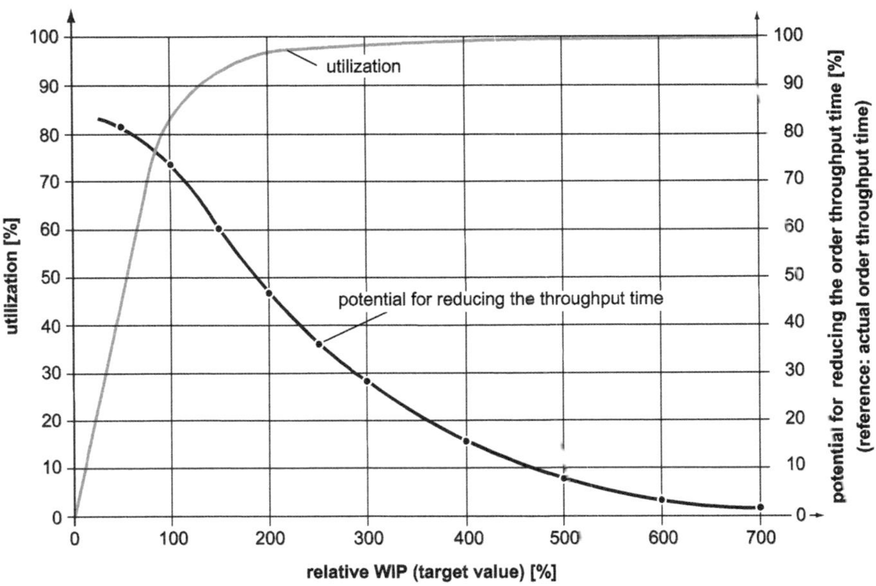

Fig. 7.22 Potential for Reducing the Order Throughput Time Solely by Reducing the WIP (Circuit Board Insertion)

7.3.3 *Deriving and Implementing Workstation Specific Measures*

7.3.3.1 Manual Insertion Workstation

On the manual insertion workstation, the circuit boards' components are mounted by hand. The station consists of 63 manual workplaces, 49 of which were occupied during the evaluation period. The results of the conducted throughput time and WIP analyses are summarized in Fig. 7.23.

As can be seen in the Throughout Diagram (Fig. 7.23a), the workstation had an almost constant WIP level during the evaluation period. The input thus corresponded well with the output. This was due to the fact that the workstation represented the lead capacity for the entire production area, on which among other things, the production's order input was oriented. At the same time though, it can also be inferred from the diagram that the WIP was approximately as high as the output volume for two weeks ($WIP_m = 3350$ hrs, $ROUT_m = 372$ hrs/SCD). Therefore, the resulting range R_m is 9.0 SCD.

The Output Rate Operating Curve (Fig. 7.23b) shows that these quite high WIP and range values were appropriate for the given conditions. The WIP therefore, could not be further reduced using production control measures unless the company had been willing to simultaneously accept a greater output loss. Since the throughput time was not sustainably influenced by the manufacturing process, the mean throughput time TTP_m with 7.5 SCD was also set through the existing struc-

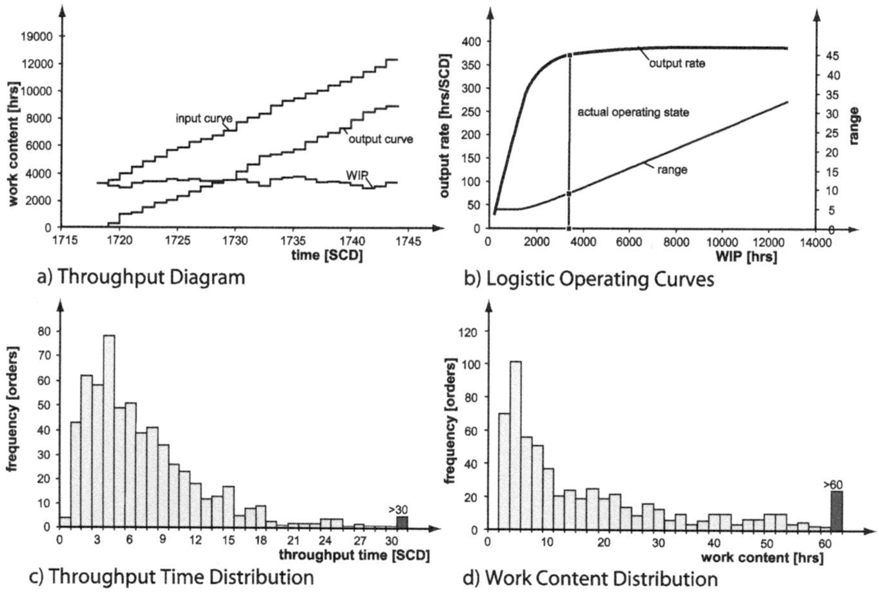

Fig. 7.23 Overview of the Analysis Results for the Manual Insertion Workstation (First Analysis)

tural conditions. The calculated mean throughput time $TTP_{m,calc}$ was 7.0 SCD. The highly variable throughput time (Fig. 7.23c) however, primarily resulted from the orders' extremely large and variable work content (Fig. 7.23d). With an output rate of approximately 7.6 hrs/SCD per individual workplace, the operation time for one operation with 40 hrs of work content is already more than 5 SCD.

Based on the Logistic Operating Curves Theory, it then follows that the existing work content structure and especially its large variance generally represents the cause of the required high throughput time and WIP level. Therefore, it stands to reason that the orders with a high work content should be reduced in order to harmonize the work content.

Upon completing the Bottleneck Oriented Logistic Analysis management decided that the following measures were to be implemented:

- All orders with a work content greater than 50 hours are to be divided when the demands are bundled from different orders.
- All orders with a work content greater than 30 hours are to be processed parallel by two or more employees. (This localized and limited division of orders prevents all the other workstations from having to also work with smaller lot sizes, which in turn would entail increased setup costs).
- The WIP level is to be adjusted to the newly achieved potential.

The first two measures mentioned, which concerned approx. 18% of the operations and approximately 55% of the entire work content, lead to a reduction of the WIP by 35%.

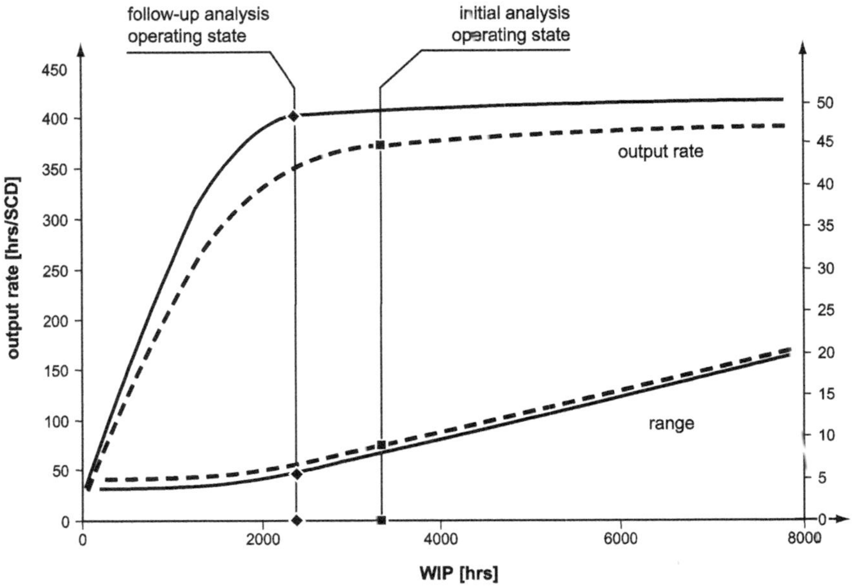

Fig. 7.24 Comparison of Logistic Operating Curves for both Analyses (Manual Insertion Workstation)

The success of these measures could be verified in a follow-up analysis nine months later (Fig. 7.24). This follow-up analysis also covered a period of 6 weeks with 29 SCD, the determined key figures from both analyses can therefore be compared directly.

The most important key figures for the manual insertion system from both analyses are presented in comparison to one another below:

• WIP	from	3350	to	2370	hrs	(−30%)
• output rate	from	372	to	402	hrs/SCD	(+8%)
• range	from	9.0	to	5.9	SCD	(−34%)
• throughput time	from	7.5	to	5.1	SCD	(−32%)
• setup time	from	25.4	to	32.3	hrs/SCD	(+27%)
• percent setup time	from	6.8	to	8.0	%	

It can easily be seen that by harmonizing the work content, the company also succeeded in converting the reduced ideal WIP for the most part into a lower WIP level and shorter throughput time. The effort invested in this is primarily documented by the increased setup time.

7.3.3.2 SMD Workstation

The second workstation that was analyzed consisted of two SMD pick and place automats, which according to the schedule were operated in three shifts (SMD:

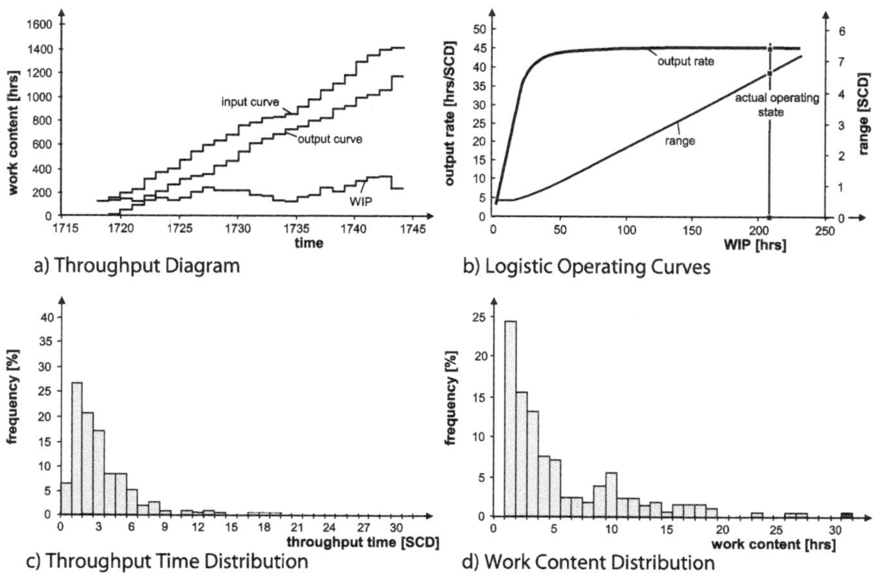

Fig. 7.25 Overview of the Analysis Results for the SMD Pick-and-Place Workstation (Initial Analysis)

Surface Mounting Device; a system for attaching electronic components to the surface).

The Throughput Diagram in Fig. 7.25a shows that the WIP strongly fluctuated during the evaluation period. The input rate was not adjusted to the output. By investigating further it became clear that the throughput time was extremely variable due to order sequence interchanges (Fig. 7.25c). In general the results indicated that the WIP with approx. 200 hrs and the range with 4.5 SCD were clearly higher than the level required to maintain the output rate. This conclusion was confirmed by the Logistic Operating Curves: The actual operating state was clearly in the overload zone (Fig. 7.25b). The LOC also indicated that there is also a reliably high utilization with a WIP of 60 hrs and a range of 1.5 SCD given. Although the work content varied quite strongly it was not the cause of the existing WIP level.

Through the follow-up analysis the conclusions regarding the potential logistic improvement were confirmed. The manufacturing process represented in Fig. 7.26 shows that the mean WIP on this station was reduced on average by 94 hours. With a slightly increased daily output rate (47.6 hrs/SCD instead of 45.1 hrs/SCD) the range was reduced to 2.0 SCD and the throughput time to 1.9 SCD. Furthermore, it can clearly be seen in the graph that the WIP level could have been further reduced by controlling the WIP more rigorously. The predicted WIP level of 60 hrs was thus attained in the first half of the evaluation period. However, since the workstation represented a capacitive bottleneck, the WIP was elevated slightly again at the end of the interval period in order to increase the utilization's reliability.

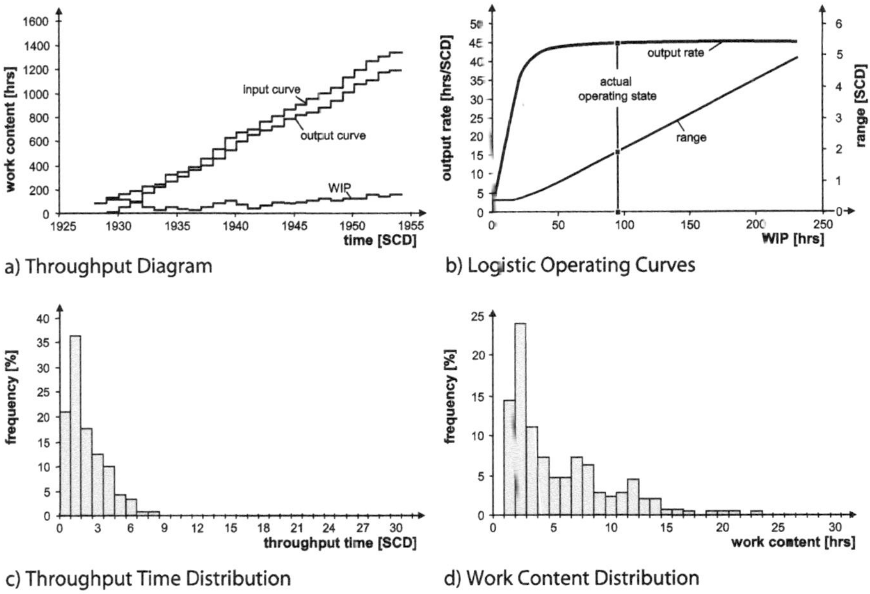

Fig. 7.26 Overview of the Results from the Follow-Up Analysis on the SMD Pick-and-Place Workstation

7.3.3.3 HF Testing Workstation

On the third investigated workstation, the functionality of the assembled circuit boards was tested. The workstation was peculiar in that a parallel conducted evaluation of failure indicators showed that the existing WIP on this workstation was blocked to a considerable degree (33%) ([Nyhu-93a], [Nyhu-93b]). The WIP is said to be 'blocked' or 'disrupted' when an order on the workstation cannot be processed despite the necessary capacity being available on the workstation. This can be caused for example by missing manufacturing tools, missing materials (in assembly processes) or closures due to technical quality defects. In this particular situation, the WIP was disrupted mainly due to a lack of test programs or in some cases where the product had been modified, because the programs and corresponding documents had not been updated.

For the workstation, two operating states were plotted on the calculated Logistic Operating Curves (Fig. 7.27). The measured operating state shows that the station was apparently working in the overload zone. However, if the WIP is reduced by the blocked portion it can be seen that the actual available WIP is just enough to avoid WIP dependent output losses.

The significance of the blocked WIP for the order throughput time was emphasized when the distributions for the throughput time and lateness (Fig. 7.28) were evaluated. The mean throughput time on this workstation was 9.5 SCD. In addition to the high mean, the strong variance of the distribution is particularly noticeable.

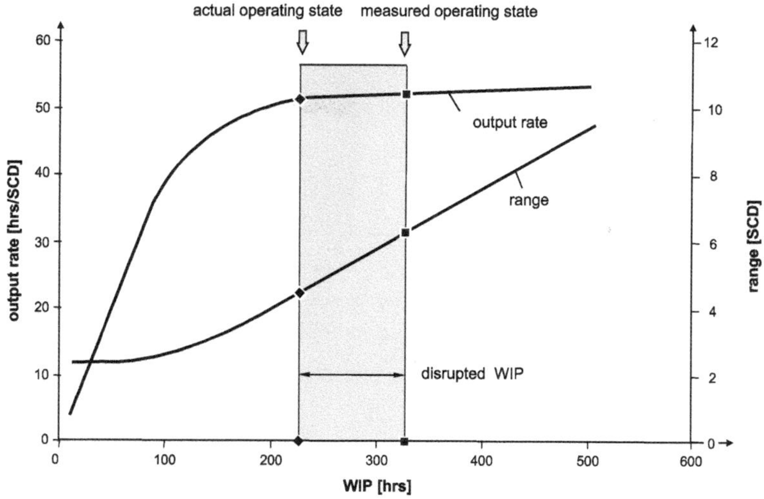

Fig. 7.27 Logistic Operating Curves for the HF Testing Workstation

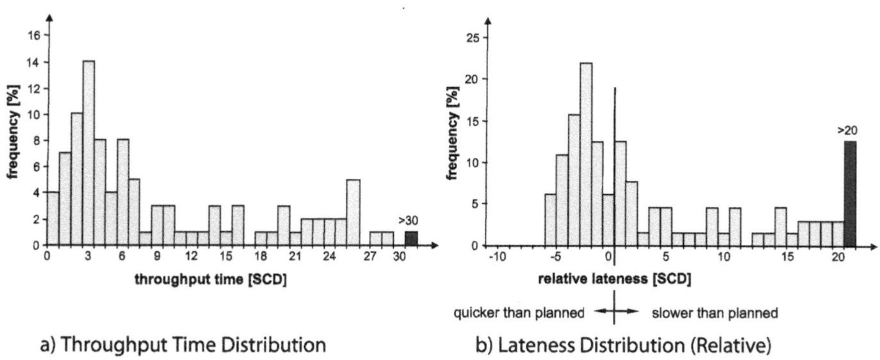

Fig. 7.28 Throughput Time and Relative Lateness Distributions for the HF Testing Center (Initial Analyses)

Such conditions make it impossible to reliably plan. This is also emphasized by the lateness distribution (actual throughput time minus the required throughput time). With a required throughput time of 5 SCD, the orders on the HF testing center were delayed on average by 4.4 SCD. The extremely large variance is also especially critical here. Whereas many orders had to wait up to a number of weeks for processing, others were processed earlier than planned in order to avoid idle time on the workstation.

As a consequence of the investigation's results, the company prioritized removing the WIP disruptions (for example by being more thorough when making product changes). The success of the implemented measures are also documented here in the results of the follow-up analyses (Fig. 7.29). Both the distributions for the throughput time and the lateness have become narrower with a correspondingly

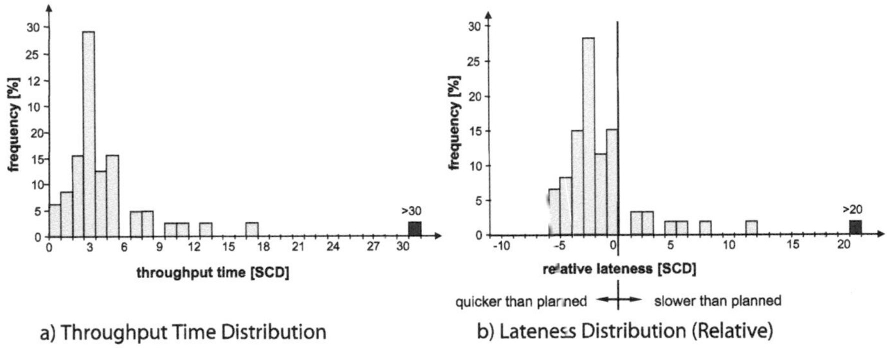

Fig. 7.29 Throughput Time and Relative Lateness for the HF Testing Center (Follow-up Analysis)

reduced mean. The effects which the company strived for with respect to increasing the schedule reliability and reducing the throughput time could thus also be attained on this workstation.

7.3.4 Summary of Application Experiences

The company was able to successfully use the potential predicted through the Logistic Operating Curves on other workstations as well. Since the results of further investigations have been reported elsewhere ([Wien-93a], [Ewal-95]), we will not go into an extensive description of them here.

The success of all the conducted measures is visibly documented in the comparison of the chosen key figures for both analyses (Fig. 7.30). Given an approximately equal output rate, the WIP, range and throughout time were clearly reduced. The changes with regards to the order throughput time though are the most noticeable. Here the fact that the adopted measures primarily concentrated on the workstations that were predominantly responsible for the order throughput time bears results and substantiates that the activities were focused on workstations relevant to the throughput time. Indeed the goal of halving the throughput time could not be totally achieved in the first attempt, however, the results gained from the follow-up analysis indicated further potential for reducing the throughput time.

The company had already implemented a production monitoring system before the first analysis. One of the essential conditions for successfully realizing the measures derived in the project was therefore already provided. Only the Logistic Operating Curves as a 'window to the process' [Ewal-93] and the merging of the different analysis techniques into the Bottleneck Oriented Logistic Analysis presented new control elements for the company. The tools applied in this project are in the meantime completely available in the firm and continue to be utilized for the targeted design and control of the manufacturing process.

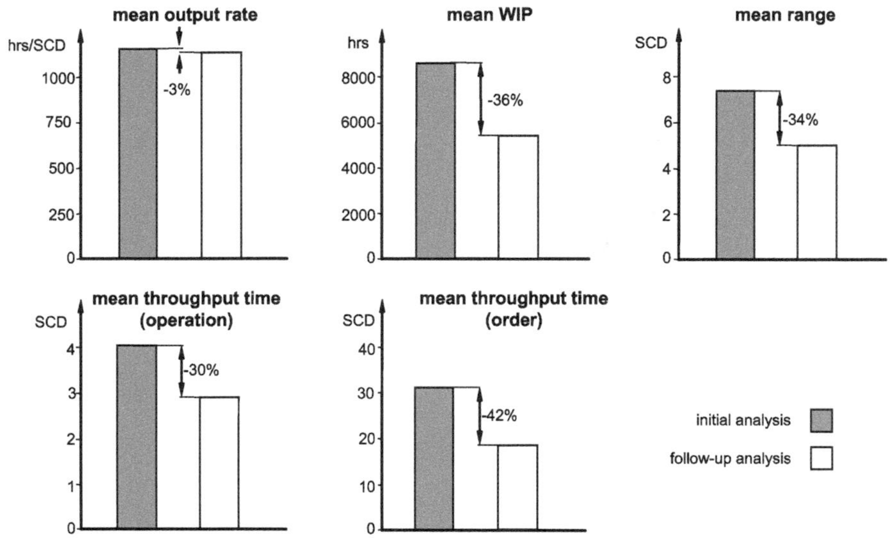

Fig. 7.30 Comparison of Results for the Throughput Time and WIP Analyses

7.4 Strategies for Implementing the Bottleneck Oriented Logistic Analysis

In order to be able to fully utilize the possibilities of a Bottleneck Oriented Logistic Analysis, it should become a natural tool and be continuously applied within the context of logistic production controls. Nonetheless, it is strongly recommended that it be applied in specific steps (Fig. 7.31). First, an initial analysis should be conducted for a limited time. Based on experience, extensive methods for improving the manufacturing process will already result from this. In addition, problems in the feedback and quality of the planning data will also become evident. Repairing these problems is an essential prerequisite for implementing the BOLA as part of a permanent production control. Once the BOLA is integrated, it serves as a continuous improvement process, which should include verifying and improving the production structure and PPC system. As we already demonstrated, improving the production structures is first based on reducing the inter-operation time between two operations by grouping workstations into segments, cells or fractals according to the products. Furthermore, with regards to the PPC system, it has to be verified that the planned throughput times are set consistently and aligned with the actual throughput time. Moreover, the currentness of the planned capacity values are frequently a weak point.

In order to apply the Bottleneck Oriented Logistic Analysis however, employees first require a systematic introduction into its fundamental principles, preferably in conjunction with training, in which concrete examples from the company

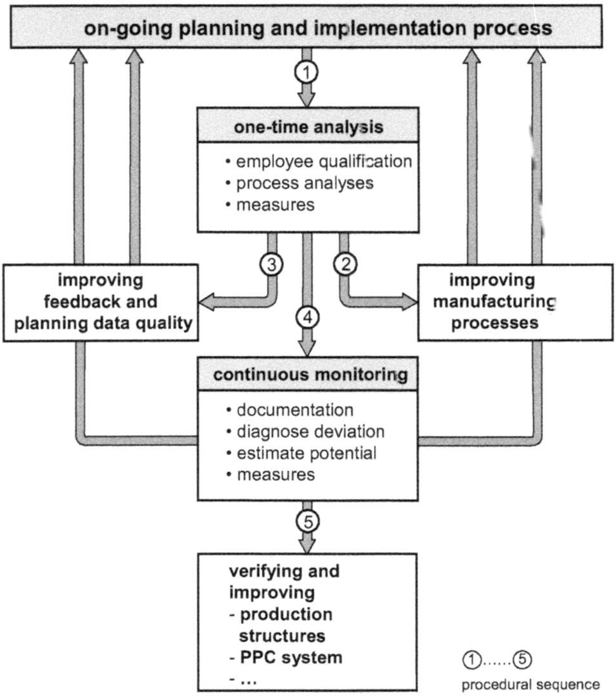

Fig. 7.31 Steps for Implementing a Bottleneck Oriented Logistic Analysis

can be observed. Since the necessary measures often run counter to ingrained habits and experience, implementing them is frequently found to be uncomfortable. Last but not least, no sustainable success is to be expected without managerial support.

Chapter 8
Applying the Logistic Operating Curves Theory to Storage Processes

While the depth of manufacturing has been significantly reduced and the demands on the production's flexibility has increased, the costs and punctual supply of materials from outside has become one of the key corporate success factors in the manufacturing and assembly industry. *Procurement and inventory management* takes on the critical role of balancing the tension between the customer's need for flexibility and the supplier's logistic performance in order to ensure an economical, high and stable delivery of semi-finished products and additionally purchased parts for production. In doing so, properly dimensioning the amount of stock to be maintained is vital. Due to its high cost impact, the stock level is an important logistic objective in procurement. The stock has to provide a buffer for fluctuating requisitions (from customers, distributors or the manufacturer's own production) as well as deviations in their suppliers (internal or external) delivery. This illustrates the traditional *dilemma of inventory management*: A balance has to be found between a high service level on the one hand, and a low stock level on the other hand.

In order to efficiently manage the procurement and inventory it is necessary to describe the interdependencies between these objectives in consideration of the operating constraints and the possibilities of influencing them. Only then can a company position themselves between the competing factors and when necessary specifically influence the procurement and inventory management processes. From the procurement and inventory control perspective, the following questions are central to the abovementioned logistic objectives:

- What levels of stock are required in order to ensure a sufficient delivery capability?
- How do disruptions to the store input and output influence the delivery capability?
- Which measures are appropriate for increasing the service level and what impact do these measures have on individual concrete cases?

It is noticeable that quite similar questions were the motivation for developing the Logistic Operating Curves (see Sect. 2.4). It thus seems obvious to derive a similar method for describing the stock management process in order to obtain effective support in answering these questions. Such a method is presented by the *Storage Operating Curves* (SOC). Storage Operating Curves illustrate the interdependencies between the stock level and the service level depending on different basic conditions. In the next chapter we will introduce the basis of these Storage Operating Curves as well as the Storage Operating Curves Theory (SOC Theory) which was derived similarly to the Logistic Operating Curves Theory. In doing so we will not only discuss the various possibilities and limitations of applying them, we will also present a prototypical example of an application.

8.1 Throughput Diagram as a Model for the Logistic Procurement Process Chain

A common method for describing the planned and realized stock management processes is the general storage model. With the help of this model (Fig. 8.1a) the chronological behavior of the stock level that results from the difference between the store input and store output is illustrated. From this model, key logistic procurement figures can be derived, which can then be used to evaluate the processes e. g., the mean stock level, the actual stock range and other key figures. Nevertheless, this model is limited with regards to being applied as a permanent monitor for the logistic procurement process. Gläßner [Gläß-95] identified the following basic deficits:

- The entire logistic procurement process chain cannot be modeled.
- The simultaneous representation of the planned and realized processes in order to compare the target and actual values is difficult to overlook.
- Individual sub-processes cannot be isolated from one another, thus the individual dynamics cannot be depicted.
- Not all of the logistic objectives can be represented.

Therefore, a process model was developed in order to implement a logistic oriented controls method for the entire logistic procurement processing chain. This new model eliminated the weakness of the general storage model, without losing the simplicity and clarity of the original [Gläß-95]. The *Storage Throughput Diagram* (Fig. 8.1b), which can be directly derived from the general storage model in a basic form, serves as the foundation for it. The principle idea of the Storage Throughput Diagram is to be able to represent the usually independent sub-processes, the store input and store output, separately from one another. In order to do so the complete input and output quantities of an item or item group are cumulatively plotted over the time for the relevant feedback date. Whereas the store input curve begins at the point of origin, the starting point of the store output curve results from the initial stock of the evaluation period. The corresponding actual stock level can be continuously determined from the vertical distance between the two curves.

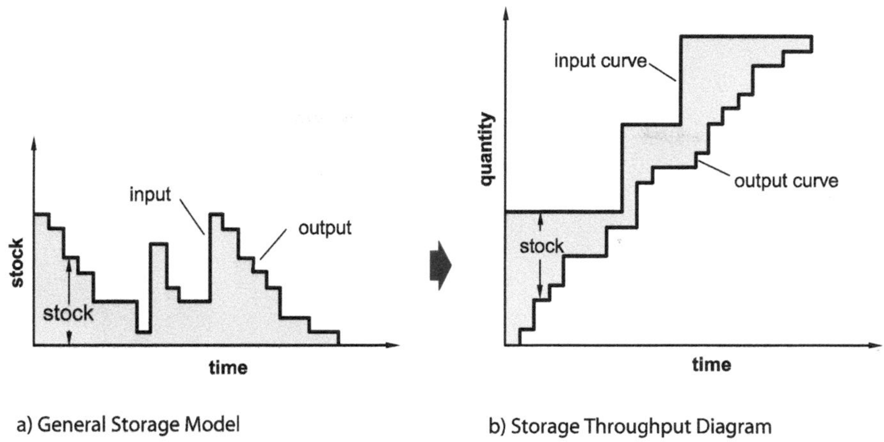

Fig. 8.1 Development of the Storage Throughput Diagram Based on the General Storage Model

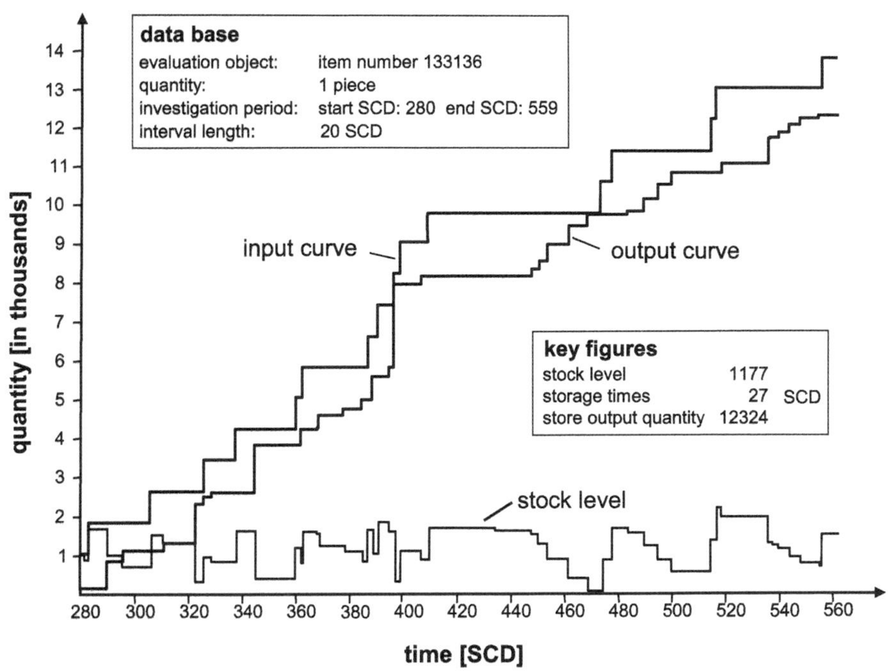

Fig. 8.2 Throughput Diagram of a Storage Process

In a further particularization step, this modeling approach can also be extended to the procurement sub-processes, whereby the ordering process is also understood as a measuring point and plotted in the Throughput Diagram. In Fig. 8.2, a Throughput Diagram for a single item is presented as an example. In addition to the visual representation of the input and output curves, a number of key performance figures have been inserted.

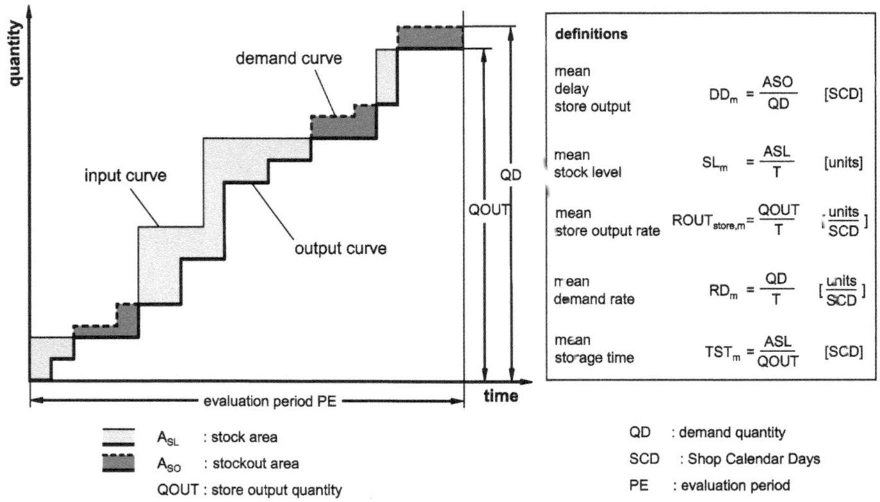

Fig. 8.3 Calculating the Basic Logistic Figures Using a Storage Throughput Diagram (Gläßner)

The stock level parameter results from the vertical distance between the input curve and the output curve. Similarly, the storage time parameter is determined by the horizontal distance between the curves. The amount of throughput during the evaluation period is described by the slope of the input and output curves.

The *delivery delay* is a measurement for quantifying the schedule adherence objective. It describes the time related delay (expressed in shop calendar days) caused by a stockout situation in the demanding production area. In order to determine the delivery delay, the stockout quantity that occurs is multiplied by the stockout time. The sum of these values are then set in relation to the required total quantity of the item being considered during the evaluation period. The delivery delay is thus a parameter which refers to both amounts and time. It is represented in a Throughput Diagram by comparing the trend of the production area's demand curve with the relevant actual output curve. In Fig. 8.3 (above) the formulas for the thus defined delivery delay as well as for additional basic key logistic figures are represented together with a Storage Throughput Diagram.

8.2 Storage Operating Curves

Employing Storage Throughput Diagrams and the key figures derived from them makes it possible to comparatively evaluate storage processes in detail based on standard operating data. Applying the discussed model however, is not adequate for comprehensively describing the conflicts between the objectives. As was the case with the Logistic Operating Curves, describing the interactions between the storage process' objectives in the form of a Storage Operating Curve has proven to be much more practical.

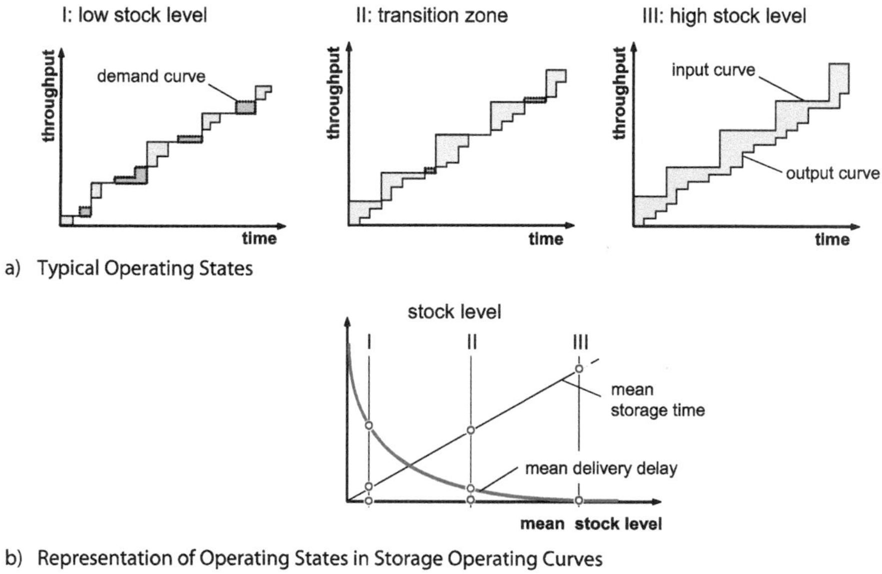

Fig. 8.4 Deriving Storage Operating Curves from the Storage Throughput Diagram

Figure 8.4 depicts such a Storage Operating Curve as well as the basic procedure for deriving this type of LOC from the Storage Throughput Diagram for a procured item [Gläß-95]. In Fig. 8.4a, three store operating states, which differ with regards to the stock and delivery delay, are reproduced in Throughput Diagrams.

Operating state III is characterized by a high stock level: All of the production requests can be immediately satisfied and there are no delivery delays in the store output. If the stock level is reduced to the point that the input and output curves just touch one another, as in operating state II, then the stock level is clearly lower. As can be seen in how the demand and output curves only negligibly diverge, there are still no significant supply shortages. If the stock level is reduced even more, the input and output curves fall together over longer durations (operating state I). During these times, there is no stock and production requisitions can therefore not be satisfied. This is visualized in the graph by the increasing stockout areas (the difference between the demand and store output curves).

If the mean delivery delay is then plotted over the mean stock level for these three processing states it results in a Storage Operating Curve for the delivery delay. The SOC describes the relation between the logistic procurement objectives, stock level and delivery delay. Furthermore, the diagonal line in the diagram indicates the proportionality of the mean stock level and mean storage time. Based on the relation between these two it is clear that given a constant store output, the mean delivery delay can also be plotted over the mean storage time, without the basic shape of the SOC changing.

The shape of the Storage Operating Curves – and therefore the achievable levels of material availability and stock – can generally be influenced by a number of different elements [Gläß-95]. These influences can be classified into three groups (Fig. 8.5): On the input side, factors which determine the source's i. e., the supplier's or in-house manufacturers' logistic performance should be considered relevant. In particular these include the delivery quantity reliability, the due-date reliability and the replenishment time. Similarly, on the output side, the demand quantities, as well as the demand quantity reliability and the due-date reliability of the demand are significant. Finally the shape of the Storage Operating Curve is also influenced by the method chosen for determining the lot size as well as by the quality of the procurement and inventory planning.

Since the logistically relevant measures impact the position and shape of the Storage Operating Curves the logistic potential of different planning measures can be determined with the help of the Logistic Operating Curves technique. Based on a fictional example, the general procedure for this is depicted in the upper part of Fig. 8.5. The Storage Operating Curves are shown for an initial state as well as two differently structured alternatives. Through the shape of the curves it is obvious that from a logistic point of view, variant 1 is better and that in contrast, variant 2 indicates worse logistic properties than the original state. Based on a recorded or defined operating state the logistic potential of the alternative plans can now be determined. Assuming that the stock level should remain unchanged, the expected improvement or deterioration of the stores delivery capability for the item being investigated can be derived from the vertical distance between the Storage Operating Curves. However, if the delivery delay should remain constant the resulting

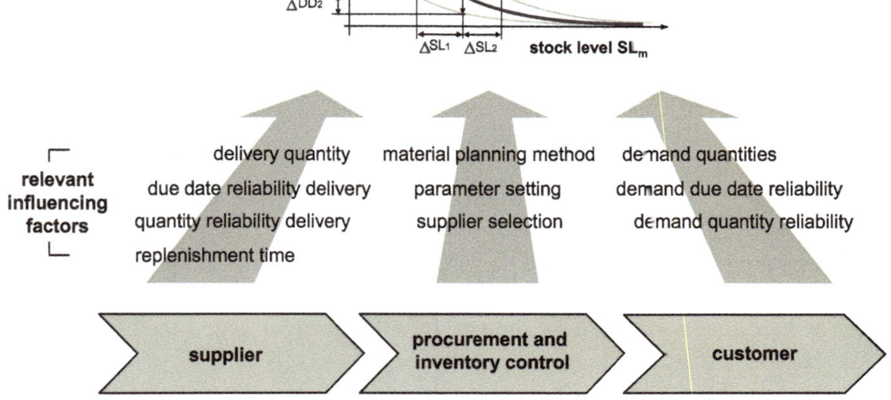

Fig. 8.5 Applying Storage Operating Curves for Describing the Logistic Potential in a Storage Process

potential for reducing the stock in variant 1 or the required stock increase in variant 2 can be determined from the horizontal distance between the curves.

8.3 Determining Storage Operating Curves Using Simulations

Generally, simulations can be used to determined Storage Operating Curves. As was already discussed in our explanation about establishing Logistic Operating Curves with simulations (see Sect. 3.2.2), the significance of chosen simulation variables for a store's key logistic figures can be investigated by specifically varying them. In the following, we will present exemplary results determined using a simulation tool developed for this purpose. The simulation system, which is based on a relational database, makes it possible to specifically manipulate the relevant factors named in Fig. 8.5 and thus to analyze their significance for the position and shape of the resulting Storage Operating Curves.

In Fig. 8.6 two chosen Storage Throughput Diagrams that were created for an item with the help of simulation are presented. Both cases were based on a mean input lot size of 60 pieces/lot, an output lot size of 1 piece/lot and a mean demand rate of 3 pieces/SCD. These parameters were however subject to considerable variation: a delivery quantity deviation of ±50%, a due-date deviation of ±9 SCD, a replenishment time of 10 SCD and a daily requirement of a maximal 27 pieces/SCD were permissible. The store input as well as the demand behavior were identical in the individual simulation trials. Only the initial stock was changed and thus also the resulting mean stock level.

The graph shows that in the selected simulation run 3 (mean stock level) stock-out situations occurred only sporadically. However, if the stock level was clearly reduced, as in simulation run 5, then shortages occurred increasingly and lasted for longer durations. After conducting further simulations runs, the determined key figures (stock level, delivery delay and storage time) can be aggregated into a Storage Operating Curve (SOC) (Fig. 8.7).

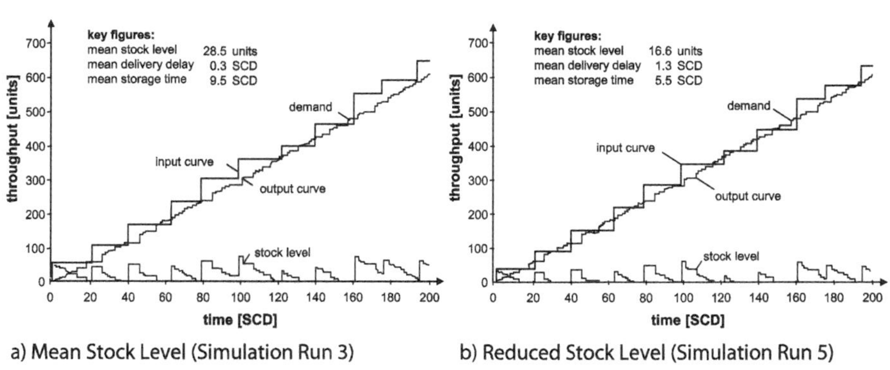

Fig. 8.6 Simulatively Determined Storage Throughput Diagrams

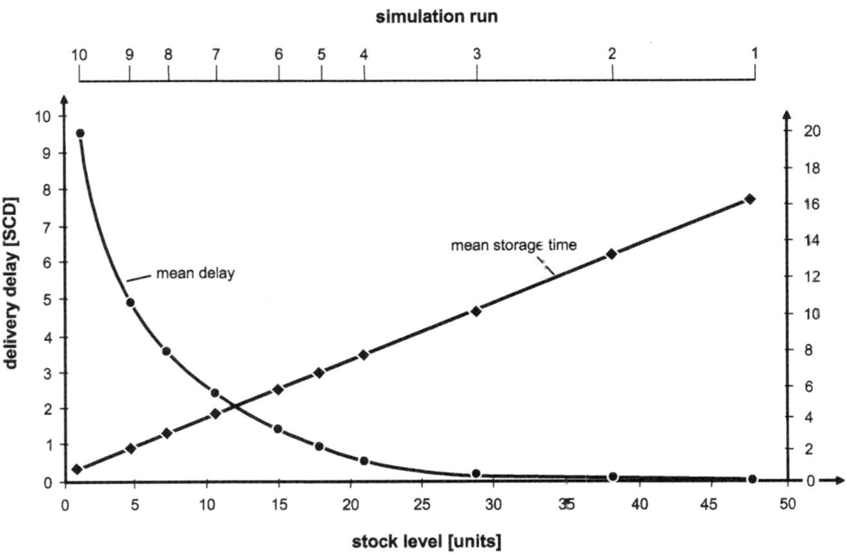

Fig. 8.7 Simulated Storage Operating Curve

Despite the extensive potential for applying simulatively generated Storage Operating Curves, it seems more improbable that these will be employed in the field. Based on experiences up until now, the effort required solely for providing realistic data can rarely be justified. Furthermore, only a large number of events can ensure a level of confidence with regards to the information derived from the SOC. This in turn requires a long evaluation period or a high level of aggregation (e. g., observing entire groups of items). It thus seems apparent, that an approximation equation for the Storage Operating Curves needs to be developed. Such an equation would facilitate a mathematical description of the relations between the stock level and delivery delay as well as the abovementioned influences while requiring less data.

8.4 Determining Storage Operating Curves Using an Approximation Equation

Similar to the Logistic Operating Curves Theory the Storage Operating Curve Theory introduced here is based on a deductive-experimental process model. The following steps provide the foundation for the modeling:

- Deriving the ideal Storage Operating Curve by analyzing the idealized process in storage.
- Expanding the model in consideration of the actual process flow or where applicable of plan deviations by integrating a generic function that can be parameterized.
- Adapting the parameters by analyzing the simulation results.

8.4.1 Ideal Storage Operating Curve

An ideal state of a storage process is characterized by the following conditions:

- The store output is continuous with a steady output rate. In the Storage Throughput Diagram the output curve can therefore be approximated as a straight line.
- Determining the economic order lot size should be considered an essential aspect of the procurement. Therefore, also in the idealized case, the input should occur discretely and in finite lot sizes.
- There are no process disruptions.

A state such as the one described here can be illustratively represented in the general storage model (Fig. 8.8a). It becomes clear that having safety stock ensures availability at all times and that shortages do not occur. The mean stock results from the sum of the safety stock level and half of the store input quantity.

Under idealized conditions, shortages also do not occur even when there is no safety stock. The situation however, changes when the mean stock level is further reduced, for example through a lower initial stock or a delayed material supply. The lower the mean stock level is, the greater the extent of the resulting shortages is.

Gläßner [Gläß-95] derived a function for the conditions described above with which the correlation between the stock level and the delivery delay in the store output can be described given the idealized conditions:

$$DD_m = \frac{1}{RD_m} \cdot \left(\frac{QIN_m}{2} - \sqrt{2 \cdot SL_m \cdot QIN_m} + SL_m \right) \tag{8.6}$$

where DD_m mean delivery delay in the store output [SCD],
SL_m mean stock level [units],
RD_m mean rate of demand [units/SCD],
QIN_m mean store input quantity (lot size) per input event [units],
$0 \leq SL_m \leq QIN_m/2$ (function domain).

This function, represented in Fig. 8.8b, is equal to zero for:

$$SL_0 = \frac{Q_{in,m}}{2} \tag{8.7}$$

where SL_0 lot stock level [units],
QIN_m mean store input quantity (lot size) per input event [units].

Here, the graph touches the abscissa with the slope zero. This lot stock level SL_0, caused by the (mean) store input's lot size, thus represents the limit at which, assuming the described ideal conditions, there is just barely no delivery delay.

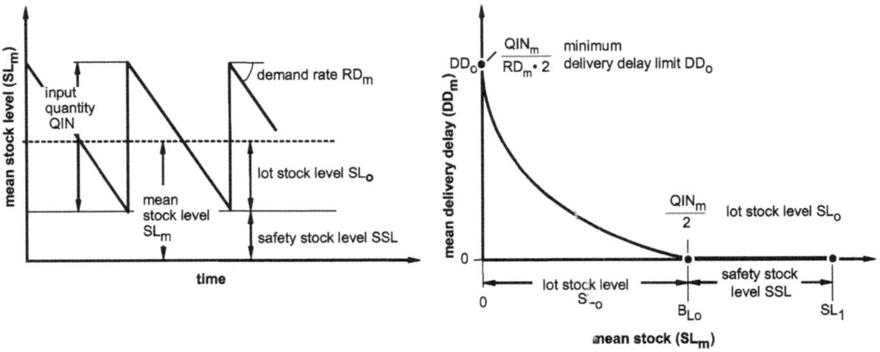

a) General Storage Model b) Ideal Storage Operating Curve

Fig. 8.8 General Storage Model as a Basis for the Ideal Storage Operating Curve

The maximal value of the function within the domain is thus defined:

$$DD_0 = \frac{QIN_m}{RD_m \cdot 2} \tag{8.8}$$

where DD_0 minimal delivery delay limit [SCD],
QIN_m mean store input quantity (lot size) per input event [units],
RD_m mean rate of demand [units/SCD].

As expected the delivery delay limit DD_0 occurs at a zero stock level. A zero stock level means that every incoming delivery is immediately forwarded completely onto the successive operation in order to counterbalance existing shortages. Shifting the store input curve further in comparison to the store output curve would in turn increase the delivery delay.

The graph of the function represented in Fig. 8.8b – derived based on the description of the idealized state – is to be considered the *ideal case* or *minimal threshold case* for a real process. Similar to the Logistic Operating Curves Theory, this function will be identified in the following as the *ideal Storage Operating Curve.*

The ideal Storage Operating Curve assumes that there are no plan deviations. A real process however is influenced by a number of different disruptions which sustainably deteriorate the stores' delivery reliability. Therefore, in a further step, the model is expanded so that the variance in the process as well as the fluctuating store output and input rates can also be included. First, in order to do so, it was decided to use an alternative way of describing the 'ideal' Storage Operating Curve.

The basis for transforming the function is once again the C_{norm} function that we already employed in deriving the Logistic Operating Curves Theory (see Sects. 4.2.1 and 4.2.2). Stretching the first quadrant of the C_{norm} function in the positive x-direction by SL_0 and stretching it in the y-direction by DD_0 results in the following basic form of an ideal Storage Operating Curve:

$$\left.\begin{aligned} SL_m(t) &= SL_0 \cdot t \\ DD_m(t) &= DD_0 \cdot \sqrt[C]{1-t^C} \end{aligned}\right\} \tag{8.9}$$

where $SL_m(t)$ mean stock level [units],
$DD_m(t)$ mean delivery delay [SCD],
SL_0 lot stock level [units],
DD_0 minimal delivery delay limit [SCD],
$0 \le t \le 1$ running variable,
C C value.

With regards to the results, this equation is for $C = 0.5$ identical to Eq. 8.6, as can easily be proven by converting Eqs. 8.7, 8.8 and 8.9. By applying Eq. 8.7, the first part of Eq. 8.9 can therefore be solved for t:

$$t = \frac{SL_m \cdot 2}{Q_{in,m}} \tag{8.10}$$

where $SL_m(t)$ mean stock level [units],
$QIN_{in,m}$ mean store input quantity (lot size) per input event [units],
t running variable.

If a similar transformation of Eq. 8.8 is now undertaken, where C is set at 0.5 and the expression for t (Eq. 8.10) is adopted for the lower part of Eq. 8.9, it results in the equation developed by Gläßner for the ideal Storage Operating Curve.

The advantage of this new formulation (Eq. 8.9) is that this function can be easily modeled by the parameters contained within it. Thus, real processes including plan deviations can be taken into consideration in the mathematical model. The possibilities for expanding the equation described in the following are not applicable to the initial Eq. 8.6.

8.4.2 Integrating Plan Deviations

The idealized state, which was used as a foundation for the ideal Storage Operating Curve, does not take into consideration any process disruptions e. g., quantity or scheduling deviations in the input or output. However, it is especially these plan deviations which characterize real stock behavior. Therefore, safety stocks have to be maintained in order to ensure the stores high service level. How large this safety stock has to be, will first be described separately for every single plan deviation.

Quantity and due-date deviations can be traced back to both an unsatisfactory logistic delivery capability as well as to internal deficits in the procurement planning. Often, important planning parameters such as the replenishment time are based only on estimated values and are not regularly verified or adapted. In order to integrate the resulting due-date deviations as well as the possible quantity deviations in the equation for calculating the Storage Operating Curves, their influence on the stock and delivery delay limits should first be analyzed.

The lot stock level SL_0 represents the value above which there is no longer delivery delays. With due-date deviations in the store input the stock has to be increased up to the point where even the worse case state – i. e., the maximal delay – can still just be buffered. These deviations reduce the stock in comparison to the originally planned level.

In Fig. 8.9a this case is depicted using the general storage model. Given possible due-date deviations, the safety stock has to be at least as high as the product of the mean demand rate and the maximum due-date deviation. Only then can it be ensured that zero stock levels for the observed item do not occur. This safety stock level, abbreviated in the following as $SSL_{1,min}$ in order to differentiate between it and other stock components, is calculated as:

$$SSL_{1,min} = \left| L^{+}_{max} \cdot RD_m \right| \tag{8.11}$$

where $SSL_{1,min}$ safety stock level for due-date deviations [units],
L^{+}_{max} maximum positive lateness (delays) [SCD],
RD_m mean rate of demand [units/SCD].

In order to avoid making errors with the algebraic signs, the deviations investigated here will be generally described by the absolute value.

When there are only quantity deviations (Fig. 8.9b) the stock can usually be increased just to the point where the maximal partial delivery (caused also by defective items) does not yet cause delivery delays. As a consequence, the required safety stock level $SSL_{2,min}$ corresponds to the maximal partial delivery:

$$SSL_{2,min} = \left| Q^{-}_{max} \right| \tag{8.12}$$

where $SSL_{2,min}$ safety stock level for delivery quantity deviations [units],
Q^{-}_{max} maximum negative quantity deviations (partial delivery) [units].

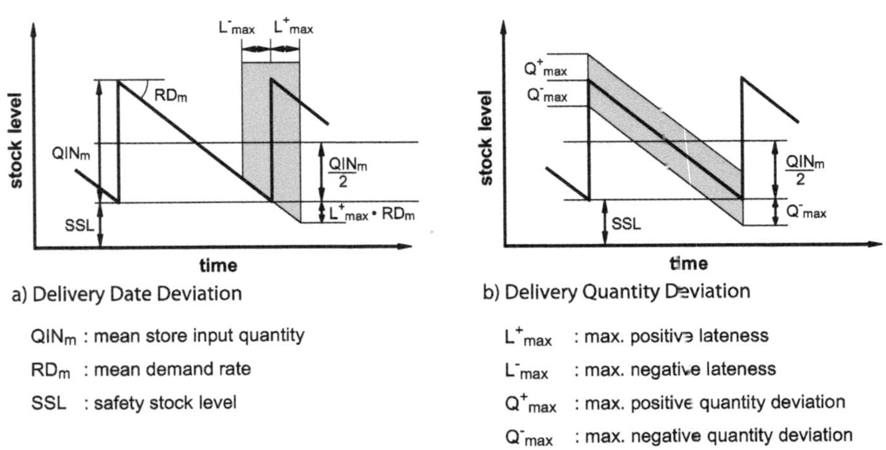

Fig. 8.9 Causes and Effects of Deviations from the Planned Stores Input

Quantity deviations only need to be considered to the extent that they cannot be counter-balanced by pulling forward the next delivery. This is the case when the replenishment time is greater than the time span between the two deliveries. Otherwise, the delivery quantity deviations are not critical, since when they occur, they can be countered through a corresponding planning intervention. Safety stock is therefore not required for plan deviations such as these.

Shifting the demand dates or off-setting quantities is also possible from the side of the store output. In contrast to the results on the store input side, it is usually the demand behavior during the replenishment time that is significant for the problems in the store output and not a single event. Shifting a single demand does not generally lead to supply problems, since possible plan deviations can be buffered by the existing lot stock. It is only with the last of the requisitions that are supposed to be covered by an input lot that the delivery capability is not guaranteed without safety stock.

If however, the rate of demand during the replenishment time can generally change, then safety stocks need to be maintained in order to avoid lasting effects on the delivery capability (Fig. 8.10a). This safety stock level ($SSL_{3,min}$) can be determined as:

$$SSL_{3,min} = (RD_{max} - RD_m) \cdot TRP \tag{8.13}$$

where $SSL_{3,min}$ safety stock level for demand fluctuations [units],
RD_{max} maximal rate of demand [units/SCD],
RD_m mean rate of demand [units/SCD],
TRP replenishment time [SCD].

An evaluation period equal to that of the replenishment time should be used for determining the maximum and minimal rate of demand.

The previously mentioned causes of deviations can occur individually or in combinations, so that the abovementioned effects can add up in a critical case.

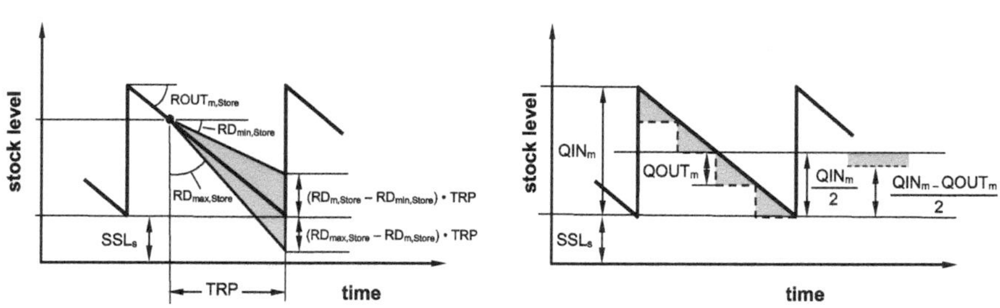

a) Demand Deviation

b) Discrete Storage Output

$RD_{m,store}$: mean store output rate
$RD_{max,store}$: maximum store output rate
$RD_{min,store}$: minimum store output rate
TRP : replenishment time

QIN_m : mean store input quantity
$QOUT_m$: mean store output quantity
SSL_S : safety stock level

Fig. 8.10 Causes and Effects of Deviations from the Planned Store Output

However, insofar that a stochastic independence can be assumed, the individual effects cannot only be reinforced, but can also compensate for each other. Based on statistics the total necessary minimal safety stock level SSL_{min} can be determined according to (see [Sach-84]):

$$SSL_{min} = \sqrt{SSL_{1,min}^2 + SSL_{2,min}^2 + SSL_{3,min}^2} \tag{8.14}$$

where SSL_{min} minimal safety stock level [units],
$SSL_{1,min}$ safety stock level for due-date deviations [units],
$SSL_{2,min}$ safety stock level for quantity delivery deviations [units],
$SSL_{3,min}$ safety stock level for demand fluctuations [units].

After substituting Eqs. 8.11, 8.12 and 8.13 in 8.14:

$$SSL_{min} = \sqrt{\left(L_{max}^+ \cdot RD_m\right)^2 + \left(Q_{max}^-\right)^2 + \left(\left(RD_{max} - RD_m\right) \cdot TRP\right)^2} \tag{8.15}$$

where SSL_{min} minimal safety stock level [units],
L_{max}^+ maximum positive lateness (delay)[SCD],
Q_{max}^- maximum negative quantity deviation (partial delivery) [units],
RD_{max} maximal rate of demand [units/SCD],
RD_m mean rate of demand [units/SCD],
TRP replenishment time [SCD].

Here, it is assumed that the individual plan deviations are statistically independent from one another. The safety stock level determined with Eq. 8.15 is the minimal amount required in order to ensure the service level at all times, despite possible quantity and schedule deviations.

In order to evaluate real stock situations it finally has to be considered that the store output generally does not occur continuously as was assumed when deriving the ideal Storage Operating Curve, but rather also in lots. Through the discrete stock withdrawal the mean stock level, which without considering the emergence of disruptions is required in any case to maintain the delivery capability, is reduced (Fig. 8.10b):

$$SL_0 = \frac{QIN_m - QOUT_m}{2} \tag{8.16}$$

where SL_0 lot stock level [units] (when considering discrete stock withdrawal),
QIN_m mean store input quantity (lot size) per input event [units],
$QOUT_m$ mean store output quantity (lot size) per output event [units].

Taking into consideration these circumstances as well as the necessary safety stock caused by disruptions, results in an adjusted (or practical) minimal stock level SL_1 with which the stores delivery capability can be ensured at anytime:

$$SL_1 = \frac{QIN_m - QOUT_m}{2} + \sqrt{\left(L_{max}^+ \cdot RD_m\right)^2 + \left(Q_{max}^-\right)^2 + \left(\left(RD_{max} - RD_m\right) \cdot TRP\right)^2} \tag{8.17}$$

where SL_l practical minimal stock level [units],
QIN_m mean store input quantity (lot size) per input event [units],
$QOUT_m$ mean store output quantity (lot size) per output event [units],
L^+_{max} maximum positive lateness (delay) [SCD],
Q^-_{max} maximum negative quantity deviation (partial delivery) [units],
RD_{max} maximal rate of demand [units/SCD],
RD_m mean rate of demand [units/SCD],
TRP replenishment time [SCD].

An adjusted minimal delivery delay limit DD_l can be derived along the same lines. Too early deliveries from the suppliers lead to irregular stockpiling, just as over-deliveries or low rates of demand do. If the mean stock level in the store nevertheless runs toward zero (a state which can only be due to blatant planning errors), it then means that the delivery delay due to these disruptions is reduced in comparison to the actual situation, but that it has to be greater in comparison to the state that is free from these disruptions:

$$DD_l = \frac{QIN_m - QOUT_m}{2 \cdot RD_m} + \sqrt{\left(L^-_{max}\right)^2 + \left(\frac{Q^+_{max}}{RD_m}\right)^2 + \left(\frac{(RD_m - RD_{min}) \cdot TRP}{RD_m}\right)^2} \tag{8.18}$$

where DD_l practical minimum delivery delay limit [SCD],
QIN_m mean store input quantity (lot size) per input event [units],
$QOUT_m$ mean store output quantity (lot size) per output event [units],
L^-_{max} maximum negative lateness (premature delivery) [SCD],
Q^+_{max} maximum positive quantity deviation (over-delivery) [units],
RD_m mean rate of demand [units/SCD],
RD_{min} minimal rate of demand [units/SCD],
TRP replenishment time [SCD].

Using the limits determined here, we can calculate a Storage Operating Curve in which the previously described disruptions are depicted and can be evaluated with regards to logistic aspects. In order to do so, the practicl minimal stock level (Eq. 8.17) and delivery delay (Eq. 8.18) need only be substituted in Eq. 8.9

$$\left.\begin{aligned} SL_m(t) &= SL_l \cdot t \\ DD_m(t) &= DD_l \cdot \sqrt[C]{1 - t^C} \end{aligned}\right\} \tag{8.19}$$

where $SL_m(t)$ mean lot stock level (as a function of t) [units],
$DD_m(t)$ mean delivery delay (as a function of t) [SCD],
SL_l practical minimal stock level [units],
DD_l practical minimal delivery delay limit [SCD],
$0 \leq t \leq 1$ running variable,
C C value.

As a result of this equation the calculated function's base point in comparison to the ideal Storage Operating Curve, moves further away from the point of origin depending on the extreme values of the existing disruptions (Fig. 8.11). If there

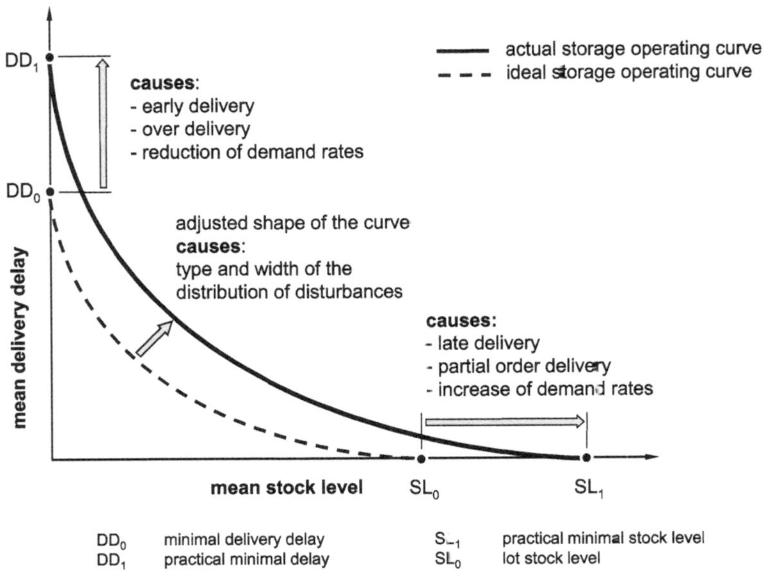

Fig. 8.11 Factors which Influence Actual Storage Operating Curves

are plan deviations which lead to stock levels lower than originally planned, then corresponding safety stocks need to be maintained in order to counteract possible delivery delays. This case is described in the SOC by the stretching of the function in the positive direction of the x-axis. If on the other hand, despite too early deliveries or other general deviations that cause stock increases there is a zero stock level, the resulting delivery delay increases. The Storage Operating Curve is thus stretched along the positive y-axis. How the shape of the function further changes between the limits when there are other deviations is essentially determined by the type and range of the disturbances' distribution.

We can clarify this correlation with a mental experiment: Imagine that two distributions of deviations with identical extremes are given. The limit SL_1 and DD_1 thus change in the same way in both cases. However, the distributions differ in their shape. In the first case, the distribution is highly populated in the region surrounding the mean. In contrast, in the second case, it is assumed that schedule deviations generally occur around the extremes.

In the first, very narrow distribution, strong deviations from the ideal process occur only seldom. The Storage Operating Curve will thus change more in its ends and not so much in between. The real SOC has to therefore be quite close to the ideal Storage Operating Curve. On the other hand, in the second case due to the broader distribution of the disturbances that was assumed, the resulting Storage Operating Curve clearly eases away from the curve of the lower limit, since the deviations from the ideal process (despite the same limits) emerge with varying degrees of intensity.

In order to be able to map the descriptions of the distributions shape's influence onto the real Storage Operating Curve Eq. 8.19 still has to be parametrized.

8.4.3 Parametrizing the Approximation Equation

Modeling the calculated Storage Operating Curve is undertaken in Eq. 8.19 by changing the value of C for the C_{norm} function. The smaller the chosen value of C is, the more the function first approximates the ideal Storage Operating Curve and thereafter the coordinate system's point of origin (see also Fig. 4.6).

In order to obtain a typical value for the C_{norm} parameter, with which the Storage Operating Curve can be appropriately modeled according to the demands, an additional base point should first be defined. The point at which the co-ordinate's system's bisector angel intersects with the C_{norm} function is well suited for this. For this intersection (X_{is}, Y_{is}) it follows:

$$\left.\begin{aligned} X_{is} &= Y_{is} \\ Y_{is} &= \sqrt[C]{\frac{1}{2}} \end{aligned}\right\} \tag{8.20}$$

For the ideal Storage Operating Curve C is 0.5. The coordinates of the defined point of intersection for the C_{norm} function are therefore:

$$X_{is} = Y_{is} = \sqrt[\frac{1}{2}]{\frac{1}{2}} = \left(\frac{1}{2}\right)^2 = \frac{1}{4} \tag{8.21}$$

In order to calculate the ideal Storage Operating Curve the C_{norm} function is stretched by the limits SL_0 and DD_0. This stretching also covers the defined base point, therefore the values of SL_{is} and DD_{is} for the ideal Storage Operating Curve are derived as:

$$\left.\begin{aligned} SL_{is} &= \frac{1}{4} \cdot SSL_0 \\ DD_{is} &= \frac{1}{4} \cdot DD_0 \end{aligned}\right\} \tag{8.22}$$

In order to further expand the modeling of the function and take into consideration plan deviations, the parameters SL_1 und DD_1 are determined. Through these parameters the Storage Operating Curve's limits are stretched. Now however, as was demonstrated in the abovementioned mental experiment, the function also has to be constructed to reflect the distribution shape of possible plan deviations (Fig. 8.12, right).

A first guess for C can be derived from the preceeding mental experiment as well as from the coordinates for the support points that were previously calculated for the function. We already discussed how the Storage Operating Curve in the transition range strongly approximated the lower limiting curve when the distribution of the plan deviations was very narrow. It can therefore be assumed that the Storage Operating Curve touches the lower limiting curve at the above calculated base point SL_{is}, DD_{is}. Assuming this, a typical initial value for C can then be cal-

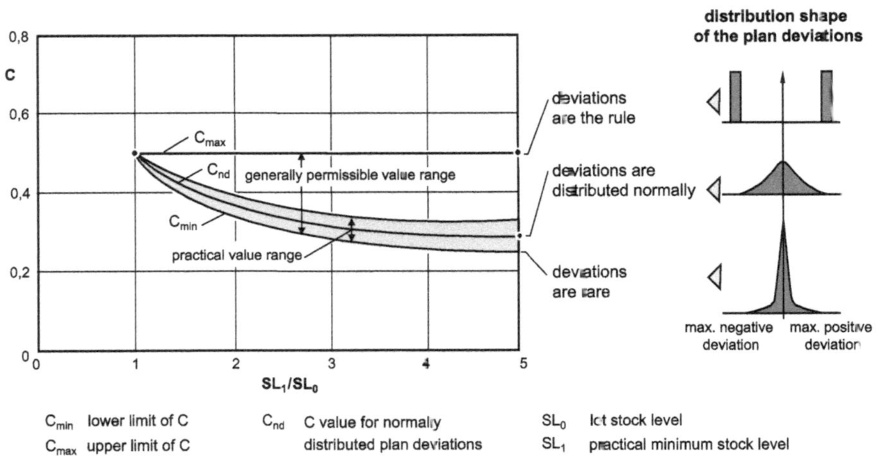

Fig. 8.12 Permissible and Practical Ranges for the Value of C in Calculated Storage Operating Curves

culated, which will be called the lower limit C_{min}. If the values of SL_{is} and DD_{is} are then substituted in Eq. 8.19 for $SL_m(t)$ and $DD_m(t)$, then:

$$\left(\frac{DD_0}{4 \cdot DD_1}\right)^{C_{min}} + \left(\frac{SL_0}{4 \cdot SL_1}\right)^{C_{min}} = 1 \tag{8.23}$$

where SL_0 lot stock level [units],
DD_0 minimal delivery delay limit [SCD],
SL_1 practical minimal stock level [units],
DD_1 practical minimal delivery delay limit [SCD],
C_{min} lower limit for C.

If,

$$\frac{DD_0}{DD_1} \cong \frac{SL_0}{SL_1} \tag{8.24}$$

(this is at least applicable when the distributions for the planning deviations are symmetric), then:

$$C_{min} \cong \frac{\ln\left(\frac{1}{2}\right)}{\ln\left(\frac{SL_0}{4 \cdot SL_1}\right)} \tag{8.25}$$

where SL_0 lot stock level [units],
SL_1 practical minimal stock level [units],
C_{min} lower limit for C.

If for example, the practical minimal stock level SL_1 is double the size of the lot SL_0, then C_{min} is calculated at 0.33 – when triple the size then $C_{min} = 0.279$. When there are no plan deviations then C_{min} is 0.5.

A Storage Operating Curve, which is calculated with a thus determined C_{min}, touches the ideal Storage Operating Curve at the additionally defined base point (SL_{is}, DD_{is}). This is however at best permissible with a very narrow distribution of plan deviations. The more broadly distributed the deviations are, the more strongly the calculated Storage Operating Curve has to detach from the ideal SOC. The value for C therefore has to also become larger. Independent of the absolute value of the plan deviations, the maximal permissible value for C is generally 0.5.

In Fig. 8.12 (above) the discussion surrounding the C value of the C_{norm} function is illustrated. In addition to the total permissible value range for the C, a range which is pertinent for practical situations and a function for a normal distribution of plan deviations that was empirically derived from simulation results, are also inserted in the graph. This function can also be numerically approximated with:

$$C_{nd} \cong 0.15 \cdot (C_{max} - C_{min}) + C_{min} \qquad (8.26)$$

where C_{nd} value of C for a normal distribution of schedule deviations,
C_{max} maximal permissible value of C ($C_{max} = 0.5$),
C_{min} lower limit for C.

It can be shown that Eq. 8.26 also generally applies to other forms of the plan deviations distribution. However, in individual cases the factor standing before the bracketed expression (here, 0.15) should be adjusted. This factor should be reduced when the distributions are narrower and increased when the distributions are broader. Concrete conclusions nevertheless, can only be made based on a further series of tests.

Determining the practically relevant range depicted in Fig. 8.12 stems from the consideration that store inputs generally have the correct quantity and are on time or occur with only minimal plan deviations. Therefore in the initial approximation it can be assumed that the distribution of the deviations tends to be normal or at least that significant plan deviations are not representative.

8.4.4 Verifying Storage Operating Curves Using Simulations

The approximation equation introduced here was verified based on a comparison of calculated Storage Operating Curves with the results of simulation studies. In the following we present two examples of this simulation based model evaluation.

In the first example, the Storage Operating Curve that was determined using simulations and presented in Fig. 8.7 is used. Based on the minimal or maximal value of the underlying normally distributed quantities and due-date deviations as well as the fluctuation range of the demand rate, the stock limits (SL_0 und SL_1; Eqs. 8.7 and 8.17) and the delivery delay limits (DD_0 and DD_1; Eqs. 8.8 and 8.18) can be determined. The value for C, which is to be used in the Storage Operating Curve equation, is established by the ratio of the calculated stock limits (Fig. 8.12; Eqs. 8.25 and 8.26). Using these SOC parameters, the Storage Operating Curve can then be calculated and compared to the simulation results. In Fig. 8.13 it can be seen that the simulation results correspond well with the calculated ones.

As these and other conducted trials indicate, the Storage Operating Curves can also be calculated in consideration of quantity and plan deviations in the store input as well as with regards to fluctuating demand rates. The parameters which are directly or indirectly entered in the approximation equation mostly have an elementary significance (Fig. 8.15). Both the end points for the ideal and approximated 'actual' Storage Operating Curves can be analytically determined provided that the behavior of the input and output of the observed stock items can be sufficiently enough described through the mean and extreme values. Only the value of C has to currently be established empirically. If the distributions for the schedule deviations are approximately normal, then a typical value can also be analytically determined for it, at least in an initial approximation.

If the distributions of the plan deviations strongly differ from the normal distribution, then in individual cases special simulations studies should be conducted in order to empirically support parametrizing the SOC equation. Beforehand though, it should be verified if it is necessary to know the exact shape of the Storage Operating Curve for the planned application or if for example the practical minimal stock level is a sufficient basis for making decisions.

In the second simulation run the quantity and scheduling deviations were evenly distributed and all other conditions remained the same. The results, again in comparison with the calculated Storage Operating Curves, are shown in Fig. 8.14. Due

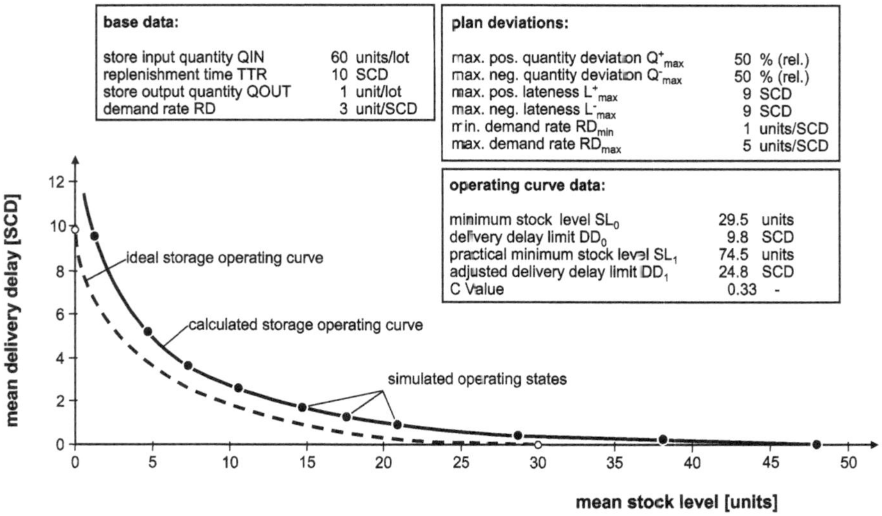

Fig. 8.13 Simulation Based Verification of Calculated Storage Operating Curves (normal distributed plan deviations)

to the identical minimal and maximal values for the distributions, the same values for the delivery delay limit and stock level limit were established for the operating curve calculation as in the first presented simulation run. Nevertheless, as expected a larger, in this case empirically determined, value for C (0.38 instead of 0.33) had to be employed in the SOC equation (see also Fig. 8.12) in order to target a high coverage between the results of the simulation and calculations.

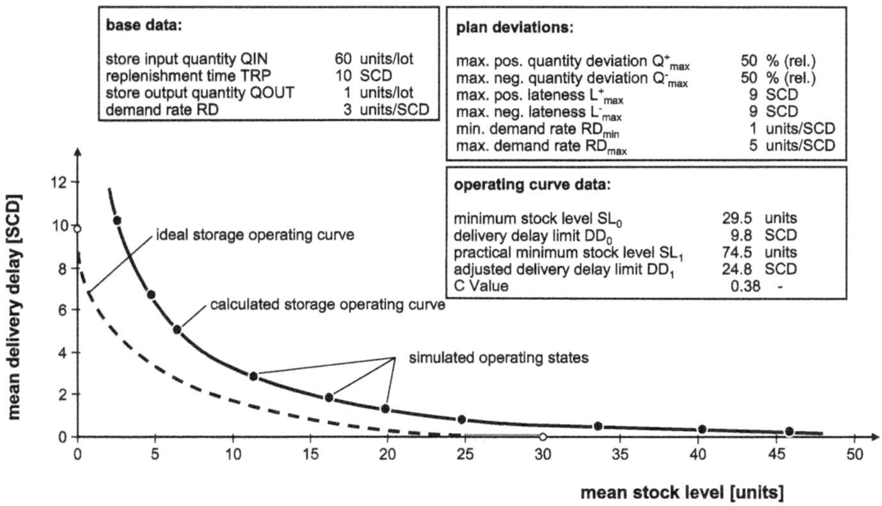

Fig. 8.14 Simulation Based Verification of Storage Operating Curves (equally distributed plan deviations)

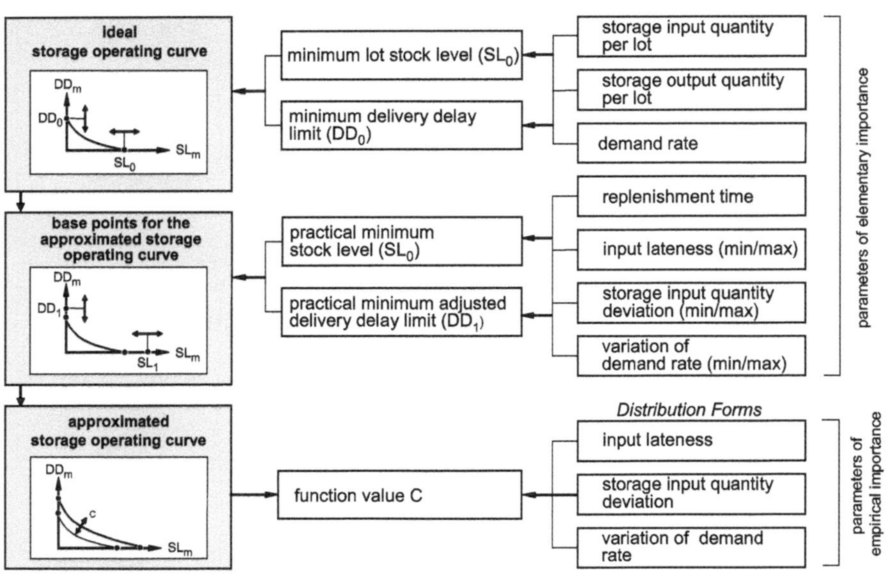

Fig. 8.15 Parameters for the Calculated Storage Operating Curves

8.5 Possible Applications

Storage Operating Curves are a promising method for mathematically describing both the logistic objectives of stock management as well as the interdependencies between the objectives. The range of application for Storage Operating Curves extends from setting targets through to evaluating processes and determining parameters up to analyzing potential (Fig. 8.16).

Storage Operating Curves thus facilitate a qualitative and quantitative description of the relations between the mean delivery delay and the mean stock level. This technique is thus predestined to support Logistic Positionings and target settings within this dilemma. Depending on the basic conditions (value, replenishment time, strategic importance) of the specific item, target values for the delivery delay and stock level can be agreed upon in consideration of these interdependencies.

Provided that the targets are quantified, the *actual state can be evaluated* from a logistic perspective. In order to do so, it is necessary to continuously record the actual state and the relevant key figures needs to be calculated when monitoring the procurement and inventory.

Planning target values also facilitates determining the material planning parameter, safety stock, in consideration of the targets. An important tool for maintaining this parameter is thus provided.

Finally, the Storage Operating Curves can be used to analyze *potential for changes*. Provided that during the target setting process it is determined that a primary goal (e. g., a given stock level) cannot be realized without negatively impacting the second target, alternative measures concerning the supplier, the ordering process and/or the users requisition behavior can be evaluated. In the

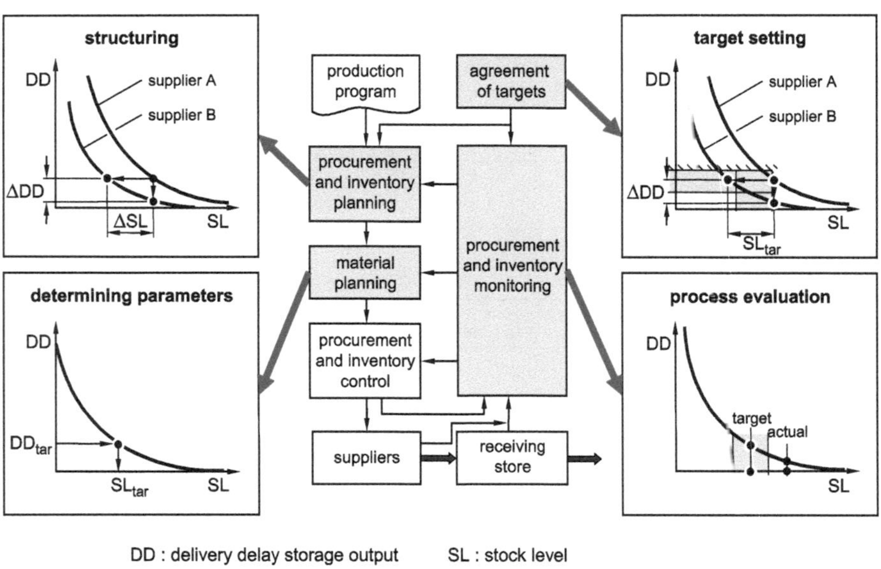

Fig. 8.16 Applications of Storage Operating Curves in Procurement and Inventory Management (Gläßner)

upper left hand part of Fig. 8.16, this is illustrated using the example of a *logistic oriented supplier evaluation*. Here, the differing SOC result from the supplier's varying delivery behavior. Quantifying the potential for improvement with regards to the stock level and supply reliability of the production based on this, allows logistic criteria as well as price and quality aspects to be drawn into the decision making process [Gläß-95].

8.6 Fields and Limits of Application

As the above discussion demonstrates, Storage Operating Curves can generally simplify the decision making process and develop a broader base in all of the essential areas of stock management. However, the previously introduced model is based on statistical methods and calculating means, therefore a sufficiently large number of events are required in order to ensure the statistic validity of the model. Consequently, Storage Operating Curves can only be limitedly employed within the operative planning and process monitoring and are instead best suited for a long term and more strategically oriented processing design.

Furthermore, applying the Storage Operating Curve technique to all of the items in stock, generally does not seem possible or practical. The mathematical model introduced here for calculating SOC requires far less effort than the corresponding simulation investigation. Nonetheless, due to the large variety of parts it would be impossible for a material planner to calculate the Storage Operating Curves for each of the articles he/she is responsible for in order to make the decisions pertinent to them. Since the material demands are frequently diverse and complex, it is often necessary to classify the spectrum of items with regards to their individual importance, in order to appropriately determine the amount of planning and monitoring effort each of the items should be allocated. The best known method for segmenting the article spectrum according to the value/quantity relation [Bult-88] is referred to as *ABC Classification* (Fig. 8.17, upper left).

Other classification methods can also be helpful in order to concentrate on the most important or critical item segments. Thus, it is possible to further differentiate by grouping the article spectrum according to the suppliers' delivery reliability. This is referred to as *UVW Classification*. In the upper, right section of Fig. 8.17 the principle of this type of classification is visually represented. The variability of the delivery date deviations is mapped over the relative percentage of items (or optionally the suppliers). Provided that deliveries which are too early are assessed as uncritical, an additional parameter can be integrated instead of the variability of the plan deviation. U items (or U suppliers) distinguish themselves through predominantly punctual deliveries, whereas in segment W, it has to be assumed that the availability is constantly endangered.

On the store output side, the demand rate's consistency is a relevant criterion for differentiating and can be described using *RIS Classification* (Fig. 8.17, lower left) (RIS stands for: regular, irregular and/or sporadic, see also [Kugl-95]). The

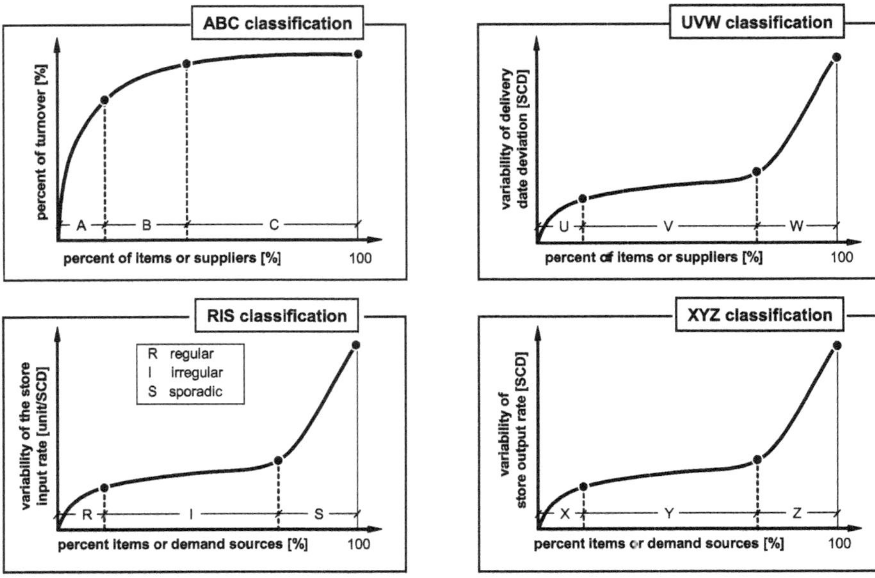

Fig. 8.17 Possibilities for Logistically Segmenting Spectrums of Items

second important factor is the predictability of the demand (i. e., its scheduling reliability), which is quantified through the variability of the demand lateness. Segmenting the items according to the predictability of the demand is referred to as *XYZ Classification* (Fig. 8.17, lower right). In XYZ classification, the amount of time required for replenishment is also included: The less time required, the more precise the demands can be planned. Categorizing the items in separate segments is, therefore, also determined by the suppliers delivery capability.

Each of the abovementioned segmenting techniques can be applied separately in order to derive planning measures for improving the process. In particular though, an appropriate combination of the techniques help to concentrate the material planning activities on segments critical to success.

Figure 8.18 depicts a possible arrangement of procurement and material planning strategies for special item segments. The basis for the *multi-dimensional segmenting* employed here is a combined RIS/XYZ analysis with which the regularity of the store output and its schedule reliability is described. After, the results of the ABC and UVW analyses can be allocated within this matrix. It can also be shown in which fields the Storage Operating Curves can be practically employed.

Independent of the demand regularity, the delivery certainty is basically determined by the stores input behavior and thus the suppliers' due-date and quantity reliability or that of the supplying manufacturer. Therefore the safety stock level can also primarily be dimensioned as a function of this delivery reliability. If Storage Operating Curves are applied in doing so, then the Logistic Positioning can be supported through it. It can thus be determined – when necessary in connection with the results of an ABC classification – which stock dependent delivery delays should be accepted for the observed item. It can also be practical to logistically

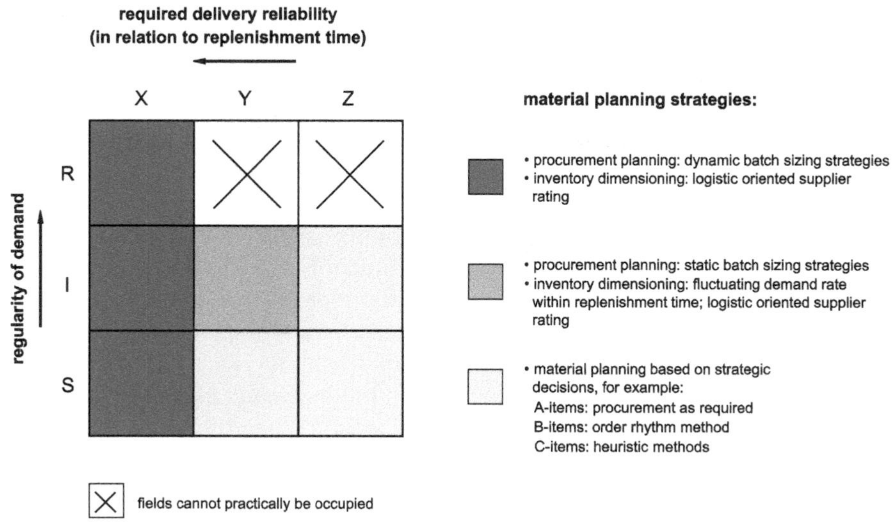

Fig. 8.18 Deriving Problem Oriented Material Planning Strategies (Example)

evaluate the suppliers, in order to quantify how they influence the required safety stocks. In the following section, we will present an example of how to do this.

The procurement quantity for X items can be determined with dynamic batch sizing methods (see for example [Fort-77*], [Nyhu-91]). The aim of these methods, is to minimize the costs which are dependent on the purchased quantities. The costs for the purchase process as well as the tied-up capital costs for the process of maintaining stock are included in the calculation method. With respect to the capital costs, the chronological trend of the demands is considered. In this way, the dynamic batch sizing methods integrate both the RIS and ABC classification through the underlying processing logic. In cases where the demand is regular (R/X items), static batch sizing methods (e. g., batch sizing according to Andler or Harris) also lead to comparable results. These methods assume a steady demand trend and are much simpler than dynamic approaches with regards to the required data and mathematical methods. Finally, when demands are sporadic, the batch sizing may result in a decision to procure items on demand.

If the uncertainty of the planning increases, then the use of dynamic batch sizing should also be reduced. Whereas statistical methods can still be employed with UY items, the applicable material planning strategies become even more difficult when the predictability or reliability of the demand continues to worsen. If the demand is very sporadic, the material planning strategies are mainly dependent on strategic decisions, which in turn can for example be oriented on an ABC classification. The possibilities for applying the Storage Operating Curves also decrease as the demands become more unpredictable. In addition to the above described fields of application, the sizing of the stock can only be supported by the Storage Operating Curves still in the UY segments, whereby in addition to the disruptions caused by the supplier, the demand fluctuations should also be taken into consideration.

8.7 Examples of Applying Storage Operating Curves in order to Evaluate Suppliers

Storage Operating Curves have been employed for example in an industry based project in order to evaluate suppliers with regards to logistics. The firm manufactured a large, diverse group of complex, electronic products and consequently had to control an extremely heterogeneous spectrum of purchased materials. In the following section, we will present an example from this investigation.

Despite a large stock of purchased parts, the company could not always ensure that the production's demands were met. The cause for this quite typical problem was seen as the suppliers insufficient delivery reliability [Kear-92]. In the first stage of the project, the delivery reliability of 1180 incoming deliveries during an evaluation period of 18 weeks was analyzed in order to quantify the situation. The parameter delivery reliability is defined here as the difference in time between the actual delivery date and that promised by the supplier. In Fig. 8.19 it is evident that of the 1180 evaluated deliveries, only 45% were delivered according to schedule. A further 37% the deliveries occurred with to some degree considerable earliness, where as the remaining 18% were delivered late.

When interpreting the distribution reproduced in Fig. 8.19 two circumstances need to be taken into consideration: First, the scheduled replenishment time was on average 65 work days – this relativizes the range of the delivery date deviation. Second, due to the multiple utilization of an item in different orders, a single pending delivery can block a number of orders. Therefore the delayed deliveries, even when they were only 18% of the recorded deliveries, were assessed as particularly critical.

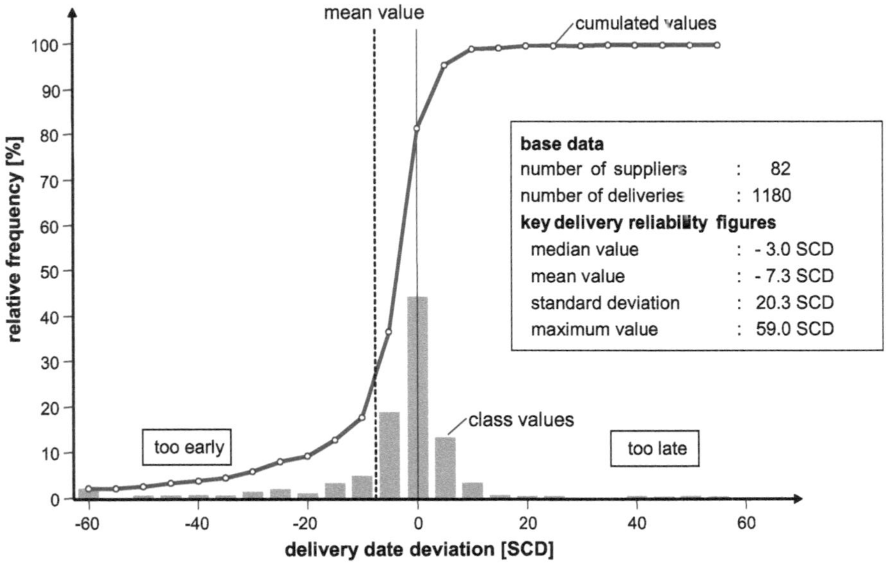

Fig. 8.19 Suppliers' Delivery Reliability (Practical Example)

It was further investigated if and to what degree individual suppliers were particularly responsible for the delivery date deviations. In order to do this, the suppliers were segmented according to the UVW classification. Figure 8.20 illustrates the results whereby the maximal lateness is represented over the relative number of suppliers. After consulting with the company, only suppliers that had performed at least five deliveries during the evaluation period were included in the evaluation (37 from 82 suppliers). The maximal lateness was chosen as a classification characteristic for the UVW analysis, because the main goal here was applying the Storage Operating Curves in order to evaluate the supply reliability of their own production. The classification limits were set at five days for the U suppliers and fifteen for the V suppliers. These limits can generally be chosen freely, but nevertheless should be oriented on the existing conditions.

The results of the analysis show that not all of the suppliers were equally responsible for the delivery date deviations. Over 50% of them could generally maintain the schedule they had promised. Nevertheless, approximately 15% of them had exceeded due dates by 15 days or more. The unsatisfactory logistic delivery capabilities of a few suppliers generally caused considerable planning uncertainty, which from the perspective of the production could only be buffered through high safety stock levels.

These circumstances can also be described quantitatively with the help of the Storage Operating Curves so that a logistic oriented assessment of the suppliers can be built on it.

The mean store input quantity (procurement lot size) as well as the mean demand rate, which are relevant for determining the ideal Storage Operating Curves, were then established for an exemplary chosen electronic item (an AD/DA voltage transformer). The ideal Storage Operating Curve calculated according to Eq. 8.9 is depicted as the lower limiting curve in Fig. 8.21. The plotted stock level SL_0 of

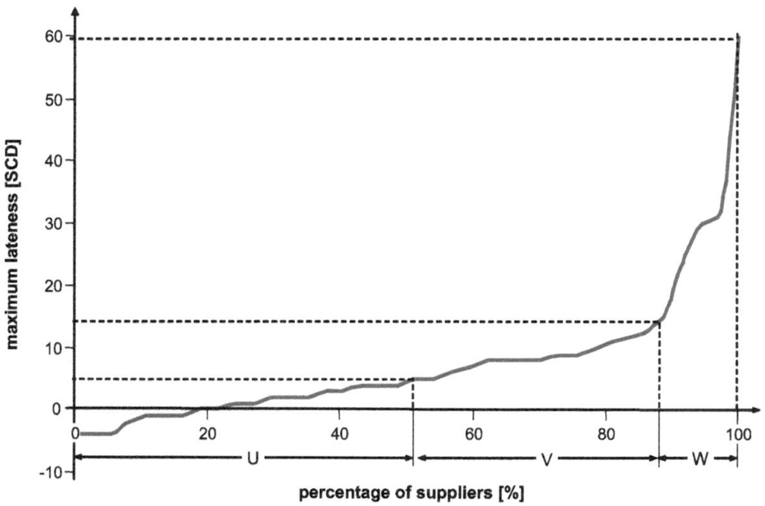

Fig. 8.20 UVW Classification of Suppliers

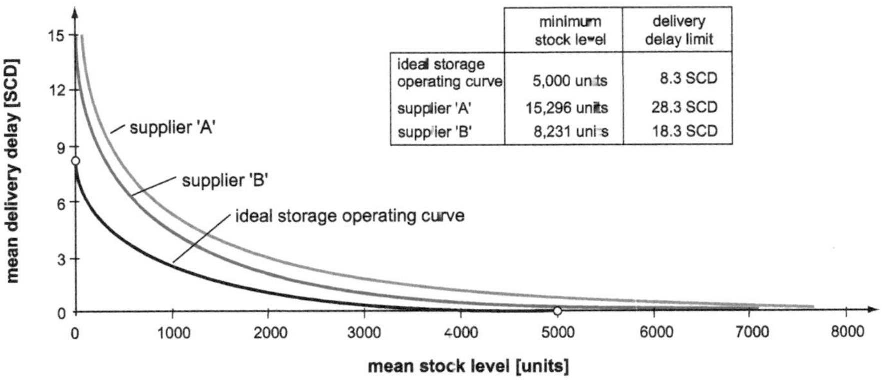

Fig. 8.21 Evaluating Suppliers by Means of Calculated Storage Operating Curves (Example: AD/DA Transformer)

5000 pieces was caused exclusively by the procurement lot size and thus through the material planning. We will consciously not take into consideration the store output lot sizes here (i. e., we will assume a steady store output), since we are only analyzing the problem of the delivery date deviations.

Furthermore, the delivery behavior of the corresponding supplier (here identified as supplier A) was evaluated. During the evaluation period, there were single cases of quantity deviations (partial deliveries) of up to 50%. Both positive and negative lateness (+ 15/– 20 SCD) were determined. Additionally, it should be noted that the behavior of all of the deliveries completed by the supplier were evaluated in order to be able to draw upon a larger base for the key statistic data (in the absence of a longer data collection period). Practical applications should be limited to a group of items or even a special item in order to, when necessary, consider the specific conditions of an item.

The limits for SL_1 and DD_1, which are calculated based on the thus described supplier behavior (Eqs. 8.17 and 8.18) as well as the Storage Operating Curve that was calculated from that (Eq. 8.19) are also displayed in Fig. 8.21. Due to the supplier's behavior a mean stock level of approximately 15,300 pieces is required in order to reliably prevent delivery delays. The difference between it and the limit SL_0 of 10,300 pieces is to be interpreted as the required safety stock component due only to this suppliers schedule behavior.

In addition, the delivery behavior of a second supplier that served a similar spectrum of articles, was analyzed (supplier B). Generally, it could be established that it was much more reliable with regards to the delivery quantity and schedule reliability. Accordingly a lower limit SL_1 resulted and thus the SOC was also shaped more favorably. The safety stock due to this supplier's behavior is almost 70% lower than with supplier A.

This example demonstrates how Storage Operating Curves can in a very simple way support both evaluating suppliers with regards to logistics and analyzing their potential as well as designing planning parameters (e. g., the safety stock).

Through the qualitative and quantitative description of the interdependencies between the stock level and the delivery delay it is possible to employ Storage Operating Curves for Logistic Positionings and for setting targets within this field of tension. Based on the statistically oriented method of modeling, they are best applied particularly in long term and strongly strategic oriented process designing.

It remains to be emphasized that the Storage Operating Curves are not an independent storage model, but are rather a valuable complement to already existing models and methods.

The goal of future studies should be to expand the Storage Operating Curves Theory so that not only the mean values, but also other statistical parameters (e. g., maximal value of the delivery date deviation; proportion of delivery operations with delivery delay) can be mathematically described. The potential application field of calculated Storage Operating Curves would thus be extended considerably.

Chapter 9
Applying the Logistic Operating Curves Theory to Supply Chains

(Author: Dr.-Ing. Stefan Lutz)

Up to now we have concentrated our discussion on individually examining production or storage areas through the use of Logistic Operating Curves. Next, we will turn our attention to the interactions between the production and storage areas. By doing so, we can then consider entire supply chains and their logistic performance.

A supply chain generally consists of various manufacturing or assembly stages partially linked with one another through storage stages, which in turn serve to decouple processes. One of the challenges in designing supply chains is that the available resources as well as both the WIP and stock have to be coordinated with each other.

In the following sections, we will show which system of logistic targets exists in a supply chain (Sect. 9.1) and how the logistic objectives impact one another (Sect. 9.2). This knowledge about the interdependencies can then be used to quantify how logistic processing states impact the supply chain's overall performance (Sect. 9.3). The example of a supply chain for a tool manufacturer will serve to illustrate this.

9.1 Supply Chain Objectives

In addition to the system of targets for production logistics there is also a system of targets for stock management. Its primary goal is also profitability: On the one hand, it pursues high logistic efficiency expressed through minimal delivery delay and a high service level, and on the other hand, it pursues low logistic costs, expressed through low storage costs and low stock levels. In Fig. 9.1 both the target systems are inserted into a supply chain. This results in a system of targets for the entire supply chain, which also pursues the goals of logistic performance and logistic costs. Here, the logistic costs are expressed through an even utilization of resources. In addition, the costs are predominantly influenced by the WIP and the

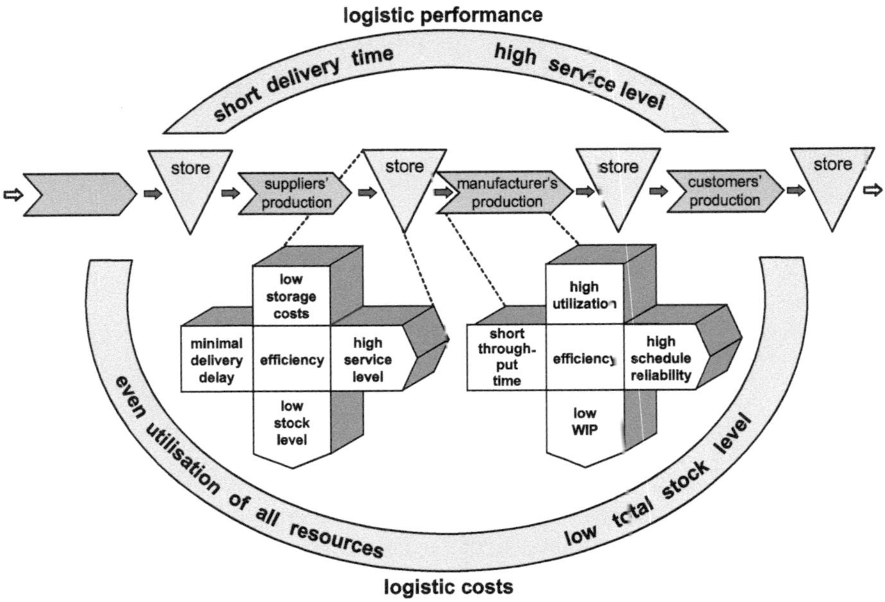

Fig. 9.1 The Target Systems in a Logistic Supply Chain

stock maintained in the various supply chain stores. The goal therefore, is to maintain a minimum stock level.

From a customer's perspective, a supply chain's logistic performance is expressed by a short delivery time and a high service level. Here, customers can be both end users as well as enterprises that are not integrated into the supply chain.

9.1.1 Weighted Service Level

One possibility of describing the logistic performance of a storage area is the key delivery delay figure that was introduced in Chap. 8. However, in practice the key service level figure is used more frequently. It describes the percentage of demands that could be covered by existing stock ([Ters-94], [Wien-97], [Silv-98], [Lucz-01]). This parameter can be differentiated as the weighted and unweighted service level. The unweighted service level (SERL) is derived from the ratio of orders served on time and with the exact quantity to the total number of demanded orders. An order's lot size is not reflected in determining this key figure as it can distort the figure for specific problems. It is therefore not obvious if a large or small quantity of demands were delivered late at a specific service level. Consequently, the unweighted service level does not allow any conclusions to be made about which percentage of the entire demand could be taken from the store on time.

In contrast, with the weighted service level ($SERL_w$) there is a weighting of the demanded orders with their corresponding lot size. This key figure results from the ratio of demanded items withdrawn on time to the total demand:

$$SERL_w = \frac{\sum_{i=1}^{m} LS_{do,i}}{\sum_{i=1}^{n} LS_{d,i}} \tag{9.1}$$

where $SERL_w$ weighted service level [-],
$LS_{do,i}$ lot size per demand served on time [units],
m number of demands served on time [-],
$LS_{d,i}$ lot size per demand [units],
n number of demands [-].

The weighted service level describes the store's logistic performance unit-wise. By weighting it with the demands' lot sizes, the significance of the different sizes of orders is taken into consideration. Therefore, a large order not served on time has a correspondingly greater influence on the weighted service level than when a smaller order is delayed.

9.1.2 An Approximation Equation for a Service Level Operating Curve

The service level, like the delivery delay, is dependent on the mean stock level. In order for an enterprise to deliberately position itself between a high service level and an as low as possible stock cost, it has to be known how the stock level and an item's resulting weighted service level influence one another. This interaction can be described by a so-called *Service Level Operating Curve (SLOC)*. The Service Level Operating Curve like the Logistic Operating Curves is based on a transformation of the C_{norm} function. How the SLOC is developed as well as how it is verified by simulations is described extensively in [Lutz-02]. The Service Level Operating Curve can be found with:

$$\left.\begin{aligned} SL_m(t) &= SL_0 \cdot \left(1 - \sqrt[C]{1 - t^C}\right) + t \cdot (SL_1 - SL_0) \\ SERL_w(t) &= 1 - \sqrt[C]{1 - t^C} \end{aligned}\right\} \tag{9.2}$$

where SL_m mean stock level [units],
SL_0 lot stock level [units],
SL_1 practical minimal stock level [units],
$SERL_w$ weighted service level [-],
C value of C for the C_{norm} function [-],
t running variable ($0 \leq t \leq 1$).

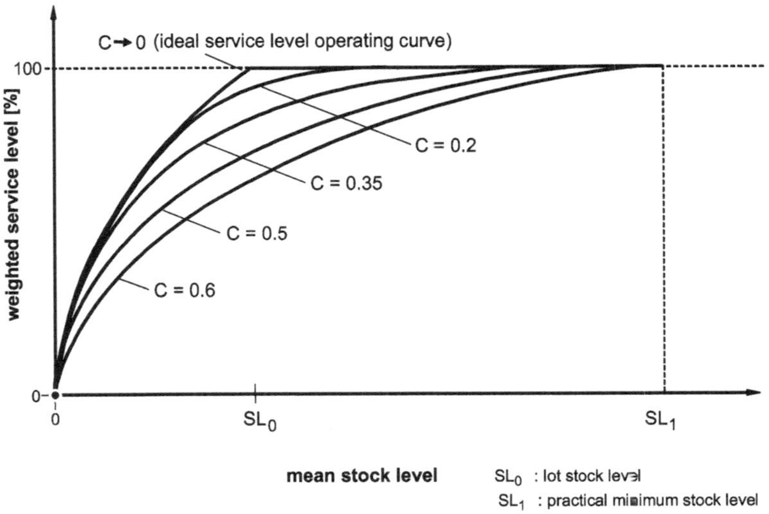

Fig. 9.2 Weighted Service Level Operating Curve of an Item for Different Values of C

The weighted Service Level Operating Curve is thus defined in sections: The value range of Eq. 9.2 results in $SL_m(t) \in [0; SL_1]$ and $SERL_w(t) \in [0; 100\%]$. For the mean stock level $SL_m > SL_1$, $SERL_w = 100\%$.

For given values of SL_0, SL_1 and C a variate pair for $SL_m(t)$ and for $SERL_w(t)$ can be calculated for every t ($0 \leq t \leq 1$) with Eq. 9.2. As an example, the shape of the Service Level Operating Curve for different values of C are displayed in Fig. 9.2.

C can be any value $C > 0$. The more C approaches 0, the closer the SLOC approximates the ideal Service Level Operating Curve. The practical minimal stock level (SL_1) forms the point on the SLOC, at which (for any value of C) a service level of 100% will be achieved. Thus in calculating the service level, the Service Level Operating Curve reflects the extremes of the plan deviations that occur. The greater the plan deviations, the greater SL_1 is. As a consequence, the function is stretched horizontally. When the mean stock level is situated above the practical minimal stock level ($SL_m > SL_1$), it results by definition in a service level of 100%. The shape of the weighted service level for stocks below the practical minimal stock level (SL_1) results from the value of C chosen for determining the Service Level Operating Curve. Thus, the greater the plan deviations are, the greater the value of C is [Lutz-02]. Determining the value of C required for calculating the SLOC is currently only empirically possible.

The Service Level Operating Curves can be applied to support decision making in a number of stock management areas (Fig. 9.3).

With regards to the objectives, the purpose of the Service Level Operating Curves is thus two-fold: First, evaluating goals with regards to how they conform to the logistic objectives and second, determining the target values. Based on the SLOC, it can be established whether or not the desired targets in consideration of

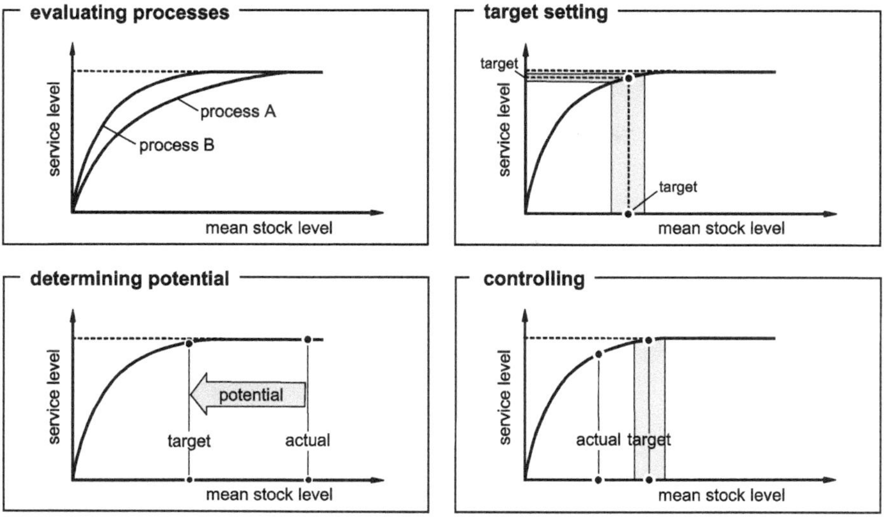

Fig. 9.3 Fields of Application for Service Level Operating Curves

the stock and service levels are at all compatible with one another. That is, if the target state is not located on the Service Level Operating Curve, it means that the desired service level cannot be achieved under the existing conditions given the targeted mean stock level.

The Service Level Operating Curves also take on particular significance when determining the logistic potential. By comparing the mean stock level of an operating state with the target value for the stock, the potential for adjusting the stock can be established. When this is completed for all the items of a storage stage, the total potential for adjusting the stock is attained. The stock can be adjusted both by reducing stores as well as by increasing them in case the actual stock level is found below the target stock level. In this case though, it should also be considered whether or not the causes for the required stock could be optimized.

9.2 Correlations between the Supply Chain's Logistic Parameters

The corresponding logistic objectives and factors which influence storage stages and production processes are subject to interdependencies. Thus for example, the logistic performance of a supply chain's finished goods store is generally influenced by the plan deviations in the stores input. These plan deviations result however from the logistic performance of the previous production whose logistic achievement is in turn influenced by their production store or input store. This causal chain can be continued and begins at the start of the supply chain, that is as a rule at the suppliers. If the effects of a process on those which follow them are known, the supply chain can be logistically analyzed, and potentials can be identified.

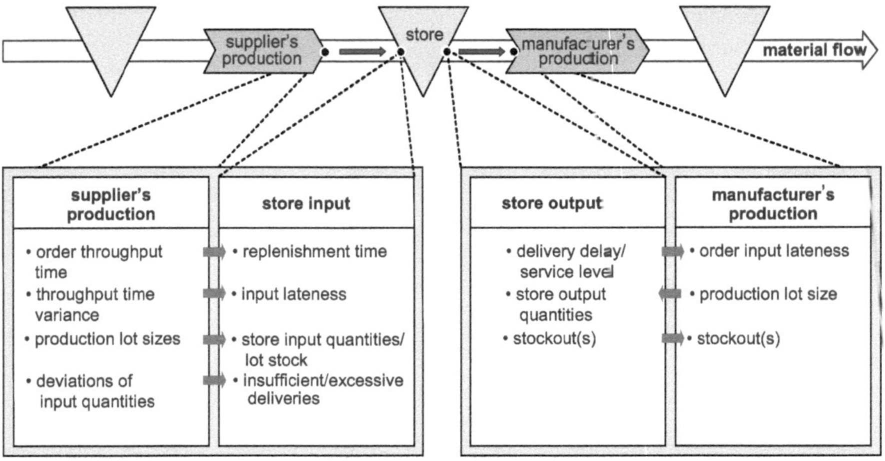

Fig. 9.4 Correlations between the Logistic Parameters of a Supply Chain (Fastabend)

The logistic output parameters of the production process are the input parameters for the storage process. In the same way, the output parameters of a storage process are the input parameters for the following production. They can thus also be referred to as coupling parameters [Fast-97]. The correlations between the logistic parameters are presented in Fig. 9.4.

The logistic characteristics of the production determine the input parameters of the stock dimensioning in the store. An exception here are the fluctuations in the demand rate induced by the following production or customer. Consequently, the replenishment time in the store corresponds with the order throughput time of the observed item in the upstream production. The input lateness of the store is determined through the production's output lateness. Yu [Yu-01] was able to show that the output lateness of a production is generally influenced by the distribution of the throughput time, whereby the throughput time distribution is directly dependent on the production's WIP level (see also Sect. 4.10). When the WIP level on the workstation is low, the throughput times are generally more narrowly distributed than when the WIP levels are high. The available capacities and the flexibility with which they can react to changes in the WIP level also influences a production's schedule deviations. Sufficient capacities can prevent backlogs – which would otherwise lead to lateness – from developing in the production.

The manufacturing lot sizes determine the quantity of the store input and therefore have a direct influence on the lot stock.

Over deliveries and partial deliveries in a storage stage represent quantity deviations in the upstream production that need to be taken into consideration when dimensioning the stock levels. Producing rejects or items with the wrong quality for example, can cause deviations in the input quantity. However, there can also be planning uncertainties that result in deviating amounts being produced.

Causal chains can be formed using the correlations described here. In Fig. 9.5 such a causal chain is explained using the example of how a decreased WIP level impacts the following store in a manufacturer.

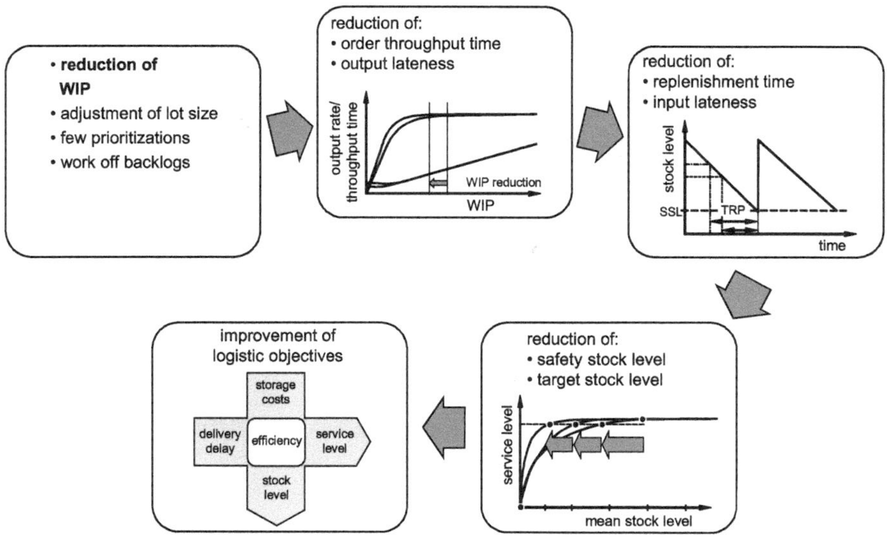

Fig. 9.5 Possible Effects of WIP Reductions on a Downstream Store

By reducing the WIP and adjusting the lot size, the throughput times in the production can be decreased. From the perspective of a downstream store, this reduces the replenishment time and minimizes fluctuations in the demand during the replenishment time. At the same time, the production's schedule reliability can be improved by decreasing the WIP level, using sequencing rules sparsely and working off the production backlog. Through that, the extremes of the input lateness are also reduced. Both measures therefore allow the replenishment time and the output lateness, two essential parameters for determining the size of the safety stock level, to be improved.

The input and output parameters are similarly linked in the store-production direction. A delivery delay or a poor service level in a storage stage can lead to an unsatisfactory availability of materials, which in turn can cause delays in starting an order. The store output quantity is generally determined by the planned production orders. Therefore the shortage in the store output corresponds to the shortage in the following production.

We have thus shown how the dependencies of the logistic objectives in the supply chain can be described using the explained interdependencies and how appropriate measures can be derived for achieving the goals.

9.3 Example of a Supply Chain Logistic Analysis

Based on the example of a tool manufacturer's finished goods store, we will now demonstrate how the changes in a supply chain can impact both the delivery capability from the customer's perspective and the WIP in the supply chain. The

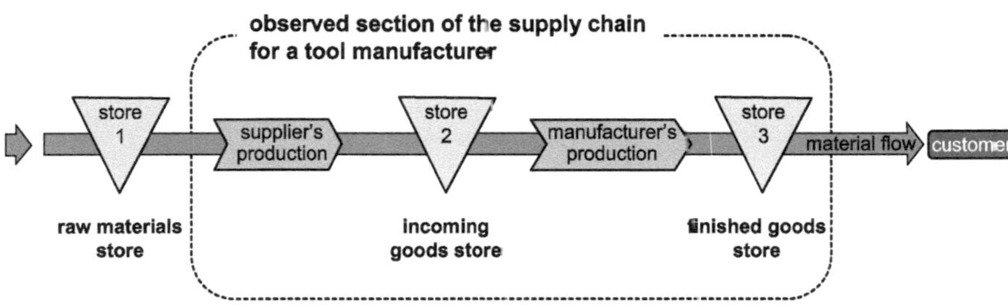

Fig. 9.6 Observed Section of a Tool Manufacturer's Supply Chain

3,161 observed items were precision parts, delivered directly to the end users (generally metal processing firms). Since the customers expected very short delivery times the items had to be sufficiently stocked in the finished goods store in order to ensure immediate availability. The manufacturer's finished goods store is at the end of the analyzed supply chain and is served by the manufacturer's production. The semi-finished goods are found in the incoming store. This store is stocked by the supplier's production (see Fig. 9.6).

The total potential of the observed section of the supply chain was determined by a partial analysis of the individual elements. The results of which were then linked to a comprehensive overview. In addition to the storage stages, the production processes were also able to be examined regarding their logistic potential.

9.3.1 Logistic Oriented Storage Analysis of the Manufacturer's Finished Goods Store

Like the Bottleneck Oriented Logistic Analysis, which identifies the logistic potential of production processes (see Chap. 7), the Logistic Oriented Storage Analysis (LOSA), which identifies the logistic potential of storage processes, was developed at the University of Hannover's Institute of Production Systems and Logistics (IFA) [Lutz-01, Lutz-02].

The goal of the Logistic Oriented Storage Analysis is to decrease the stock in store and at the same time increase the customer's level of service. The essential elements of the LOSA are calculating the potential for improvement and identifying measures that help exploit the potential.

9.3.1.1 Calculating Potential Based on Logistic Operating Curves

For each of the stored items, the potential for reducing or increasing the stock can be determined from the difference between the target stock level and the mean stock level. The total potential of the finished goods store is the sum of the potentials for the individual articles.

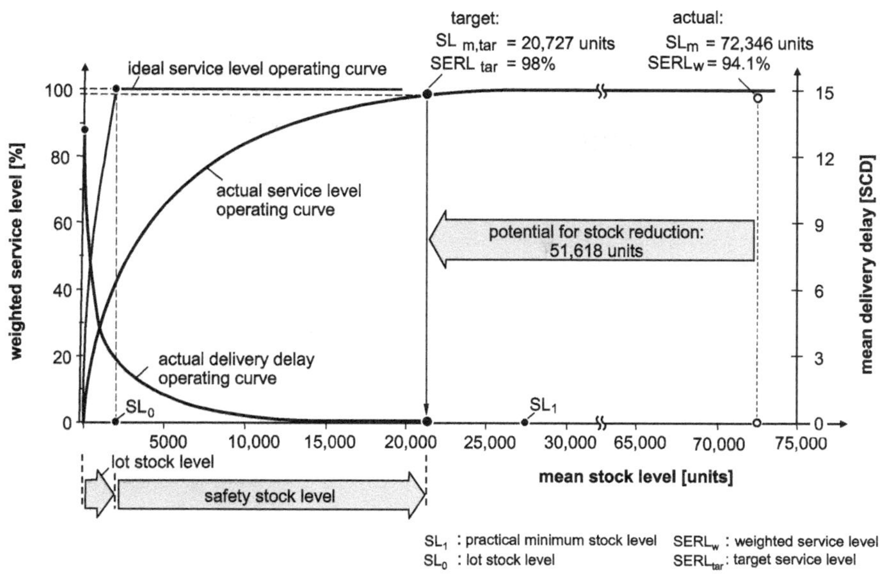

Fig. 9.7 Storage Operating Curves for Item 347121

Using item no. 347121 as an example, we will now illustrate this procedure. This item, which is responsible for 2% of the total mean stock level, represents the largest individual proportion of the store's stock. Figure 9.7 shows the calculated actual and ideal Service Level Operating Curve for this item as well as the actual Delivery Delay Operating Curve (DDOC). These were determined based on the extremes of the plan deviations (Eq. 9.2). In addition, the actual operating states were plotted. There is no stock related cause for the fact that despite the high mean stock level, the operating state is not exactly positioned on the Service Level Operating Curve. Instead, this is due to a planned delayed delivery of specific demands.

The mean stock level of the item during the evaluation period was 72,346 units. The target value of the weighted service level for this item was 98%. With the assistance of the SLOC we can determine what target stock level is required in order to attain this targeted service level. Here we can see that for this item, the target stock level was 20,727 units. The stock could thus be reduced to this target level. Through this measure alone the stock would be reduced by 51,618 units, and thus release a corresponding amount of capital.

Figure 9.8 shows an excerpt from a list of the items ranked according to their stock potential. The stock levels for 71% of the items had the potential to be reduced, whereas 29% had a negative stock potential i. e., the stock had to be slightly increased. With regards to the number of individual units, the potential for decreasing them was considerable. Whereas, there was the possibility to reduce the stock by approx. 1.7 million units, the necessary stock increases amounted to only 0.3 million units. Consolidated, the total potential for reducing the stock in the finished goods store was 1.44 million units. The total stock could thus be re-

rank	item number	ABC (turnover)	mean stock level [units]	target stock level [units]	potential for stock reduction [units]	relative potential for stock reduction [%]
1	347121	A	72345.7	20727	51613	71.3
2	366826	A	57527.6	16137	41391	71.9
3	366255	A	36484.2	11453	25032	68.6
4	366827	A	32455.1	12732	19723	60.8
5	[illegible]	A	19699.3	5767	13933	70.7
			[illegible]	4712	[illegible]	
2243	337953	C	[illegible]			
2244	324221	C	301	301	0	0,0
2245	338158	C	233	233	-1	-0.4
2246	332129	C	353	354	-1	-0.3
				445	-1	
[illegible]	339158	A				
3160	367071	B	872	4897	-4026	-461.8
3161	347129	A	1386	8139	-6753	-487.2
		total:	**3538030**	**2094875**	**1443155**	**40.8**

reduced stock level

increased stock level

Fig. 9.8 List of Items Ranked According to Stock Potential

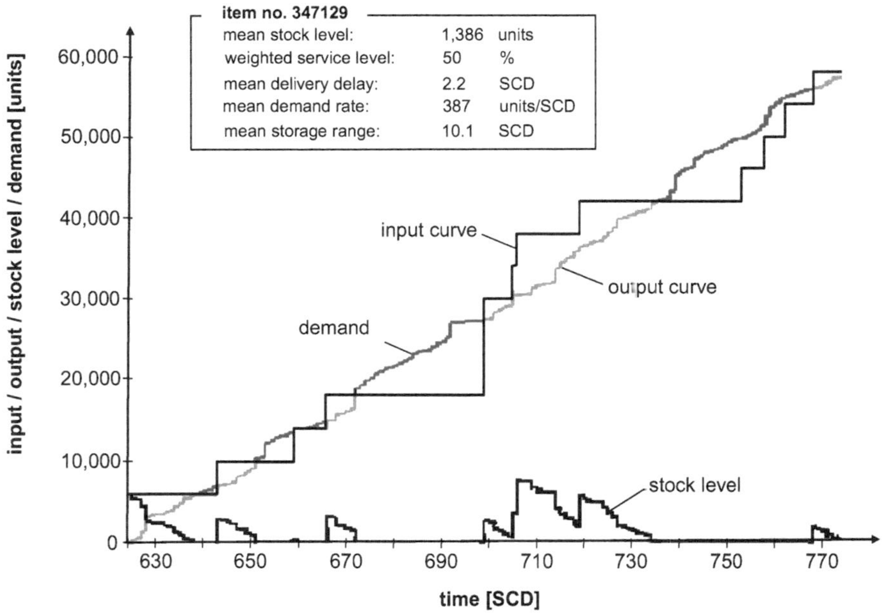

Fig. 9.9 Storage Throughput Diagram for Item No. 347129

duced by this amount, without the stock related loss of service level falling below the target value.

Item no. 347129 illustrates why both increasing and decreasing stock levels is necessary (see Fig. 9.9). The store for this item has the greatest need to be increased (see Fig. 9.8, last row); as can be seen in the corresponding Storage Throughput Diagram (Fig. 9.9) there are frequently shortfalls.

Indeed, over extended periods of time the stock level is completely depleted. Whereas, the mean stock level was 1,386 units during the evaluation period the

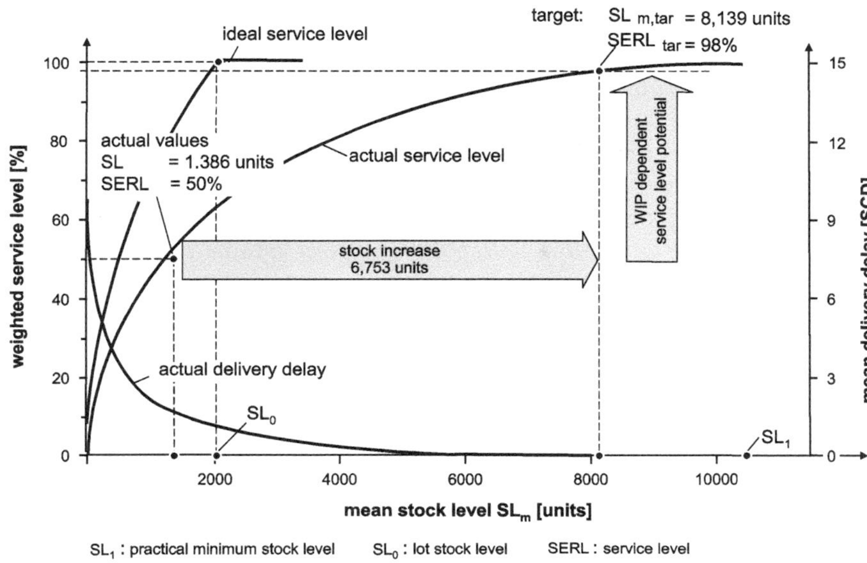

Fig. 9.10 Storage Operating Curves for Item No. 347129

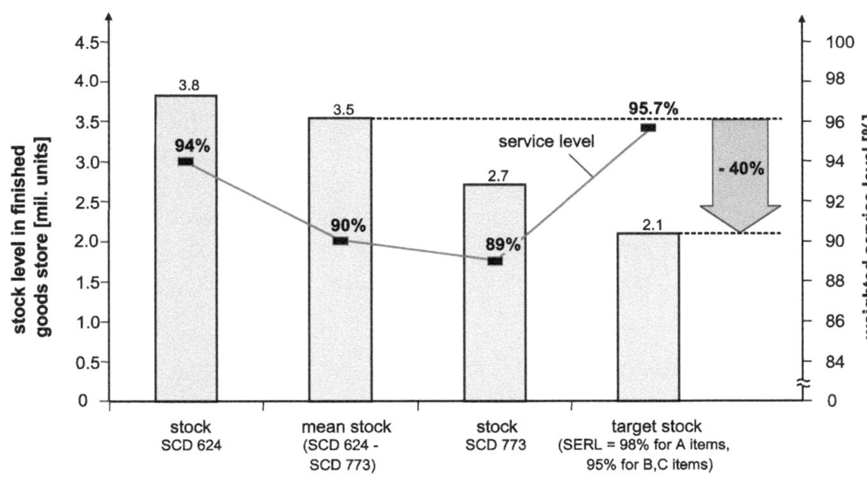

Fig. 9.11 Potential for Stock and Service Levels within the Finished Goods Store

weighted service level was just 50%. Taking into consideration the actual deviations from the plan that occurred, the target level was determined to be 8,139 units. This can be seen in the Storage Operating Curve represented in Fig. 9.10.

In general, the stock for item no. 347129 was too low to achieve the target service level. It was therefore necessary to significantly increase the stock level in order to attain the desired service level.

Figure 9.11 provides an overview of the total stock level and the total target stock level for the finished goods store during a time period of 149 SCD. The input's maximal lateness was 20 SCD and the replenishment time was 32.5 SCD. Comparing the stock levels of all the items at the beginning and end of the evaluation period shows that the stock level had noticeably decreased during the investigation. The cause of this was the temporary input reduction during a production changeover. However, as the stock was reduced, the service level also fell considerably below the target value. The stock reduction was completely uncontrolled. Through a targeted, problem oriented dimensioning of the stock using the Service Level Operating Curves the stock level could be further reduced while at the same time increasing the service level. Comparing the mean stock level during the evaluation period with the target stock level, shows that the stock could potentially be reduced by approx. 40%.

Assessing the stock through the manufacturing costs of the various items revealed that the value of the mean stock level during the evaluation period was approx. 5.9 million euros. The value of the stock could be reduced to approx. 3.3 million euros. A potential of approx. 2.6 million euros means that assuming storage costs of 20%, approximately 520,000 euros can be saved per year. This tied-up capital is then released and available for investments, for example in order to reduce the setup costs of the upstream production. Through that, smaller lot sizes could be economically manufactured, which in turn would mean smaller lot sizes in the finished goods store.

The decisive control lever for determining the size of the stock is setting the safety stock level. This is accomplished with the Service Level Operating Curves so that only the actual occurring planning uncertainty will be buffered. Previously, safety stock levels were calculated based exclusively on the monthly demand. This explains the excessive stock level. The safety stock level was significantly reduced for 83% of the items through the new calculation based on the Service Level Operating Curve. Using the Storage Operating Curves, the required stock can be assigned to the stored items depending on the problems. The stock in the finished goods store is then adjusted to the plan deviations that may arise.

9.3.1.2 Deriving Measures

In order to derive appropriate measures, it has to be determined which factors are predominantly responsible for the target stock level (comprised of the lots and safety stock levels). Based on the percentage of the practical minimal stock level (SL_l) that the individual plan deviations are responsible for, we can determine their respective proportion of the target stock [Lutz-02]. The analysis of the target stock level of the finished goods store resulted in the distribution of stock illustrated in Fig. 9.12. The target stock level of the finished goods store consists of 43.1% lot stock and 56.9% necessary safety stock.

The breakdown in Fig. 9.12 shows that in particular the safety stock which is required due to fluctuations in the demand rate during the replenishment time (36.7%) is a very large component of the target stock level. It represented 64.5% of the safety stock. The delivery lateness is the second largest influence factor on the total safety stock. Therefore, measures for reducing the stock should be ap-

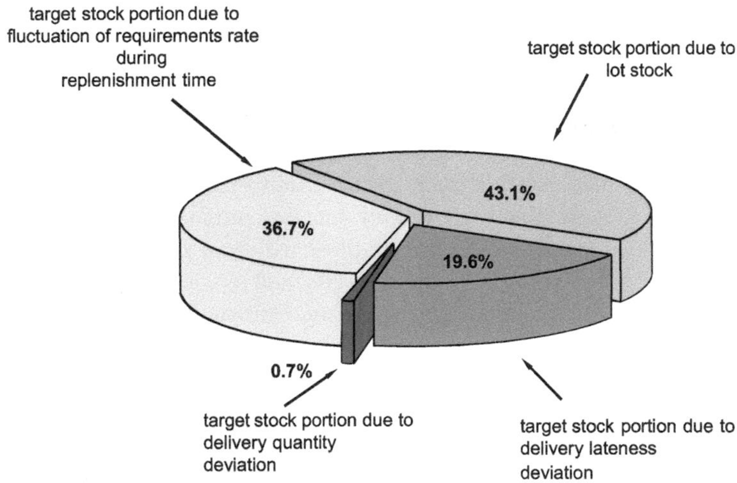

Fig. 9.12 Target Stock Distribution (Example)

plied to these two causes and not to the delivery quantity deviations, which at 0.7% of the total target stock is an almost negligible component. Decreasing the delivery quantity deviations would only effect the total stock marginally.

In comparison narrowing the distribution of the demand rate during the replenishment time would have a large impact. This factor is comprised of both the rate of demand and replenishment time components. The rate of demand can barely be influenced by the manufacturer. Nevertheless, there is still the possibility to control the demand through appropriate marketing measures, such as sales actions or progressive discounts. In the end though, the customer decides whether there is a demand. Within the company the replenishment time can be influenced through targeted measures such as shortening the throughput time in the upstream production. The finished goods store's safety stock levels can also begin to be decreased by reducing the delivery lateness by improving the planning and decreasing lateness in the production.

If the plan deviations caused by the stock are noticeably reduced, further attempts should still be made in order to decrease the input lot sizes and instead increase the frequency of the deliveries to the stores. In this way, the lot stock can be reduced further.

9.3.2 Conducting a Bottleneck Oriented Logistic Analysis of the Manufacturer's Production

Since the logistic output parameters are also the input parameters (plan deviations) for determining the size of the finished goods store, the next step in examining the supply chain is thus a Bottleneck Oriented Logistic Analysis (see Chap. 7) of the manufacturer's production. Here, the correlations between the production's logistic objectives are examined and the potentials for improving the logistic performance are derived. Based on these results, it can be estimated to which degree the stock in the finished goods store can be adjusted to changing boundary conditions.

How the logistic objectives output rate, WIP and throughput time interact on a workstation can be quantitatively described using the Logistic Operating Curves. If the target WIP levels are fixed for all of the workstations in a production, the corresponding throughput times for the workstations can be determined from there. Based on the individual workstation's throughput times the order's total throughput time in the production can be found. With these results, and in consideration of the material flow, the production's total potential can be estimated with regards to decreasing the WIP and throughput time. Figure 9.13 illustrates the outcomes of the Bottleneck Oriented Logistic Analysis for the observed manufacturer.

With the help of the Logistic Operating Curves a Logistic Positioning was conducted for all of the workstations. By comparing the actual and target values for the mean WIP, output rate and throughout time of the workstations, the logistic potential of the workstation could be estimated. Furthermore, target values for the logistic objectives could be determined, which in turn influence the sizing of the next store.

The Bottleneck Oriented Logistic Analysis of the investigated tool manufacturer's production showed that the mean order throughput time could be reduced from 32.5 SCD to 15.5 SCD by decreasing the WIP by approximately 50%. This decreased throughput time can be attained just by reducing the WIP on the workstations in the production. The production throughput time corresponded to the replenishment time with which the stock level in the finished goods store was dimensioned. The production throughput time and therefore the replenishment time can now be reduced by approx. 50%, this thus has a direct influence on the

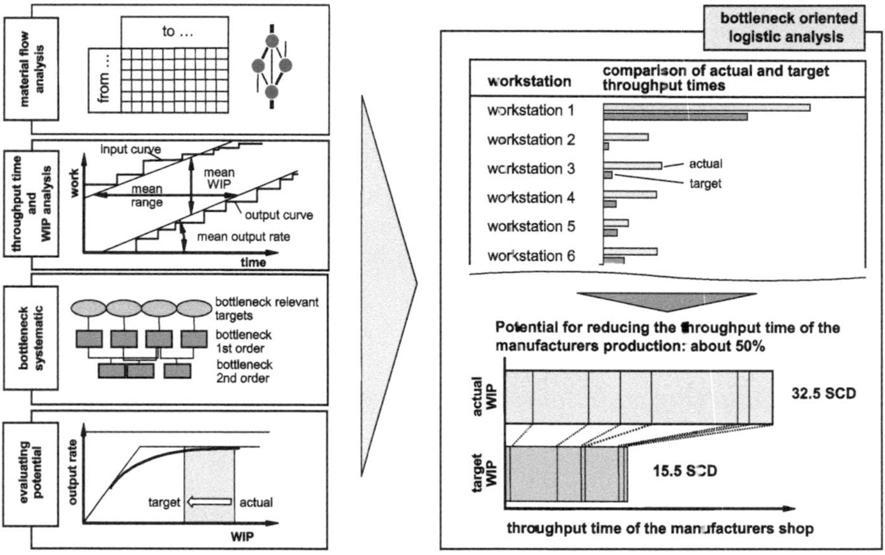

Fig. 9.13 Elements of a Bottleneck Oriented Logistic Analysis for Determining the Logistic Potential in a Production (Based on [Wind-01])

part of the safety stock caused by the fluctuating demand rate during the replenishment time.

In the observed production, the throughput times were also highly variable. As Yu [Yu-01] has shown, the throughput time distribution and consequently the output lateness decrease in the course of reducing the WIP. In this way, a further factor that influences the finished goods store's safety stock level is positively impacted. If both the replenishment time and the input lateness can be reduced, then two influence factors which determine the stock certainty are decreased so that the safety stock levels can be further reduced. Figure 9.14 illustrates the impact of these two measures.

In order to improve the production's scheduling reliability, it was assumed that the reliability would continue to improve as the WIP was reduced. Furthermore, considerably decreasing the production backlog continued to be a goal. This measure would also positively affect the productions scheduling reliability thus the tolerance range for the production output lateness was set at 10 SCD. In addition, the management decided that since larger schedule deviations could only result from considerable planning errors or unusual load peaks, they should not be satisfied by stock in the finished goods store.

With the reduced replenishment time and input lateness values in the finished goods store, new safety stock or target stock levels could be determined using the Service Level Operating Curves. The smaller plan deviations will result in a more favorable shape for the SLOC, as is for example illustrated in the lower left part of

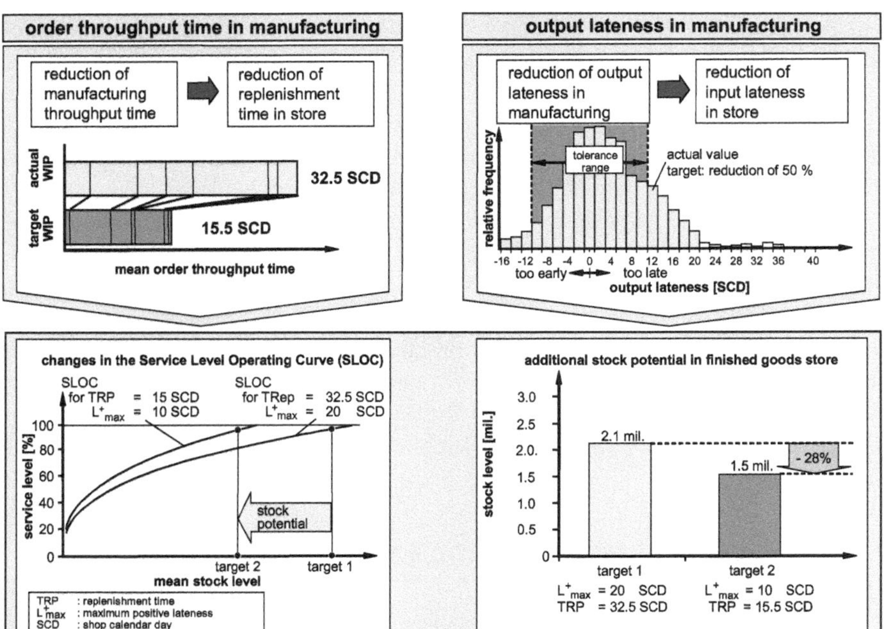

Fig. 9.14 Impact of Optimizing the Manufacturer's Production Logistics on the Finished Goods Store

Fig. 9.14. Therefore, a lower total stock also results. The examination of the entire spectrum of items shows that by improving the logistic objectives of the manufacturer's production the stock in the finished goods store could be further reduced by 28% without decreasing the service level.

9.3.3 Logistic Oriented Storage Analysis of the Manufacturer's Input Stores

The manufacturer's production scheduling reliability is, as far as they are concerned, dependent on the availability of materials in the input stores. When there is a lack of materials, the order's production start is delayed and in turn they are not punctually available in the finished goods store. The mean deviation from the planned input in the manufacturer's production was 1.2 SCD. This corresponds to the mean delivery delay with which the item can be taken from the input store. The mean deviation from the planned input of the supplier into the stores was 6.4 SCD. In order to at least partially buffer the schedule uncertainty of the supplier, the manufacturer maintained high safety stock levels in the input stores. Nevertheless, as can be derived from the delivery delays, there were frequent problems with material supply.

Another reason for the very high stock level in the input store stems from the fact that the supplier's and manufacturer's productions were based on different lot sizes. Due to high setup costs and long setup times the supplier produced lot sizes which, in particular for B and C items, are many times larger than the manufacturer's actual demand. For these items, there is a correspondingly long storage time and high stock level. Since there was a history of frequent changes to the product specifications, there was also a high risk that many of the items would have to be scrapped. In fact, approximately 11% of the stock in the input stores showed no stock movement at all during the evaluation period

Next, through the Logistic Oriented Storage Analysis of the manufacturer's input stores, the items were classified according to the demand for them (ABC Classification). Due to the importance of A items, the safety stock level was sized so that it ensured a service level of 100%. In order to guarantee the immediate availability of materials, the lower limit of the safety stock level was set so that at least one production sized lot could be withdrawn. By examining the factors which caused the stock level, it became clear that the predominant component (51%) of the safety stock level resulted from the supplier's lateness.

9.3.4 Bottleneck Oriented Logistic Analysis of the Supplier's Production

The results of the Bottleneck Oriented Logistic Analysis which was conducted for the supplier resulted in a potential for reducing the throughput time by approximately a further 20%. It was also possible to reduce the WIP by a similar magnitude. By planning measures such as an improved prognosis of the manufacturer's demands, the goal was to reduce the maximal output lateness of the supplier's production corresponding to the maximal input lateness of the input stores, to a maximal of 10 SCD. With these entry values, the target stock level in the manufacturer's input store can be set at 0.7 million items.

For the B and C items, which were in weak demand, the lot sizes were drastically reduced by the supplier in order to reduce the lot stock. Since the demand for these items was very irregular and in small quantities, the stockpiling of the safety stock lead to disproportionately high costs. It was therefore, decided to forego safety stock for B and C items. Instead, a part of the supplier's capacity was reserved for B and C items, so that they could be produced punctually in the lot size demanded by the manufacturer.

9.3.5 Supply Chain's Total Potential

Through the described measures, the total stock in the finished goods store could be reduced by 72%. Value-wise it resulted in a potential of 76% while at the same time clearly increasing the service level.

If having a service level of 100% in finished goods store results in practically no further input lateness for the manufacturing orders, then the output lateness is also reduced. During the evaluation period, the order input lateness for the production was on average 1.2 SCD. The start of approx. 30% of the orders was thus delayed. If this value is now reduced to zero, it can be assumed that the production's output lateness would also be reduced by this amount. From the point of view of the finished goods store, this means that the mean input lateness can be further reduced by 1.2 SCD. Through this measure, the resulting target level for the finished goods store is a further 8% below the previous target level.

The measures for improving the logistics which have been described so far are summarized in Fig. 9.15. First, the mean stock levels in the finished goods store can be adjusted by determining the necessary level of safety stock. Afterwards this target level can be further reduced by logistically optimizing the upstream productions. So that the most extensively possible improvements can be targeted in the manufacturer's production the input store's stock has to be adjusted in the next step. The organization of the supplier's production is also crucial here. The stock potential of the entire supply chain, illustrated in Fig. 9.16, results from a combination of the above discussed measures.

If all of the measures were implemented with the expected results, there would be a total potential for reducing the stock in the finished goods store by approx. 60% of the mean stock level found during the evaluation period. The stock in the entire supply chain during the evaluation period was approximately 9.3 million units at various processing levels. Through the measures that improve the logistic performance of the individual supply chain elements, there is a potential, which when realized would reduce the total stock in the supply chain by approximately 55% to 4.1 million units. At the same time the supply chain's weighted service level in relation to the customer, can be increased to the target level of 98% for A items and 95% for B and C items.

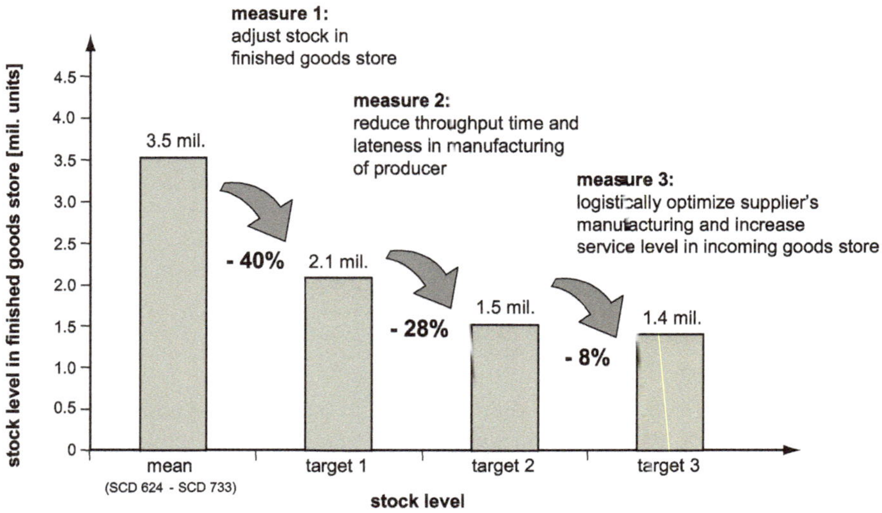

Fig. 9.15 Sequence of Measures and Stock Potential in the Finished Goods Store

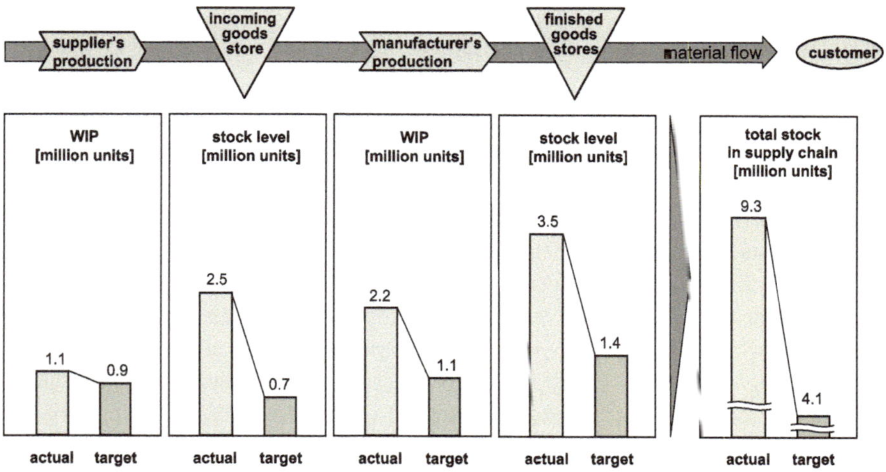

Fig. 9.16 Sequence of Measures and Stock Potential in the Finished Goods Store

9.4 Summary of Applying Operating Curves to the Supply Chain

In the preceding we have described the interactions of the logistic objectives in the stores with the help of the Storage and Service Level Operating Curves, and in interplay with the Logistic Operating Curves, for the production. If both approaches are linked with one another, it is possible to analyze the logistic interactions within the supply chain. This makes it possible to determine the available potential with respect to the service and stock levels over a number of progressive steps, as well as to aggregate the total potential.

By combining the Logistic Oriented Storage Analysis with the Bottleneck Oriented Logistic Analysis for the production area, there is a consistent analysis method available, which makes it possible to quantify how the logistic objectives interact with another both in the storage stages as well as in the entire supply chain.

Since the required stockpiles can be allocated to the logistic disruptions in the supply chain which cause them, the necessary measures can be derived. Such measures can then be evaluated with regards to how they impact the logistic goals and obtaining them. In this way, support is provided for the often problematic process of selecting measures. At the same time, the awareness of problems pertinent to attaining goals is sharpened through a cost-oriented evaluation of the stock.

Chapter 10
Conclusions

A company can only be successful when its production fulfils the market's complex and quickly changing demands. Designing and controlling production are thus key functions for a company and need to be rigorously oriented on the company's goals. Further developing the organization as well as the planning and controls is necessary and occurs on a wide scale, because many companies are strongly motivated due to disappointing logistic performance figures.

The methods chosen for this within the industry are quite diverse. However, they are mainly oriented on reducing the production complexity by simplifying the structure in order to create manageable and highly responsive factories. Due to the dynamic of the changes and the multiplicity of external and internal influence factors though, these design measures are also difficult to manage. Moreover, even small production units experience the *dilemma of operations planning*: The logistic objectives are contradictory. Ensuring a high utilization requires a high WIP, which leads to long throughput times and thus often to poor scheduling adherence. This is further aggravated by the fact that changes in the firm's environment can strongly shift the ranking of sub-goals.

In order to make the dilemma of operations planning controllable through an adapted process and production design, it is necessary to have knowledge of the interdependencies between the logistic objectives and the possibilities of manipulating them. Due to the diversity of the processes as well as the internal and external influences, these interdependencies are not easily detectable or describable. Designing and controlling the production therefore has to be supported by appropriate models.

Models describe planned or realized systems abstractly, but very concentratedly. Generally, they do not deliver an exact duplicate of the reality, but instead are meant to reproduce the relevant properties of a system with sufficient enough accuracy, tailored to a specific application. Due to practicality, the more routinely the model is to be applied, the lower the effort required to employ it has to be.

Models are also used to describe the dilemma of operations planning. Here, It has proven to be quite helpful to represent the logistic objectives in the form of

Logistic Operating Curves. These make it possible to describe the objective's interdependencies also quantitatively and thus allow a Logistic Positioning within the field of tension created by their innate competitiveness.

In order to develop Logistic Operating Curves, three different modeling approaches are generally appropriate:

- Simulation as an experimental modeling approach.
- Queuing theory as a deductive modeling approach.
- Logistic Operating Curves Theory as a deductive-experimental modeling approach.

The advantages and disadvantages of the three modeling approaches are comparatively evaluated and illustrated in Fig. 10.1. The relatively minimal effort required to develop the simulation model distinguishes it from the others. It is therefore easy to make adjustments to changing conditions. Furthermore, single events can be described. Thus for example, the flow of individual orders through the production can also be described. This model is quite widely accepted in particular due to the possibility of the graphical animation. However, the simulation has to be newly setup and evaluated for each application case and the results are generally not transferable. Therefore, applying simulations is also connected with a great deal of effort. With respect to today's pressure to make decisions the time required to obtain useful results is too long. Despite the extensive possibilities for implementing them, Logistic Operating Curves created by simulations are thus not common in the industry.

Queuing theory has also not been able to establish itself in production companies with meshed structures. Although queuing models require only a minimal amount of effort there are still assumptions and conditions that need to be met and these

modeling approach	disadvantages	advantages
simulation 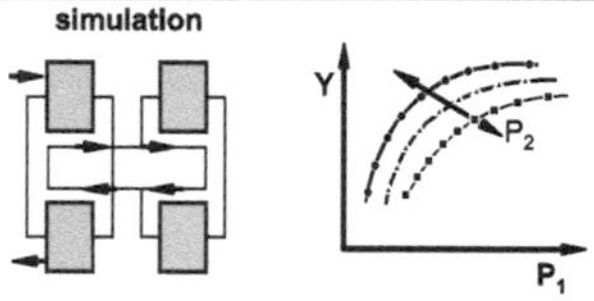	• model validity can only be proven in points • necessary to re-build model for every application • results cannot be generally transferred • greater application effort	• minimal development effort • possible to adapt model • can describe single events • highly accepted
queuing model 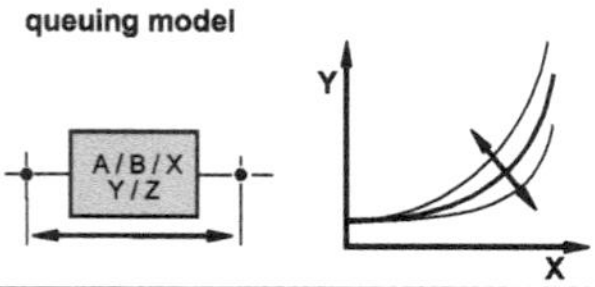	• great development effort (non-recurrent) • model can only be limitedly adapted, if at all • parameters to some extent do not meet practical conditions • limited to resources perspective	• model structure and parameters are derived from basic laws • minimal application effort
Logistic Operating Curve Theory 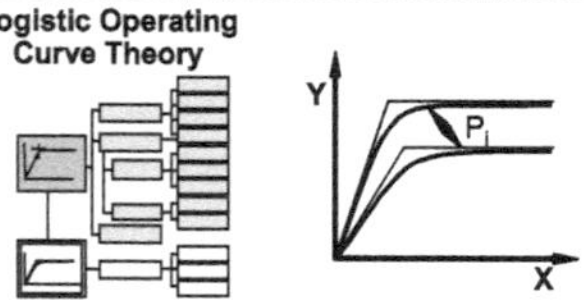	• greater development effort (but non-recurrent) • limited to resources perspective	• model structure and parameters are mostly derived from basic laws • minimal application effort • model can be adapted within defined limits

Fig. 10.1 Comparative Evaluation of Alternative Approaches to Modeling Production Processes. (P: Model Parameters)

typically do not exist in real manufacturing processes. Adjusting a model to specific conditions is also usually not possible, or requires the quite costly development of a generally new model. Moreover, the descriptive parameters required for applying the model are often not practical, especially with more complex models.

The Logistic Operating Curves Theory is an alternative approach which has already proven itself frequently. With the LOC Theory the dependencies between the production processes' logistic objectives and the ways in which they can be manipulated can be described with a relatively simple mathematic model. We have shown through illustrative practical applications how production processes can be designed and evaluated from a logistic perspective based on operational feedback data.

It has to be emphasized here that one of the particular advantages of the Logistic Operating Curves Theory is that the model can usually be adapted to changes in conditions just through the parameterization. Furthermore, the LOC Theory stands out because the model's structures and parameters were almost completely derived from basic laws. Therefore, when the prerequisites for application are met, basic correlations between the logistic objectives can be described independent of a specific case. We were thus able to derive the basic laws of production logistics, which promote the fundamental understanding of the static and dynamic behavior of production systems.

One of the principal disadvantages of the Logistic Operating Curves Theory is that it is restricted to a resource perspective. In order to describe the orders flow through the production with regards to logistic aspects, a combination of these modeling approaches with other methods is thus required. One such possibility is the *Bottleneck Oriented Logistic Analysis* with which the LOC Theory is implemented in conjunction with analyses of the material flow, throughput time and WIP. Based on two practical applications, we demonstrated not only how this method can clearly represent production processes, but also how it can indicate logistic bottlenecks in the material flow and develop and evaluate measures to eliminate them. Furthermore, we have also illustrated how to calculate the impact on the delivery behavior.

The possible applications for the Output Rate and Throughput Operating Curves that we have presented here, demonstrate that the Logistic Operating Curves Theory can successfully describe the dilemma of operations planning qualitatively, quantitatively and with consideration to specific operational conditions. The LOC Theory can thus be applied to specifically influence the production processes.

The Storage Operating Curves and Service Level Operating Curve are important additions to the Logistic Operating Curves. The SOC and SERL make it possible to depict the interdependencies between the logistic objectives within a store. Finally, by combining the Storage Operating Curves and the Logistic Operating Curves the logistic interactions within a supply chain can be analyzed. The available potential for the delivery performance and inventory can therefore be determined over a number of value-adding stages.

Appendix: Software Documentation

In order to more easily understand the theory for calculating Logistic Operating Curves as well as the application possibilities that have been introduced in this book, a number of demo programs have been developed. Included on the accompanying CD, these software programs also provide readers with the opportunity to test them in concrete examples:

- **LOPROM.xls**
  Content: analyze operational feedback data; determine key figures; create work content and throughput time distributions as well as Throughput Diagrams; create Logistic Operating Curves; provide assistance in determining potential by varying the WIP; compare Logistic Operating Curves with changed structural parameters
  Due to the complexity of the LOPROM.xls software, extensive documentation has been developed.
- **SROC.xls**
  Content: develop Schedule Reliability Operating Curves
- **SOC.xls**
  Content: develop Storage Operating Curves; compare Storage Operating Curves with changed structural parameters
- **TOLS.xls**
  Content: calculate economically optimal manufacturing lot sizes based on Throughput Oriented Lot Sizing; present cost and value-adding trends

The latest versions of these software programs can also be downloaded from the institute's website: **http://www.ifa.uni-hannover.de/software**.

System Requirements

The software was developed using Microsoft® Excel and Visual Basic®. The following system requirements are thus necessary:

- IBM personal computer or compatible system
- Pentium 233 or greater; at least 16 MB RAM
- At least Microsoft® Windows XP
- Microsoft® Excel 97 – 2007; for LOPROM at least Excel 2003, Service Pack 3
- Display resolution of at least 800 x 600 pixels (or even better 1024 x 768)
- Mouse

A1 LOPROM – Software Documentation

Introduction

The *LOPROM.xls* software offers the possibility to transparently represent the production flow on a workstation based on operational data with the assistance of a Throughput Diagram, throughput time and work content distributions as well as with Logistic Operating Curves. This in turn allows weak points to be easily identified and measures to be derived and evaluated.

The software contains demo data that can also be replaced with actual data from an enterprise.

Running the Software

The software provided uses macros. In order for the software to operate as described here, it is recommended that the user first start Microsoft Excel (Versions 97–2007) and check the macro security level **before** opening the *LOPROM.xls* workbook. In order to do so:

- Click on *Extras* in the upper menu bar, scroll to *Macro* and then click on *Security.*
- When prompted select either a middle or low security level. With respect to the *LOPROM.xls*, choosing a middle security level will provide the user with the opportunity to decide each time the file is open whether or not the macros are activated.

Once the macro security level has been chosen, the user can then open the LOPROM.xls workbook. In order to prevent the software from being accidentally overwritten the software has been provided with password protection. Thus when prompted for the password, the file should be opened with write protection.

The software's *Start* dialogue box (Fig. A.1) appears with the *Data* worksheet in the background. From the start dialogue box various information sheets (*Data Preparation*; *End User License Agreement*) can be called-up by clicking the button next to the corresponding title. In addition, the size of the window can be automatically adjusted to the computer's configuration and the size of the current worksheet using the *Automatic Zoom Adjustment.* After clicking on the *Start Program* button, the user will be presented with the *End User Agreement.* Accepting the agreement will allow the user to begin working with the program; declining the agreement will result in the software being terminated.

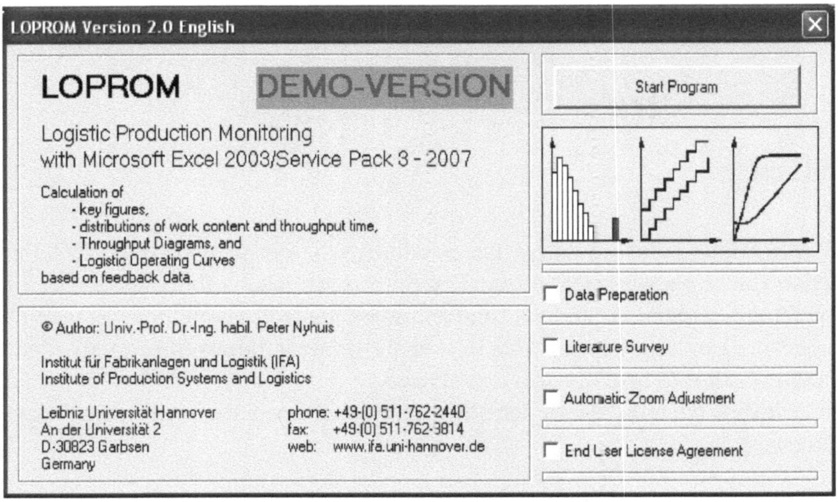

Fig. A.1 Dialogue Box *Start Sheet [opens automatically]*

Running and Utilizing the Software

After the user has accepted the End User Agreement the *Select Function* dialogue box (Fig. A.3) and the *Data* worksheet (Fig. A.2) are displayed. The *Select Function* dialogue box can be opened at any time from the *Data* worksheet, simply by clicking on the *Select Function* button at the top of the worksheet. The various functions will be discussed following the *Data* worksheet introduction.

Analyzing Feedback Data

The *Data* worksheet contains the feedback data that is to be evaluated: the input date (column B: input date), output date (column C: output date) and the respective work content (column D: work content). The input and output dates should be entered in shop calendar days and the work content in standard hours. Feedback dates (output dates) should not be keyed in for orders which have entered the workstation, but which have not yet left the station. The corresponding cells must be left empty as these orders form the final WIP.

The content of column A (order number) is not significant for the running of the software. The column is only for identification purposes. When necessary, the column can also remain empty or be used for another variable (e. g., item number). The throughput time variable, which is also indicated on the worksheet (column E) is calculated by the software. The feedback data can be entered either by copying it from another file (if possible with the command sequence: *Edit* >, *Paste Special...* >, *Value* or by manually entering it.

Select Function

Order Number [-]	Input Date [SCD]	Output Date [SCD]	Work Content [hrs]	Throughput Time [SCD]
30	133	137	7	4.0
29	120	136	14	16.0
28	130	135	26	5.0
27	130	133	11	3.0
26	131	132	6	1.0
25	119	131	10	12.0
24	125	130	11	5.0
23	126	129	16	3.0
22	118	128	26	10.0
21	120	126	15	6.0
20	122	125	5	3.0
19	121	124	14	3.0
17	120	123	20	3.0
18	115	123	6	8.0
16	118	122	28	4.0
15	110	120	49	10.0
14	117	117	1	0.0
13	116	116	5	0.0
12	110	116	7	6.0
11	109	116	30	7.0
10	104	113	15	9.0
9	103	111	21	8.0
8	106	109	17	3.0
6	105	108	26	3.0
7	105	108	2	3.0
5	102	106	11	4.0
4	101	105	23	4.0
3	101	103	19	2.0
2	94	102	21	8.0
1	98	99	20	1.0
33	132		25	
31	125		10	
32	120		15	

Fig. A.2 Worksheet *Data*

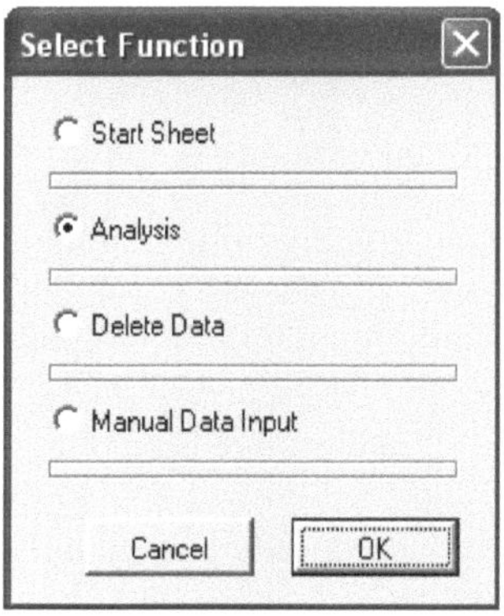

Fig. A.3 Dialogue Box *Select Function [Button Select Function]*

Once the feedback data has been entered, the different software functions can then be called up again by clicking on the *Selection Function* button at the top of the sheet. A dialogue box will then open (Fig. A.3) from which the following described functions can be selected:

Start Dialogue Box

The *Start Dialogue Box* (see Fig. A.1 above) will open again. All of the contained software functions such as the automatic zoom adjustment can thus be accessed from there.

Analysis

In analyzing the feedback data that was copied in or entered manually, the following activities are initiated internally in the software:

1. Determining the Evaluation Period: Through a dialogue box, the user will be asked what time period the analysis should be based on (Fig. A.4). The software will recommend the smallest and largest feedback output dates. These values can, however, be overwritten when necessary.
2. Sorting the Feedback Data.
3. Calculating the Throughput Time per Operation.
4. Developing and Calculating the Distributions for the Work Content and Throughput Time: This is based only on the data sets that have an output date within the evaluation period.
5. Calculating Data for the Throughput Diagram.
6. Calculating the Logistic Operating Curves: The workstation specific data that is required here will be requested through the *Workstation Details* dialogue box (Fig. A.5).

Every time the feedback data is changed, the analysis needs to be run again!

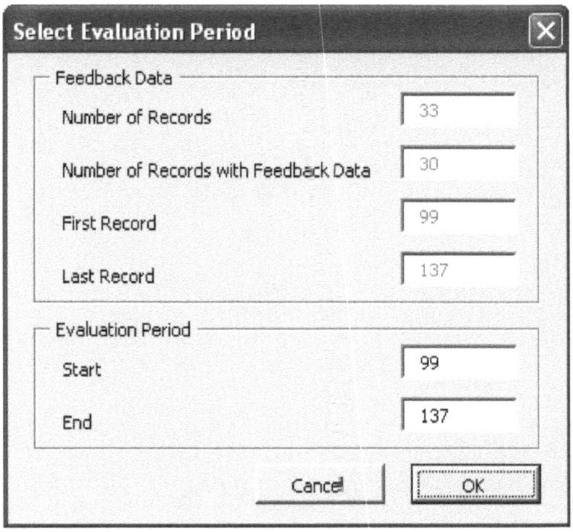

Fig. A.4 Dialogue Box *Select Evaluation Period [Selection Analysis]*

Manual Data Input

The feedback data can be entered using an Excel worksheet. On the right hand side of the dialogue box it is possible to search for specific data sets in order to modify or delete them.

Delete Data

The feedback data found in the *Data* worksheet will be completely deleted. This function should be used before new data is entered.

If the analysis is then run, the user will be requested to enter workstation specific data (Fig. A.5). In order to calculate the Logistic Operating Curves, it is imperative that the number of single workstations is correctly entered.

In order to control the data, it is internally verified whether or not the given capacity at least approximately conforms to the output rate reported during the evaluation period (Fig. A.6). Changes can be made as required.

After completing the analysis the program jumps to the *Throughput Diagram* worksheet (Fig. A.7). In the Throughput Diagram, the input, output, WIP and the cumulated capacities are plotted over the time. The dynamic behavior of the workstation during the evaluation period is thus described qualitatively and chronologically accurately.

The input, output and WIP data which the Throughput Diagram is based on are stored in the *Calc. Throughput Diagram* worksheet.

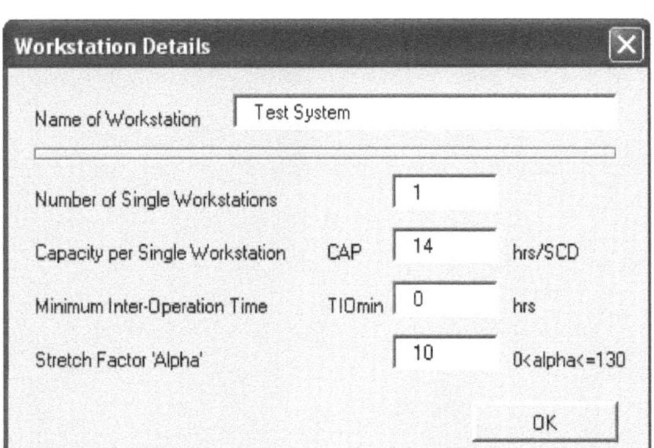

Fig. A.5 Dialogue Box *Workstation Details [Button Workstation Details]*

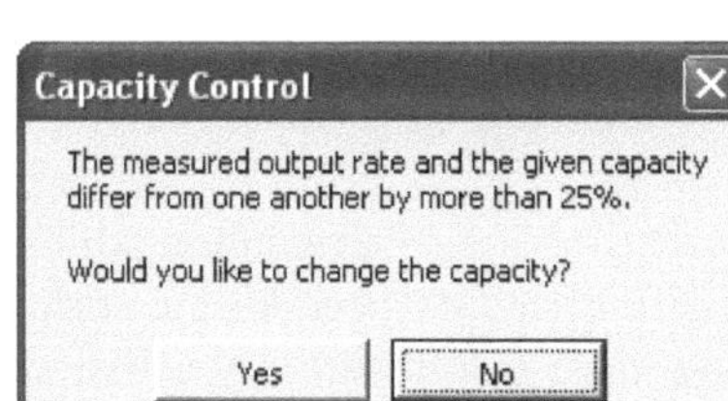

Fig. A.6 Dialogue Box *Capacity Control [Opens Automatically]*

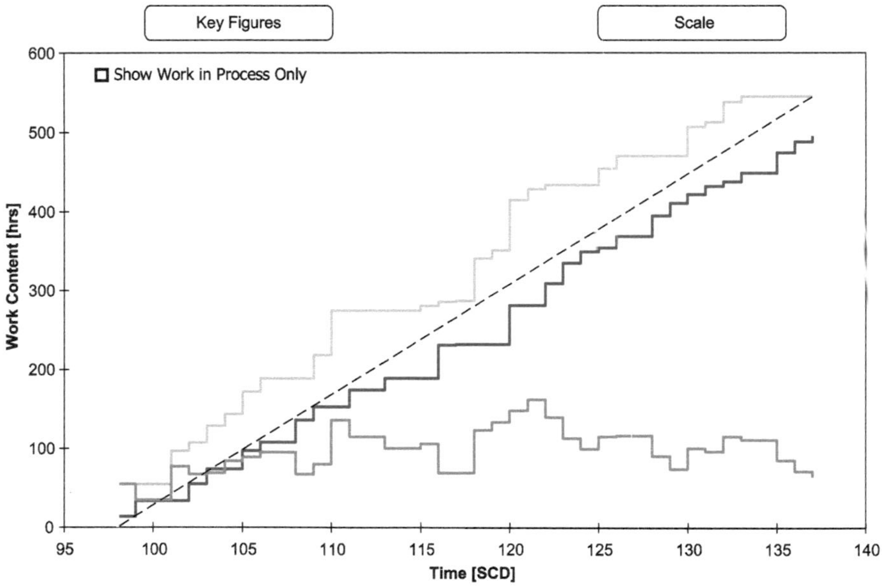

Fig. A.7 Diagram *Throughput Diagram [Worksheet Throughput Diagram]*

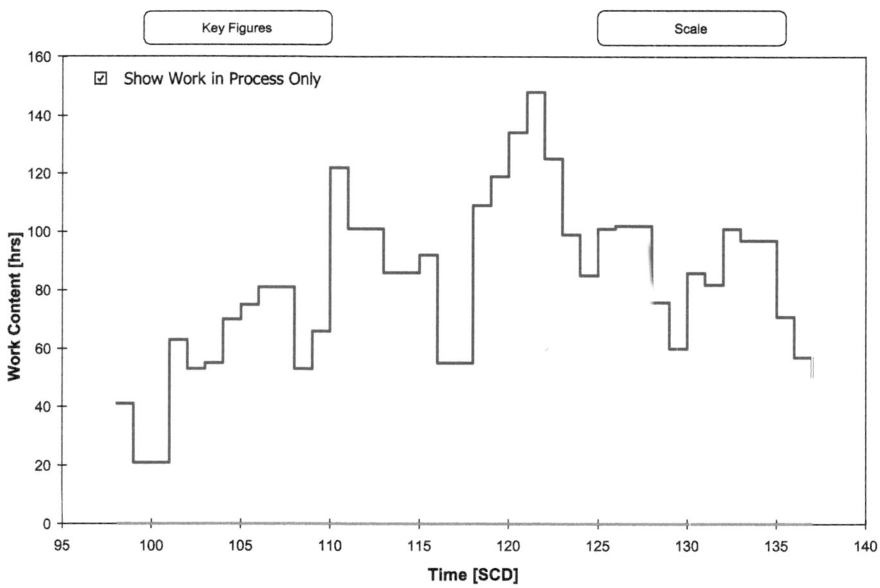

Fig. A.8 Diagram *Throughput Diagram (WIP only) [Worksheet Throughput Diagram]*

During longer evaluation periods, peculiarities in the WIP trend are not always easily identified because the scaling of the y-axis is oriented on the earliest input and latest output date. By activating the control box (*Show Work in Process Only*)

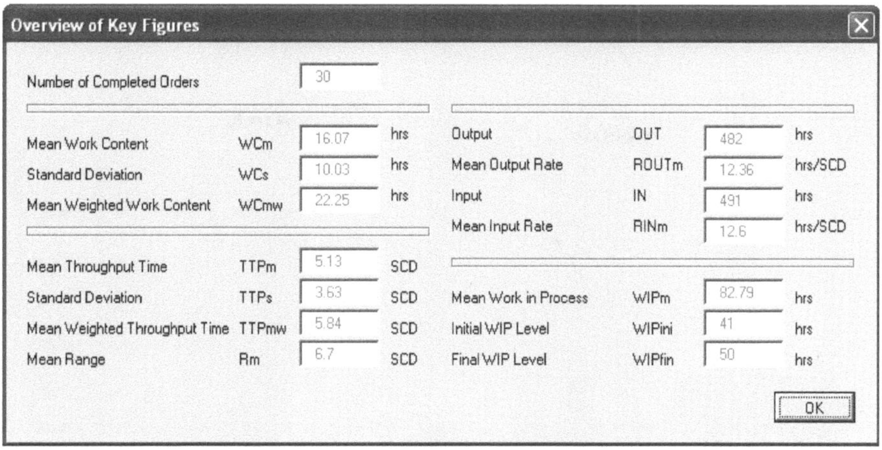

Fig. A.9 Dialogue Box *Overview of Key Figures [Button Key Figures]*

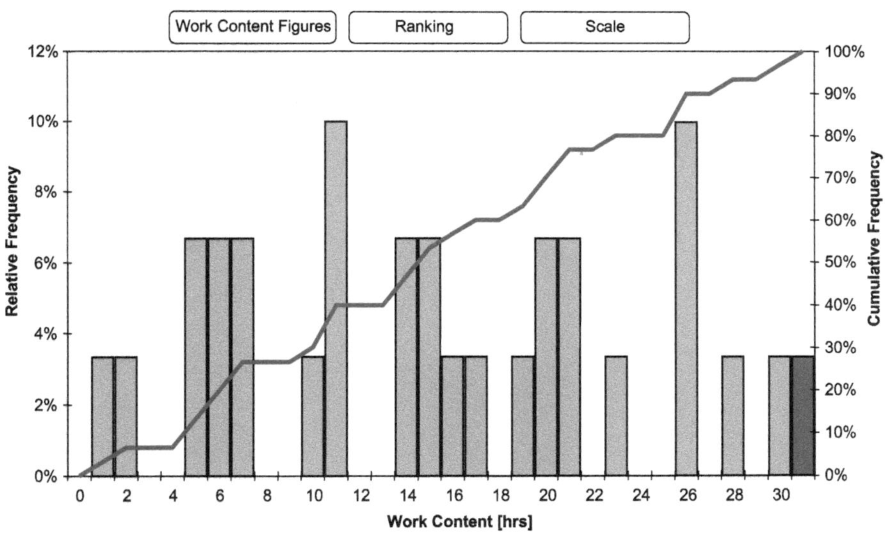

Fig. A.10 Diagram *Work Content Distribution [Worksheet WC]*

in the upper left hand of the diagram, it is possible to display the WIP trend only (Fig. A.8). In doing so the scale of the work content axis will automatically change. The most important key figures (Fig. A.9) can be displayed using the *Key Figures* button just above the graph (Fig. A.8).

Detailed results for the work content and throughput distributions can be derived from the *WC* and *TTP* worksheets (Fig. A.10 and Fig. A.11). Both graphs can be scaled using a scrollbar accessed by clicking on the *Scale* button on the respective worksheets. When required the key figures for the distributions

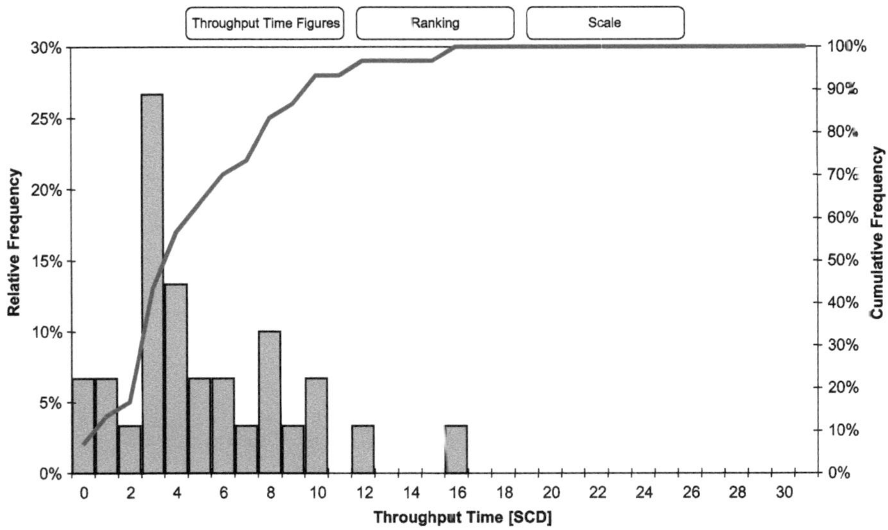

Fig. A.11 Diagram *Throughput Time Distribution [Worksheet TTP]*

Fig. A.12 Dialogue Box *Throughput Time Figures* [*Button Throughput Time Figures*]

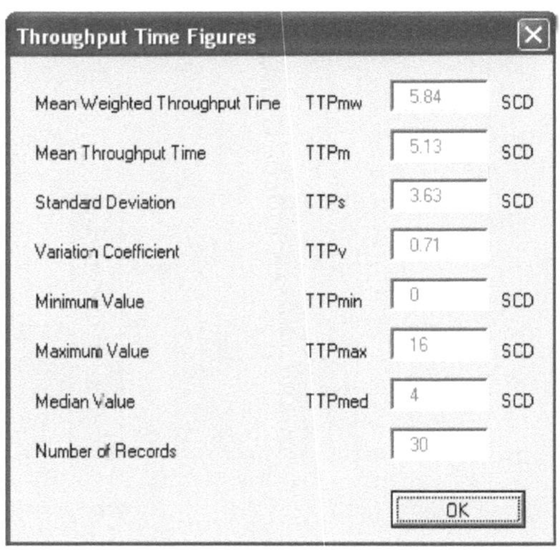

(Fig. A.12) and a list of the order numbers ranked according to the work content or throughput time can be displayed.

In the Logistic Operating Curves (*LOC* worksheet), whose values are calculated in the *Calc. LOC* worksheet, the correlations between the logistic objectives WIP, output rate, throughput time or range are represented (Fig. A.13). By default, the capacity line, the Output Rate Operating Curve and the Range Operating Curve are represented as a function of the WIP in the diagram in addition to the operating state. By clicking on the *Diagram Options* button the user can choose to display the Throughput Time Operating Curve as well as the ideal Output Rate Operating Curve. Furthermore, the user can also choose to display specific parts of the graph only (Fig. A.14).

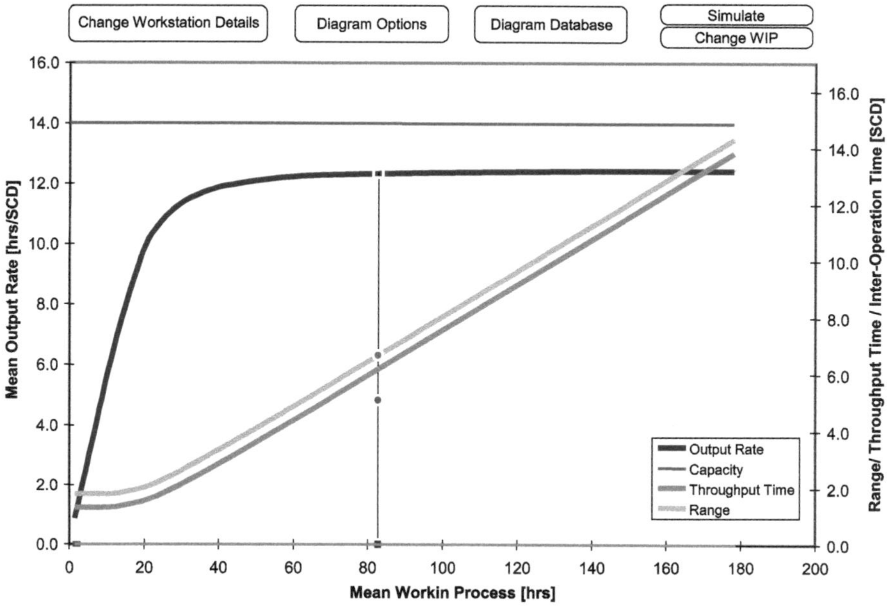

Fig. A.13 Diagram *Measured Operating Point [Worksheet LOC]*

Fig. A.14 Dialogue Box *Diagram Options [Button Diagram Options]*

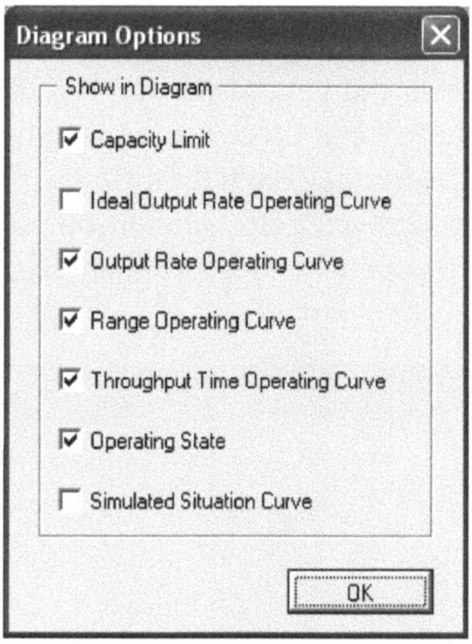

An overview of the most important data which are used in calculating the Logistic Operating Curves (Fig. A.15) can be displayed using the *Diagram Database* button. By clicking on the *Diagram Database* button, the key logistic figures that mark the measured operating state on the calculated Logistic Operating Curves will also be displayed.

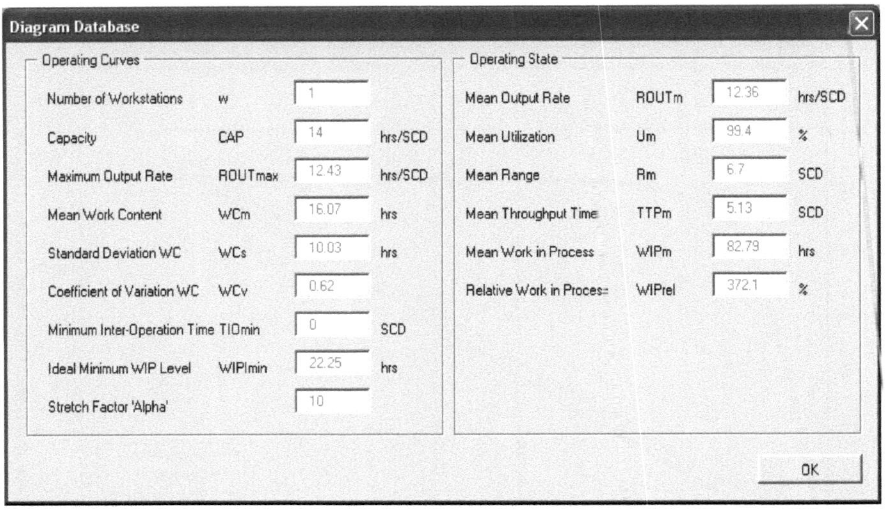

Fig. A.15 Dialogue Box *Diagram Database [Button Diagram Database]*

Determining the Potential for Reducing the Throughput Time and WIP

With the help of the Logistic Operating Curves, it can be shown which WIP and throughput times can be attained on a workstation without changing the structural conditions. The button *Change WIP*, found at the top right hand corner of the *LOC* worksheet, helps support this assessment. After clicking on the button, a dialogue box opens with a scrollbar (Fig. A.16). Using the scrollbar, a fictitious WIP can be

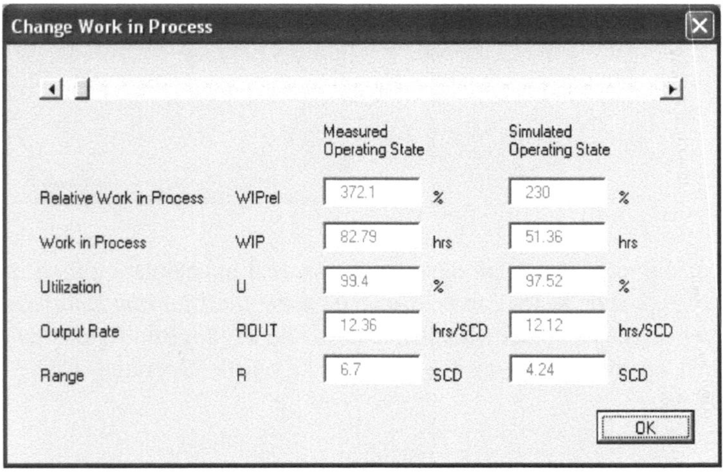

Fig. A.16 Dialogue Box *Change Work in Process [Button Change WIP]*

set for the workstation. The resulting changes for selected key figures are then provided in comparison to the measured operating state. The new operating states are then also displayed in the Logistic Operating Curves diagram (Fig. A.17).

If it appears that the desired logistic target values are not attainable with the given structural conditions, it is necessary to intervene in the work content or the work content structure in order to develop additional logistic potential. The Logistic Operating Curves Theory also provides support for evaluating possible measures. Corresponding to the possible interventions in the process the user can change the data for the capacities, work content structure, minimum inter-operation time and/or the LOC Theory's stretch factor 'alpha' by clicking on the *Simulate* button (Fig. A.18).

The probable impact of these changes is made obvious in comparison to the initial situation through a second pair of Logistic Operating Curves (Fig. A.19).

The newly calculated operating state is automatically set here so that the same output rate is reached that was set during a previous WIP change (see above). In cases where there were no WIP changes the operating state is set to the output rate measured in the initial situation. The function for changing the WIP which can be used further is now generally based on the new variant.

In the *Overview* worksheet the graphs for the Throughput Diagram, Logistic Operating Curves, throughput times distribution, and the work content times distribution are summarized in their most current state. In addition a selection of relevant key analysis figures are displayed.

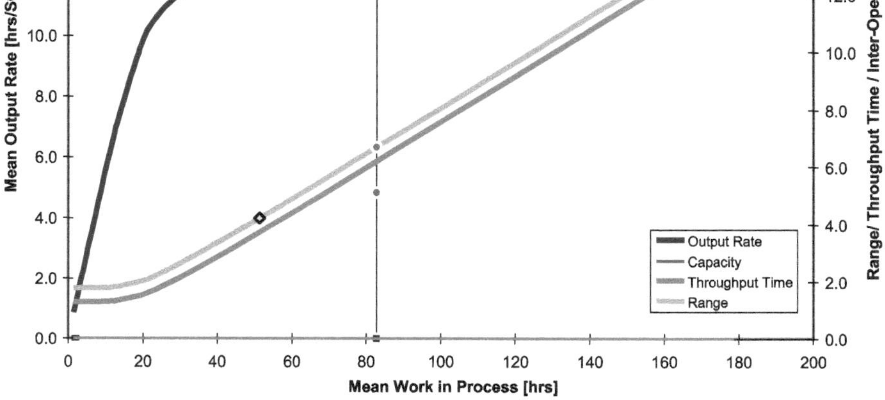

Fig. A.17 Diagram *Decreased WIP Operating Point [Worksheet LOC]*

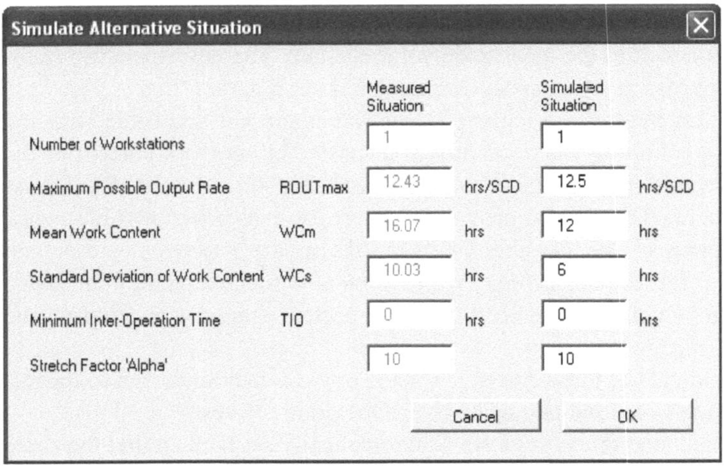

Fig. A.18 Dialogue Box *Simulate Alternative Situation [Button Simulate]*

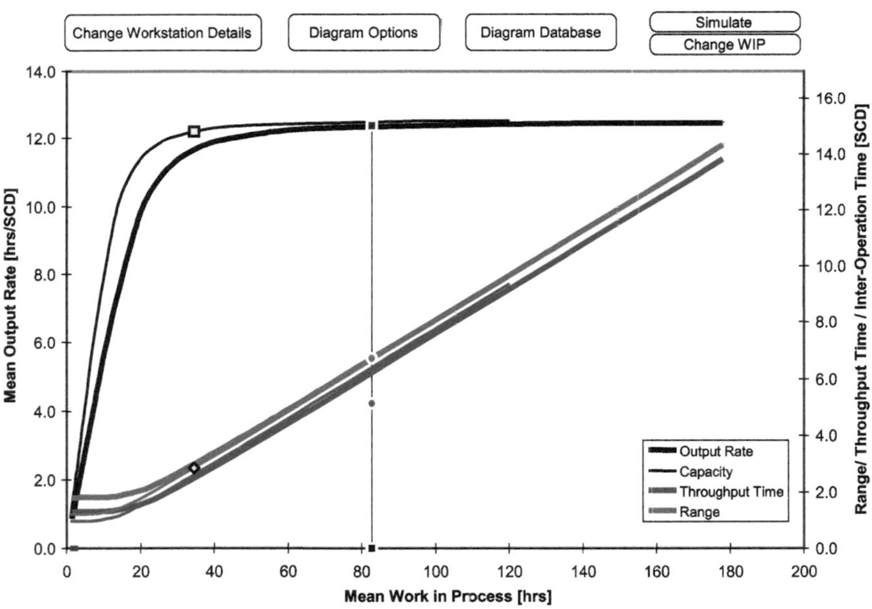

Fig. A.19 Diagram *Simulated Operating State [Worksheet LOC]*

Errors and Troubleshooting

Depending on the configuration of the computer, it is possible that when changing the data that describe the structure (see Fig. A.16) a message appears on the dialogue box that says *Invalid Values*. This is caused by invalid decimal symbols in the maximum possible output rate, mean work content and or standard deviation of the work content. In this case, the decimal symbol needs to be changed for these variables (from '. ' to ', ' or vise versa). Alternatively, it is also possible to change the symbol for the decimal overall through the systems settings (country settings).

The feedback data is checked within the software for possible data errors that could interfere with running the software. When necessary, the software will automatically be aborted with a warning about the possible cause of the error.

Nevertheless, should an error occur when running the software it is possible that this is due to one of the cells on one of the worksheets being inadmissibly overwritten. Generally, this can be resolved by running the analysis again or when necessary, by re-starting the original software. No further problems are known.

A2 SROC – Software Documentation

In addition to calculating the Schedule Reliability Operating Curves (SROC), the *SROC.xls* software explains the parameters that are involved and how they operate.

Corresponding to the explanation in Sect. 4.10 the delivery reliability is represented as a function of the WIP in Schedule Reliability Operating Curves. In order to calculate them the upper and lower limits of the tolerance range, the standard deviation of a random (to be considered exemplary) inter-operation times distribution and the mean inter-operation time as a function of the WIP are required (see Eq. 4.68). The inter-operation time can be determined and depicted as a function of the WIP using the Logistic Operating Curves Theory (see also Sect. 4.4).

In the upper part of Fig. A.20 the datasheet for the SROC software is shown. In order to use the software, the fields with the white backgrounds need to be filled in.

Based on the information provided here the software first determines a data table from which for example, the mean WIP and mean inter-operation time as a function of the running variable t can be taken (Fig. A.21). Following that the value for the delivery reliability is calculated according to Eq. 4.68. The distribution function $\phi(u)$ (referred to as 'phi' in the datasheet) is not determined through Table 4.2, but rather using a standard Excel function. The results of this calculation are presented as a graph in the lower part of Fig. A.20.

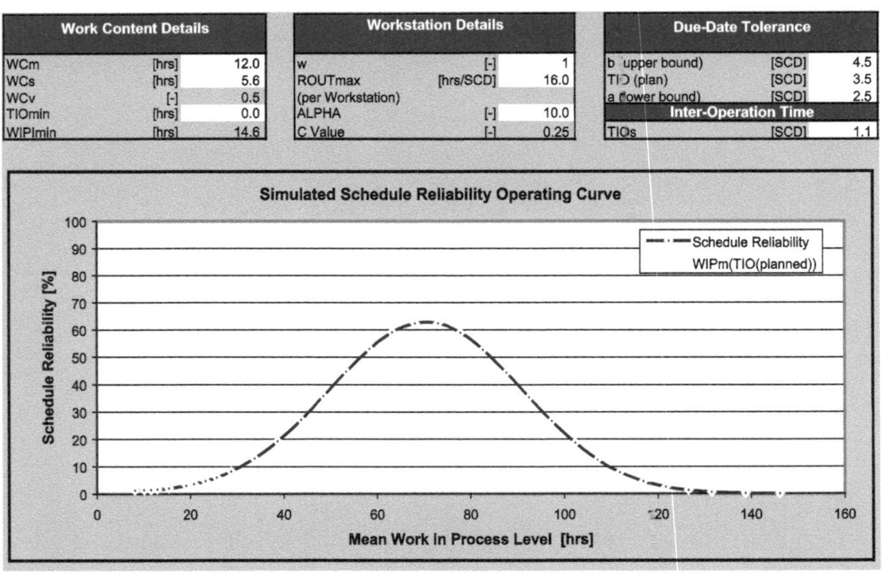

Fig. A.20 Worksheet *Data & Results (SROC)*

Calculation of Logistic Operating Curve									
t	Um	ROUTm	WIPa	WIPbuffer	WIPm	WIPrel	Rm	TTPm	TIOm
0.001	54	8.69	7.93	0.15	8.08	55	0.93	0.77	0.02
0.003	66	10.49	9.58	0.44	10.02	69	0.95	0.79	0.04
0.006	73	11.66	10.65	0.88	11.52	79	0.99	0.82	0.07
0.01	78	12.50	11.42	1.46	12.88	88	1.03	0.87	0.12
0.02	85	13.58	12.40	2.92	15.32	105	1.13	0.97	0.22
0.04	91	14.51	13.25	5.84	19.09	131	1.32	1.15	0.40
0.05	92	14.76	13.48	7.31	20.79	142	1.41	1.24	0.49
0.065	94	15.04	13.73	9.50	23.23	159	1.54	1.38	0.63
0.08	95	15.23	13.91	11.69	25.60	175	1.68	1.52	0.77
0.1	96	15.41	14.07	14.61	28.68	196	1.86	1.70	0.95
0.133	98	15.61	14.25	19.43	33.68	231	2.16	1.99	1.24
0.166	98	15.73	14.36	24.25	38.61	264	2.46	2.29	1.54
0.2	99	15.81	14.43	29.22	43.65	299	2.76	2.60	1.85
0.233	99	15.86	14.48	34.04	48.52	332	3.06	2.90	2.15
0.266	99	15.90	14.52	38.86	53.38	365	3.36	3.19	2.44
0.3	100	15.93	14.54	43.83	58.37	400	3.67	3.50	2.75
0.333	100	15.95	14.56	48.65	63.21	433	3.96	3.80	3.05
0.366	100	15.96	14.57	53.47	68.05	466	4.26	4.10	3.35
0.4	100	15.97	14.58	58.44	73.02	500	4.57	4.41	3.66
0.433	100	15.98	14.59	63.26	77.85	533	4.87	4.71	3.96
0.466	100	15.99	14.60	68.08	82.68	566	5.17	5.01	4.26
0.5	100	15.99	14.60	73.05	87.65	600	5.48	5.32	4.57
0.533	100	15.99	14.60	77.87	92.47	633	5.78	5.62	4.87
0.566	100	16.00	14.61	82.69	97.30	666	6.08	5.92	5.17
0.6	100	16.00	14.61	87.66	102.27	700	6.39	6.23	5.48
0.633	100	16.00	14.61	92.48	107.09	733	6.69	6.53	5.78
0.666	100	16.00	14.61	97.30	111.91	766	7.00	6.83	6.08
0.7	100	16.00	14.61	102.27	116.88	800	7.31	7.14	6.39
0.733	100	16.00	14.61	107.09	121.70	833	7.61	7.44	6.69
0.766	100	16.00	14.61	111.91	126.52	866	7.91	7.74	6.99
0.8	100	16.00	14.61	116.88	131.49	900	8.22	8.05	7.30
0.85	100	16.00	14.61	124.19	138.79	950	8.67	8.51	7.76
0.9	100	16.00	14.61	131.49	146.10	1000	9.13	8.97	8.22

Calculation of Schedule Reliability Operating Curve				
b-TIOm(WIP(t))	Phi(b)	a-TIOm(WIP(t))	Phi(a)	**RelS[%]**
4.48	1.000	2.48	0.987	**1.327**
4.46	1.000	2.46	0.986	**1.405**
4.43	1.000	2.43	0.985	**1.515**
4.38	1.000	2.38	0.983	**1.662**
4.28	1.000	2.28	0.979	**2.060**
4.10	1.000	2.10	0.969	**3.043**
4.01	1.000	2.01	0.963	**3.651**
3.87	1.000	1.87	0.952	**4.733**
3.73	1.000	1.73	0.939	**6.048**
3.55	0.999	1.55	0.917	**8.212**
3.26	0.998	1.26	0.869	**12.940**
2.96	0.996	0.96	0.804	**19.204**
2.65	0.991	0.65	0.720	**27.137**
2.35	0.982	0.35	0.624	**35.819**
2.06	0.967	0.06	0.520	**44.690**
1.75	0.941	-0.25	0.411	**52.965**
1.45	0.902	-0.55	0.311	**59.070**
1.15	0.848	-0.85	0.224	**62.379**
0.84	0.774	-1.16	0.150	**62.328**
0.54	0.686	-1.46	0.096	**58.918**
0.24	0.585	-1.76	0.058	**52.708**
-0.07	0.476	-2.07	0.032	**44.328**
-0.37	0.371	-2.37	0.017	**35.371**
-0.67	0.275	-2.67	0.009	**26.637**
-0.98	0.191	-2.98	0.004	**18.696**
-1.28	0.126	-3.28	0.002	**12.473**
-1.58	0.079	-3.58	0.001	**7.824**
-1.89	0.046	-3.89	0.000	**4.533**
-2.19	0.025	-4.19	0.000	**2.501**
-2.49	0.013	-4.49	0.000	**1.294**
-2.80	0.006	-4.80	0.000	**0.613**
-3.26	0.002	-5.26	0.000	**0.179**
-3.72	0.000	-5.72	0.000	**0.045**

Fig. A.21 Worksheet *Calculation* (*SROC*)

A3 SOC – Software Documentation

In order to calculate the Storage Operating Curves (SOC) for an item, data about the demand, the lot sizes in the store input and store output, as well as the maximum values for the plan deviations (quantity and due date) are required.

Figure A.22 shows the *Data & Results* worksheet in the *Storage Operating Curves* software. In order to use the software, the fields with the white backgrounds need to be filled in. The Storage Operating Curves which are depicted in the resulting graph are calculated based on Eqs. 8.17, 8.18 and 8.19. These calculations are carried out in the *Calc. SOC* and *Calc. SLOC* worksheets.

One of the central parameters of the Storage Operating Curves is the *value of C for the C_{norm} Function* (see Sect. 8.4.3). The value of C determines the distribution form of the plan deviations and the ratio of the practical minimum stock level to lot stock (SL_1/SL_0) (see also Fig. 8.12). The *C Value* worksheet provides an easy support for determining the value of C and can be accessed by clicking on the *Change C Value* button. On this worksheet the distribution form of the plan deviations is set using a scroll bar. The software then determines the value of C for the ratio SL_1/SL_0 and automatically transfers this value to the Logistic Operating Curves calculation.

The resulting graphs are automatically scaled. Should it however be necessary to adjust the scaling, for example, in order to more clearly depict the critical por-

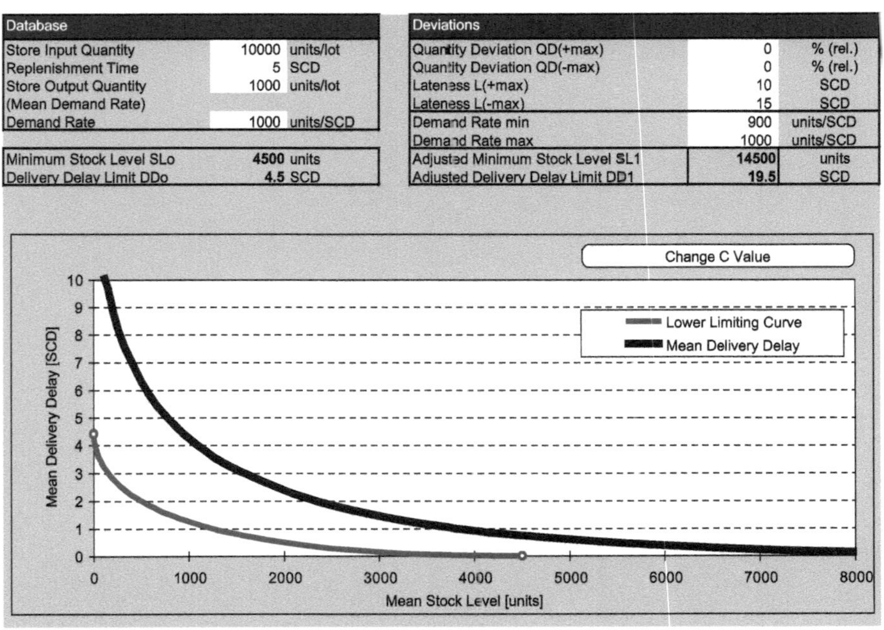

Fig. A.22 Worksheet *Data & Results (SOC)*

tion of the Storage Operating Curves, the protection for the corresponding worksheet has to be removed (Tools > Protection > Unprotect Sheet). Afterwards, the preferred axis can be activated with a double click. In the scaling register the entries

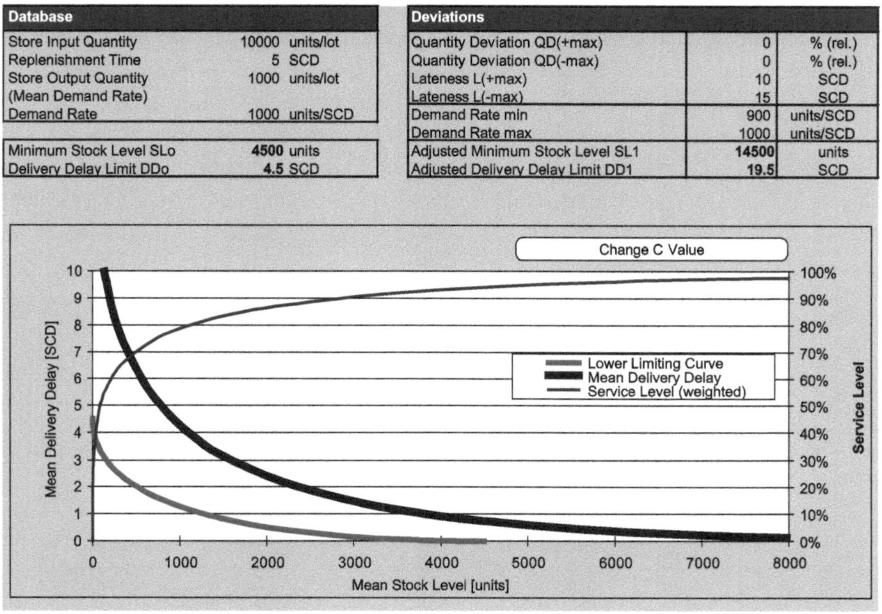

Fig. A.23 Worksheet *Data & Results (incl. SLOC)*

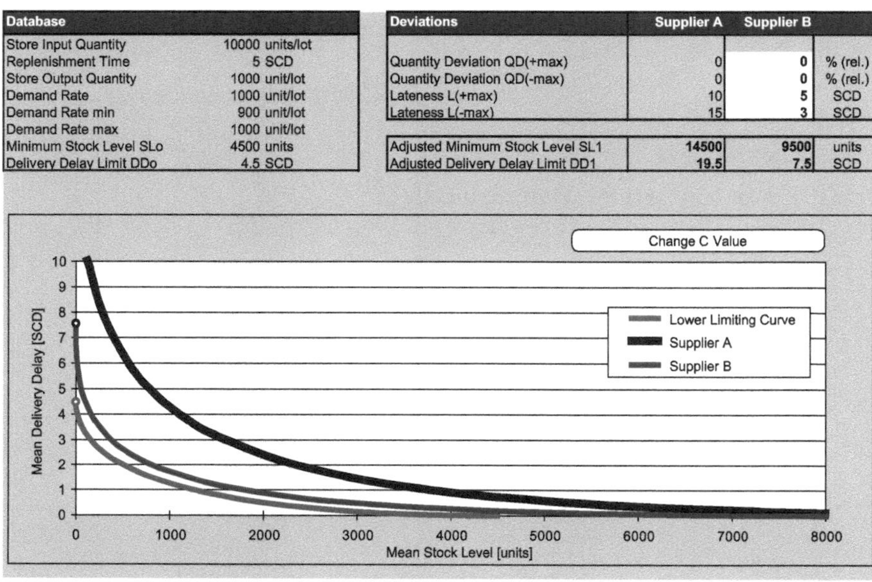

Fig. A.24 Worksheet *Supplier Comparison*

can subsequently be changed manually. Afterwards, in order to prevent the formulas that are integrated into the worksheet from being overwritten, the worksheet protection should be activated again.

In addition to depicting the *Mean Delivery Delay Operating Curve* and the *Lower Limiting Curve,* the resulting graph on the *Data & Results (incl. SLOC)* worksheet (Fig. A.23) also contains the *Service Level Operating Curve (SLOC).*

A further option comparatively evaluating the impact of different plan deviation values is offered via the *Supplier Comparison* worksheet (Fig. A.24). This can be utilized for example, to evaluate different suppliers (see also Sect. 8.7).

In order to evaluate two suppliers, the Service Level Operating Curves for both suppliers are also depicted on the *Supplier Comparison (incl. SLOC)* worksheet (Fig. A.25).

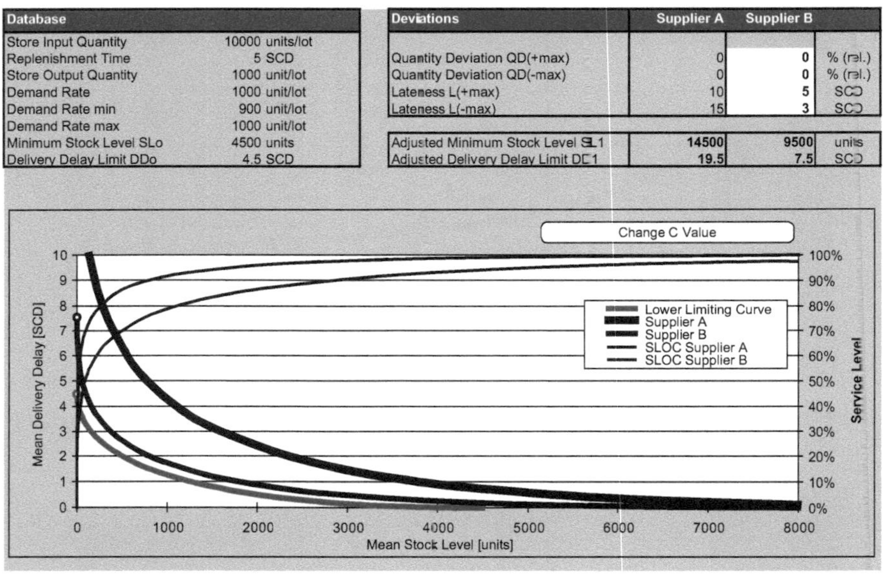

Fig. A.25 Worksheet *Supplier Comparison (incl. SLOC)*

A4 TOLS – Software Documentation

The Logistic Operating Curves make it possible to describe the correlations between the work content – and thus the lot size – and the logistic objectives. This is the basis for Throughput Oriented Lot Sizing (TOLS). In determining economically optimal lot sizes this method reflects not only the storage and order change costs, but also the capital tie-up costs in the production area (see Sect. 6.5.3.1). Extensive documentation on this method can be found in [Nyhu-91]. The TOLS software demonstrates both how this method operates as well as possible applications for it.

Figure A.26 shows the data input mask. In order to use the software, the fields with the white backgrounds need to be filled in or when necessary, deleted (e. g., when the data is related to operations). Based on the data entered, the program calculates the optimal manufacturing lot size for the base model according to Andler and according to Throughput Oriented Lot Sizing. The values in the fields with a grey background are calculated by the program and cannot be edited.

The calculated costs that are dependent on the lot size are depicted in two graphs. Figure A.27 shows the trend for the order change costs (calculated from the setup costs that were entered into the input mask) and the capital tie-up costs in storage as well as the resulting total costs curve. Their minimum is found at the optimal lot size calculated according to the Andler's base model.

Figure A.28 depicts the final costs of the capital tie-up in production. These are presented here added together with the storage costs. The calculation is made by determining the lot size dependent operation and inter-operation times. The latter

Data Input

Product		
Material Costs	€/unit	40.10
Manufacturing Costs	€/unit	81.10
Demand in Evaluation Period	units	5000
Setup Costs	€/lot	956
Shop Calendar Days in Evaluation Period	SCD	250
Tied-Up Capital Rate	%	15

Lot Size according to **ANDLER**	units/lot	**886**	Lot Range : 44 SCD
Lot Size according to **TOLS**	units/lot	**469**	Lot Range : 23 SCD

SCD : Shop Calendar Day

Details of Operations

Operation -	Workstation -	Setup Time [min]	Processing Time [min]	Hourly Machine Rate [€/hour]	Value Adding [€/unit]	Setup Cost Rate [€/hour]	Setup Costs [€]	Daily Capacity [hrs/SCD]	Planned Flow Rate -	Auxiliary Variable	Operation Time per Unit [SCD]	Inter-Operation Time per Unit [SCD]
10	4711	10	0.10	60	0.10	60	10	8	2	0.03	0.00021	0.00021
20	4712	60	1.00	80	1.33	80	80	8	2	0.25	0.00208	0.00208
30	4713	20	3.00	120	6.00	120	40	8	2	0.75	0.00625	0.00625
40	4714	240	10.00	120	20.00	120	480	24	3	1.25	0.00694	0.01389
50	4715	10	0.20	90	0.30	90	15	8	2	0.05	0.00042	0.00042
60	4716	0	1.00	60	1.00	60	0	8	2	0.25	0.00208	0.00208
70	4717	20	0.40	70	0.47	70	23	8	2	0.10	0.00083	0.00083
80	4718	15	0.40	70	0.47	70	18	8	2	0.10	0.00083	0.00083
90	4719	60	2.00	90	3.00	90	90	8	2	0.50	0.00417	0.00417
100	4720	120	5.00	100	8.33	100	200	8	3	1.88	0.01042	0.02083
		555	23.10		41.00		956			5.15	0.03424	0.05160

Fig. A.26 Worksheet *Data Input (TOLS)*

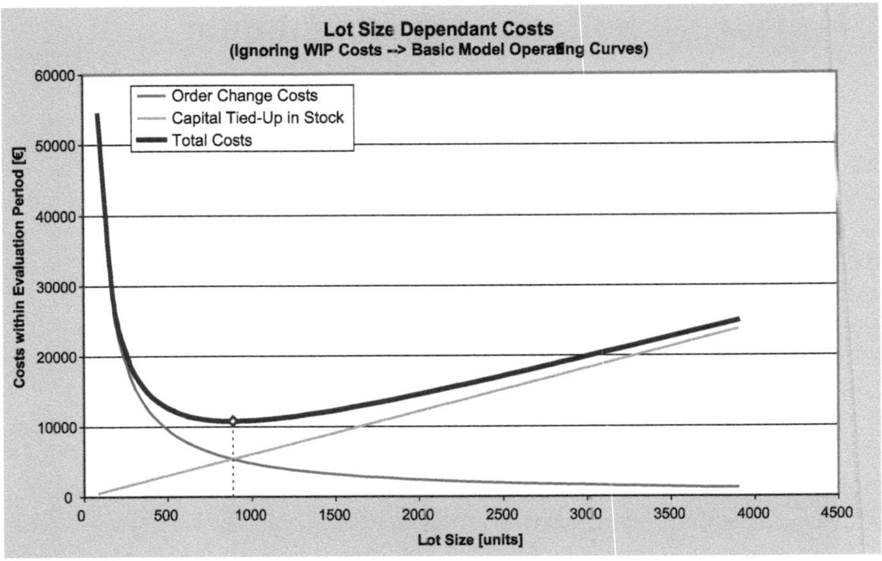

Fig. A.27 Diagram *Andler*

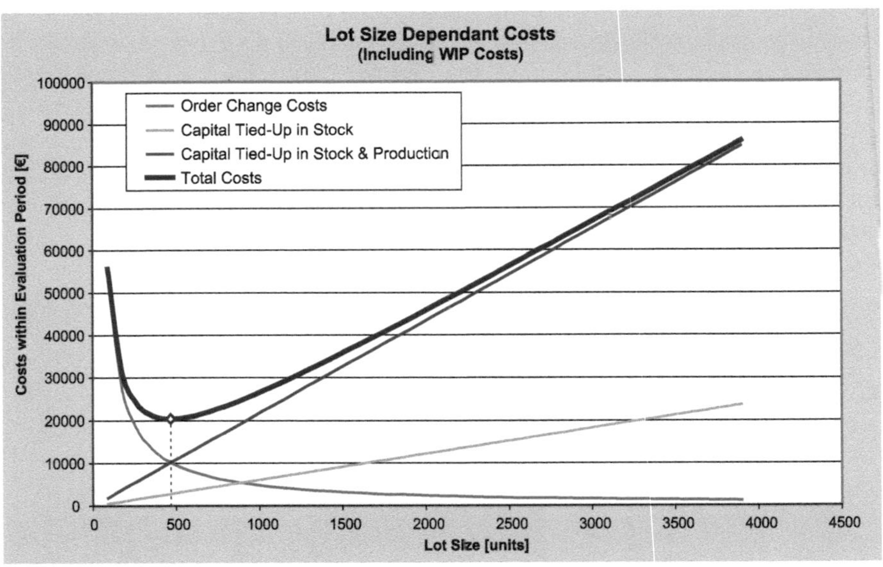

Fig. A.28 Diagram *TOLS*

are determined for each operation using the planned flow rate that was entered into the input mask. The minimum of the resulting total cost curve is found at the optimal lot size that results from applying Throughput Oriented Lot Sizing. By changing the data in the input mask, it can be seen that the higher the capital tie-up costs in production (i. e. the WIP) are, the more strongly the results of the two

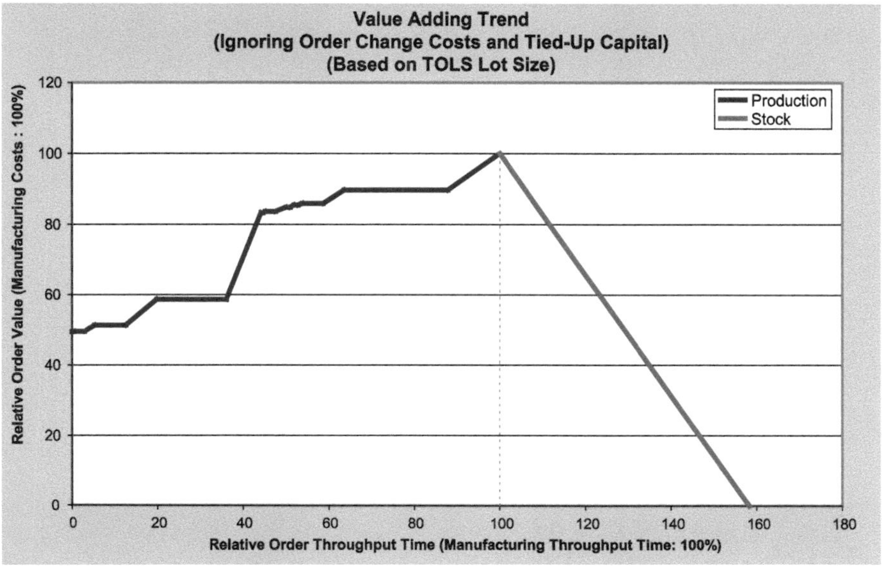

Fig. A.29 Diagram *Order Value*

methods for determining the lot size differ. The essential drivers here are the number of operations, the planned flow rate, the daily capacity per workstation, and the effort required for processing each part (described through the unit processing times).

In comparison to the capital tie-up costs in storage, the level of the resulting capital tie-up costs in the production can also be clearly inferred from the value-adding trend, depicted in Fig. A.29.

Bibliographic References

[Arno-95] Arnold, D.: Materialflußlehre. Vieweg Verlag, Braunschweig, Wiesbaden, 1995.

[Baum-93] Baumgarten, H.: Trends und Strategien in der Logistik. Technische Universität Berin, Bereich Materialflußtechnik und Logistik, 1996.

[Bech-84] Bechte, W.: Steuerung der Durchlaufzeit durch belastungsorientierte Auftragsfreigabe bei Werkstattfertigung. VDI Progress Reports, Series 2, No. 70, Düsseldorf, 1984.

[Beck-96] Beckmann, H.: Theorie einer evolutionären Logistik-Planung. Dissertation. Universität Dortmund, 1996.

[Berr-70] Berr, U.; Papendieck, A. J.: Produktionsreihenfolgen und Losgrößen in der Serienfertigung in einen Werkstattmodell. Werkstattstechnik 60 (1970), No. 4.

[Birk-95] Birkhofer, H.; Costa, C.: Auf dem Wege zur ganzheitlichen Simulation. VDI Report No. 1215, 1995.

[Brin-89] Brinkmann, M.: Dynamische Groblayout-Planung auf der Basis des Durchlaufdiagramms. Dissertation Universität Hannover, VDI Progress Reports, Series 2, No. 184, Düsseldorf 1989.

[Bron-95] Bronstein, I. N.: Taschenbuch der Mathematik. 2nd edited edition. Deutsch-Verlag, Thun, 1995.

[Bull-92] Bullinger, H.-J.: Innovative Produktionsstrukturen – Voraussetzung für ein kundenorientiertes Produktionsmanagement. In: Bullinger, H.J. (Ed.) Kundenorientierte Produktion. IOA-Forum May 1992. Springer Verlag, Munich/Vienna/Heidelberg 1992.

[Bult-88] Bultmann, D.: ABC-/XYZ-Analysen als Instrument der analytischen Logistik. In: RKW-Handbuch Logistik. 14. Lfg. XII/88, KZ 9110, Erich Schmidt Verlag, Berlin.

[Burm-97] Burmeister, M.: Auslegung der Verbrauchssteuerung bei vernetzter Produktion. Dissertation Universität Hannover, VDI Progress Reports, Series 8, No. 658. Düsseldorf, 1997.

[Conw-67*] Conway, R. W.; Maxwell, W. L.; Miller, L. W.: Theory of Scheduling. Reading, Massachusetts, Addison-Wesley Publishing Company, 1967.

[Daen-92] Daenzer, W. F.; Huber, F. (Eds.): System Engineering. Verlag Industrielle Organisation, Zürich, 1992.

[Dang-91] Dangelmaier, W.; Geck, K.: Adaptive Organisationsplanung in der Fertigung. Fertigungstechnik und Betrieb, Berlin 41 (1991) 12, pp. 635–639.

[DeBo-83*] De Bodt, M.; Gelders, L. F.; Van Wassenhove, L. N.: The Effect of Demand Uncertainty of Multilevel Lot Sizing. Part I: The Least Unit Cost. Part II: The Silver Meal Heuristic. Katholike Universiteit Leuven, 1983.

[Domb-88] Dombrowski, U.: Qualitätssicherung im Terminwesen der Werkstattfertigung. Dissertation Universität Hannover, VDI Progress Reports, Series 2, No. 159, VDI-Verlag, Düsseldorf, 1988.

[Egli-01] Egli, J.: Transportkennlinien: Ein Ansatz zur Analyse von Materialflusssystemen. Dissertation Universität Dortmund, Verlag Praxiswissen, Dortmund, 2001.

[Eide-95] Eidenmüller, B.: Die Produktion als Wettbewerbsfaktor. Verlag TÜV Rheinland, Cologne, 1995.

[Erdl-84] Erdlenbruch, B.: Grundlagen neuer Auftragssteuerungsverfahren für die Werkstattfertigung. Dissertation Universität Hannover, VDI Progress Reports, Series 2, No. 71, Düsseldorf, 1984.

[Ever-87] Eversheim, W.; Thomae, H. G.: Simulation als Hilfsmittel zur Produkt- und Produktionsplanung. ZwF 82 (1987) Issue 6, pp. 333–339.

[Ever-94] Eversheim, W.; Fuhlbrügge, M.; Dobberstein, M.: Simulation unterstützt die Produktionssystemplanung. VDI-Z 136 (1994), No. 7/8.

[Ewal-93] Ewald, H.: Einsatz von Betriebskennlinien zur Überprüfung und Sicherung der logistischen Qualität. Contribution to the seminar 'Qualitätsmanagement in der Logistik', gfmt, Stuttgart, 1993.

[Fast-94] Fastabend, H.; Großklaus, A.: Betriebskennlinien als Controllinginstrument der Produktionslogistik. ZwF 89 (1994), 7-8, pp. 373–376.

[Fast-97] Fastabend, H.: Kennliniengestützte Synchronisation von Fertigungs- und Montageprozessen. Dissertation Universität Hannover, VDI Progress Reports, Series 2, No. 452, Düsseldorf, 1997.

[From-92] Fromm, H.: Das Management von Zeit und Variabilität in Geschäftsprozessen. CIM Management 5/92, pp. 7–14.

[Forr-61*] Forrester, J.: Industrial Dynamics, MIT Press, Cambridge, USA, 1961.

[Fort-77*] Fortuin, L.: A survey of literature on reordering of stock items for production inventories. Int. Journal of Production Research 1977, 15 (1), pp. 87–105.

[Gläß-91] Gläßner, J.; Nyhuis, P.: Zielkonflikt lösen – Durchlauforientierte Losbildung im Vergleich zu konventionellen Verfahren. Maschinenmarkt, Würzburg (1991) 8, pp. 60–64.

[Gläß-95] Gläßner, J.: Modellgestütztes Controlling der beschaffungslogistischen Prozeßkette. Dissertation Universität Hannover, VDI Progress Reports, Series 2, No. 337, Düsseldorf, 1995.

[Gned-71] Gnedenko, B. W.; Kowalenko, I. N.: Einführung in die Bedientheorie. R. Oldenbourg Verlag, Munich/Vienna, 1971.

[Goeb-96] Goebel, D.: Modellbasierte Optimierung von Produktentwicklungsprozessen. VDI Progress Reports, Series 2, No. 385, Düsseldorf, 1996.

[Gott-95] Gottschalk, E.: Management logistischer Prozesse in der Praxis. In: Wiendahl, H.-P. (Ed.) Erfolgsfaktor Logistikqualität. Springer, Berlin, 1995.

[Grob-93] Grobel, T.: Analyse der Einflüsse auf die Aufbauorganisation von Produktionssystemen. Forschungsberichte aus dem Institut für Arbeitswissenschaften und Betriebsorganisation der Universität Karlsruhe (Ed.: Zülch, G.), Volume 6, 1993.

[Gros-98*] Gross, D.; Harris, C. M.: Fundamentals of Queuing Theory. 3rd Edition. John Wiley & Sons, New York, 1998.

[Groß-96] Großklaus, A.: Ablauforientierte Produktionslogistik – eine modellbasierte Analyse. Dt. Univ.-Verlag, Gabler-Verlag, Wiesbaden 1996.

[Gute-51] Gutenberg, E.: Grundlagen der Betriebswirtschaftslehre. Volume 1: Die Produktion. 1st Edition, Berlin, 1951.

[Günz-93] Günzel, U.: Entwicklung und Einsatz eines Simulationsverfahrens für operative und strategische Probleme der Produktionsplanung und -steuerung. Dissertation. Universität Erlangen-Nürnberg. Hanser-Verlag. Munich, Vienna 1993.

[Hell-78] Heller, W.-D.; Lindenberg, H.; Nuske, M.; Schriever, K.-H.: Stochastische Systeme. Markoffketten, Stochastische Prozesse, Warteschlangen. Walter de Gruyter, Berlin/New York, 1978.

[Hess-88] Hesse, R.: Simulation ermöglicht Absicherung kreativer Planungsprozesse. AV 25 (1988) 2, pp. 61–63.

[Hold-86] Holdhof, J.: Terminplanungssysteme für Werkstätten mit heterogenen Produktionsbedingungen. VDI Progress Reports, Series 2, No. 116, Düsseldorf, 1986.

[Hopp-96*] Hopp, W. J.: Spearmann, M. L.: Factory Physics. Irwin, Chicago, 1996.

[Jain-94] Jainczyk, M.: Logistikorientierte Bewertung von Investitionsvorhaben im Produktionsbereich. In: PSK-Fachtagung (Tagungsbericht). gfmt, Munich, 1994.

[Jüne-88] Jünemann, R.; Klöpper, H.-J.: Reduzieren der Durchlaufzeit als logistische Herausforderung. Logistik, May/June 1988, pp. 80–84.

[Kalu-94] Kaluza, B.: Rahmenentscheidungen zu Kapazität und Flexibilität produktionswirtschaftlicher Systeme. In: Corsten, H. (Ed.): Handbuch des Produktionsmanagement: Strategie – Führung – Technologie – Schnittstellen. Gabler-Verlag. Wiesbaden, 1994.

[Kear-92] A. T. Kearney Inc.: Produktivität und Qualität in der Logistik – Schlüsselfaktoren im europäischen Wettbewerb. Düsseldorf, 1992.

[Kern-02] Kerner, A.: Modellbasierte Beurteilung der Logistikleistung von Prozeßketten. Dissertation. Universität Hannover, 2002.

[Kett-76] Kettner, H. (Ed.): Neue Wege der Bestandsanalyse im Fertigungsbereich. Fachbericht des Arbeitsausschusses Fertigungswirtschaft (AFW) der Deutschen Gesellschaft für Betriebswirtschaft (DGfB). TU Hannover, 1976.

[Kluß-96] Klußmann, J.; Krauth, J.; Splanemann, R.: Simulation – Spielerei oder zukunftsweisende Technik? Steigerung der Analysefähigkeit in Produktion und Logistik. wt -Produktion und Management 86 (1996), pp. 361–366.

[Knol-87] Knolmayer, G.: Materialflußorientierung statt Materialbestandsoptimierung. Erschienen in: Baetge, J. (Ed.): Logistik – eine Aufgabe der Unternehmenspolitik: ein Round-table-Gespräch. Duncker u. Humblot, Berlin, 1987.

[Knol-90] Knolmayer, G.; Lemke, F.: Auswirkungen von Losgrößenreduktionen auf die Erreichung produktionswirtschaftlicher Ziele. ZfB 60 (1990) 4, pp. 423–442.

[Köni-76] König, D.; Stoyan, D.: Methoden der Bedientheorie. Verlag Vieweg, Berlin, 1976.

[Krau-92*] Krauth, J.: Comparison 2: Preliminary Evaluation. Eurosim – Simulation News Europe, Number 4, March 1992.

[Kreu-94] Kreutzfeld, J.: Planen mit Bearbeitungsalternativen in der Teilefertigung. Dissertation Universität Hannover, VDI Progress Reports, Series 16, No. 80, 1994.

[Kugl-95] Kugler, P.; Nyhuis, P.: Controllingwerkzeuge als Fenster zum Prozeß in Produktion und Beschaffung. Contribution to the seminar 'KVP in der Produktionslogistik' der Techno-Transfer GmbH, Hannover, November 2nd and 3rd 1995.

[Kuhn-92] Kuhn, A.: Modellgestützte Logistik – Methodik einer permanenten ganzheitlichen Systemgestaltung. Tagung "Rechnergestützte Fabrikplanung '92". VDI Reports, No. 949, Düsseldorf, 1992, pp. 109–136.

[Kuhn-95] Kuhn, A.: Prozeßketten in der Logistik: Entwicklungstrends und Umsetzungsstrategien. Praxiswissen Dortmund, 1995.

[Kuhn-97] Kuhn, A.: Konstruktionssystematik für Logistikmodelle. Logistik-Jahrbuch 1997, Verlagsgruppe Handelsblatt, pp. 270–274..

[Kull-87] Kulla, B.: Ergebnisse oder Erkenntnisse – liefern makroanalytische Simulationsmodelle etwas Brauchbares? In: Biethahn, J.; Schmidt, B. (Eds.): Simulation als betriebliche Entscheidungshilfe. Springer-Verlag, Berlin a.o., 1987.

[Kupr-91] Kuprat, T.: Simulationsgestützte Beurteilung der logistischen Qualität von Produktionsstrukturen. Dissertation Universität Hannover, VDI Progress Reports, Series 2, No. 228, Düsseldorf, 1991.

[Lars-92*] Larsen, E.: Effects from use of alternative routings. 24th CIRP International Seminar on Manufacturing Systems. June 11–12, 1992, Copenhagen, Denmark.

[Look-96] Looks, S.: Kennzahlen für Fertigungsinseln. io Management Zeitschrift 65 (1996), No. 3, pp. 71–74.

[Lore-84] Lorenz, W.: Entwicklung eines arbeitsstundenorientierten Warteschlangenmodells zur Prozeßabbildung der Werkstattfertigung. Dissertation Universität Hannover, VDI Progress Reports, Series 2, No. 72, Düsseldorf, 1984.

[Ludw-90] Ludwig, E.; Nyhuis, P.; Ullmann, W.: Analyseverfahren zur Diagnose der Produktion. Contribution to the seminar "PROLOG '90", 30./31.05.1990, Munich.

[Ludw-92] Ludwig, E.; Nyhuis, P.: Verbesserung der Termineinhaltung in komplexen Fertigungsbereichen durch einen neuen Ansatz zur Plandurchlaufzeitermittlung. In: Görke, W.; Rininsland, H.; Syrbe, M. (Eds.): Information als Produktionsfaktor. Springer Verlag, Berlin/Heidelberg/New York, 1992, pp. 473–483.

[Ludw-93] Ludwig, E.; Nyhuis, P.: Plandurchlaufzeitermittlung auf der Basis logistischer Grundgesetze. Contribution to the symposium 'Zeitgemäße Ablauforganisation', Techno Transfer, 17./18.05.93, Hannover.

[Ludw-95] Ludwig, E.: Modellgestützte Diagnose logistischer Produktionsabläufe. Dissertation Universität Hannover, VDI Progress Reports, Series 2, No. 362, Düsseldorf, 1995.

[Lutz-01] Lutz, S.; Lödding, H.; Wiendahl, H.-P.: Kennliniengestützte Logistische Lageranalyse. ZWF 96 (2001) 10, pp. 550–553.

[Lutz-02] Lutz, S.: Kennliniengestütztes Lagermanagement. VDI Progress Reports, Series 2, Düsseldorf 2002.

[Math-77*] Mather, H.; Plossl, G.W.: Priority Fixation versus Throughput Planning. APICS, Intern. Conference, Cleveland / Ohio (USA), 1977.

[Meie-93] Meier, K.-J.: Auslastung contra Durchlaufzeit. ZWF 88 (1993) 2, pp. 60–63.

[Meie-95] Meier, K.-J., Bestandsoptimierung durch Fertigungsablaufsimulation. ZWF 90 (1995) 4, pp. 157–160.

[Mert-96] Mertins, K.; Rabe, M.: Kontinuierliche Unternehmensentwicklung – Evolution und Revolution. ZWF 91 (1996) 5, pp. 192–194.

[Mich-97] Michaeli, W.: Mit Produkt- und Prozeßsimulation zur virtuellen kunststoffverarbeitenden Fabrik. In: Schuh, H.; Wiendahl, H.-P. (Eds.): Komplexität und Agilität. Springer Verlag, Berlin, Heidelberg, 1997.

[Milb-94] Milberg, J.: Unsere Stärken stärken – Der Weg zu Wettbewerbsfähigkeit und Standortsicherung. Municher Kolloquium '94. mi-Verlag, Munich, 1994.

[Milb-97] Milberg, J.; Reinhart, G. (Eds.): Mit Schwung zum Aufschwung. Münchner Kolloquium '97. mi-Verlag, Landsberg / Lech, 1997.

[Möll-95] Möller, J.: Logistische Beurteilung von Fabrikstrukturen mit Betriebskennlinien. In: VDI-ADB Yearbook 95/96. VDI-Verlag, Düsseldorf, 1995, pp. 99–122.

[Möll-96] Möller, J.: Kennliniengestützte Auslegung von Fabrikstrukturen. Dissertation Universität Hannover, VDI Progress Reports, Series 2, No. 389, VDI-Verlag, Düsseldorf, 1996.

[Müll-62] Müller-Merbach, H.: Sensibilitätsanalyse der Losgrößenbestimmung. Unternehmensforschung 6 (1962), pp. 79–88.

[Noch-93] Noche, B.; Lindner, W. L.: Simulation zur strategischen Unternehmensstrukturplanung. ASIM-Tagungsbericht. gfmt, Munich, 1993

[Nyhu-91] Nyhuis, P.: Durchlauforientierte Losgrößenbestimmung. Dissertation Universität Hannover, VDI Progress Reports, Series 2, No. 225, Düsseldorf, 1991.

[Nyhu-93a] Nyhuis, P.: Betriebskennlinien zur Entscheidungsunterstützung in der PPS. In: Wiendahl, H.-P. (Ed.): Neue Wege der PPS. gfmt, Munich, 1993.

[Nyhu-93b] Nyhuis, P.; Penz, T.: Das logistische Leistungsvermögen auf dem Prüfstand. Contribution to the seminar 'Qualitätsmanagement der Logistik'. gfmt, Sept. 1993, Stuttgart.

[Nyhu-95] Nyhuis. P.: Lagerkennlinien – ein Modellansatz zur Unterstützung des Beschaffungs- und Bestandscontrolling. RKW-Handbuch Logistik 22 Lfg. XII/1995.

[Nyhu-96] Nyhuis, P.: Modelle zum Reengineering und Monitoring von Prozeßketten. In: Wiendahl, H.-P. (Ed.): Dezentrale Produktionsstrukturen. Techno-Transfer, 1996.

[Oert-77] Oertli-Cajacob, P.: Praktische Wirtschaftskybernetik. Carl Hanser Verlag, Munich, Vienna, 1977.

[Papu-94] Papula, L.: Mathematik für Ingenieure und Naturwissenschaftler, Volume 3. Vieweg-Verlag, Braunschweig, Wiesbaden, 1994.

[Penz-96] Penz, T.: Wechselwirkungen technischer und logistischer Produktionsprozesse und ihre Auswirkungen auf das Qualitätsmanagement. VDI Progress Reports, Series 20, No. 232, VDI-Verlag, Düsseldorf, 1996.

[Pete-95] Petermann, D.: Modellbasierte Produktionsregelung. Dissertation Universität Hannover, VDI Progress Reports, Series 20, No. 193. VDI-Verlag, Düsseldorf, 1996.

[Port-92] Porter, M. E.: Wettbewerbsvorteile: Spitzenleistungen erreichen und behaupten. 3rd Edition. Campus Verlag, Frankfurt/Main, New York, 1992.

[Port-93] Porter, M. E.: Nationale Wettbewerbsvorteile: Erfolgreich konkurrieren auf dem Weltmarkt. Sonderausgabe. Wirtschaftsverlag Ueberreuter, Vienna 1993.

[Prit-95*] Pritschow, G.; Wiendahl, H.-P.: Application of Control Theory for Production Logistics – Results of a Joint Project. 45th CIRP General Assembly, University of Twente, Netherlands, August 21st–26th 1995.

[Prof-77] Profos, P.: Modellbildung und ihre Bedeutung in der Regelungstechnik. VDI Reports, No. 276, Düsseldorf, 1977, pp. 5–12.

[Rice-90*] Rice, W.; Gross, R.: Simulation Technique and the WIPAC Curve for Balancing Capacity against Inventory. APICS, 33rd International Conference Proceedings, New Orleans, 1990, pp. 284–287.

[Rums-95] Rumscheid, B.: Rechnergestützte Planung und Durchführung von Simulationsstudien. In: Simulation in der Praxis – neue Produkte effizienter entwickeln. VDI-Verlag, Düsseldorf, 1995.

[Sach-84] Sachs, L.: Angewandte Statistik. 6th Edition. Springer-Verlag, Berlin, 1984.

[Sain-75] Sainis, P.: Ermittlung von Durchlaufzeiten in der Werkstattfertigung aus Daten des Fertigungsprogrammes mit Hilfe der Warteschlangentheorie. Dissertation. Technische Universität Hannover, 1975.

[Scha-92] Scharf, P.; Spies, W.: Computersimulation hilft bei der Optimierung von Produktionsanlagen. VDI-Z 134 (1992), No. 10, pp. 72–79.

[Schl-95] Schlüter, K.: Einmal ist immer das erste Mal – nützliche Hinweise für Einsteiger. VDI Reports, No. 1215, 1995.

[Schn-02*] Schneider, M.; Wiendahl, H.-P.: Modelling and Controlling the Logistic Performance of Manufacturing Departments by the Use of Logistic Process Operating Curves (LPOCs). The 35th CIRP – International Seminar on Manufacturing Systems, Seoul, Korea, May 12th–15th 2002.

[Scho-94] Scholtissek, P.; Gläßner, J.: PPS auf dem Prüfstand – Simulationsgestützter Verfahrenstest. ZwF 89 (1994), Issue 3, pp. 96–98.

[Scho-96] Scholtissek, P.: Simulationsprüfstand für Logistikkonzepte. Dissertation Universität Hannover, VDI Progress Reports, Series 2, No. 377. VDI-Verlag, Düsseldorf, 1996.

[Sell-90] Sell, R.: Angewandtes Problemlösungsverhalten – Denken und Handeln in komplexen Zusammenhängen. 3rd Edition. Springer-Verlag, Berlin, Heidelberg, New York, 1990.

[Spen-96] Spengler, Th., Stemper, M, Rentz, O.: Produktionssteuerung mit Betriebskennlinien und Belastungsorientierter Auftragsfreigabe. ZwF 91 (1996) 7–8, pp. 348–351.

[Spur-94] Spur, G.; Stöferle, Th.: Handbuch der Fertigungstechnik, Volume 6, Fabrikbetrieb, Carl Hanser, Munich/Vienna 1994.

[Spri-92] Springer, G.: Simulationsgestützte Mitarbeiterausbildung am Beispiel der Fertigungssteuerung. Dissertation Universität Hannover, VDI Progress Reports, Series 2, No. 269. VDI-Verlag, Düsseldorf, 1992.

[Stro-77] Stroyan, D.: Qualitative Eigenschaften und Abschätzung stochastischer Modelle. Verlag Oldenbourg, Munich, 1977.

[Ulfe-86] Ulfers, H.-A.: Erfahrungen mit der belastungsorientierten Auftragssteuerung ABS bei Siemens. In: Wiendahl, H.-P. (Ed.): Fachbericht zum Seminar 'Praxis der belastungsorientierten Fertigungssteuerung' des Instituts für Fabrikanlagen der Universität Hannover, Hannover, 1986.

[Ullm-94] Ullmann, W.: Controlling logistischer Produktionsabläufe am Beispiel des Fertigungsbereichs. VDI Progress Reports, Series 2, No. 311, Düsseldorf, 1994.

[VDI-93] Simulation von Logistik-, Materialfluß- und Produktionssystemen. VDI – Gesellschaft Fördertechnik Materialfluß Logistik. VDI Guidelines, 3633, Sheet 1, 1993.

[Vieh-85] Viehweger, B.: Planung von Fertigungssystemen mit automatisiertem Werkzeugfluss. Dissertation. Technische Universität Berlin. HanserVerlag, Munich, 1985.

[Voig-90] Voigt, J.-P.: Veränderte Markt- und Verbraucheranforderungen an die Stahlerzeuger. Stahl und Eisen 110 (1990), Issue 4.

[Warn-93] Warnecke, H.-J.; Zeh, K.-P.: Strukturauslegung der Fertigung – ein integriertes Verfahren zur Leistungsbewertung und -optimierung. FB/IE 42 (1993) 3, pp. 112 –119.

[Webe-87] Logistikkostenrechnung. Berlin, Heidelberg 1987.

[Wede-89] Wedemeyer, H.-G. v.: Entscheidungsunterstützung in der Fertigungssteuerung mit Hilfe der Simulation. Dissertation Universität Hannover, VDI Progress Reports, Series 2, No. 176, Düsseldorf, 1989.

[Wien-87] Wiendahl, H.-P.: Belastungsorientierte Fertigungssteuerung. Hanser Verlag, Munich/Vienna 1987.

[Wien-89] Wiendahl, H.-P.: Grundgesetze der Logistik. fördern + heben, 1989, Issue 4/89, pp. 289–296.

[Wien-93a] Wiendahl, H.-P.; Nyhuis, P.: Die logistische Betriebskennlinie – ein neuer Ansatz zur Beherrschung der Produktionslogistik. In RKW (Eds.): RKW-Handbuch Logistik. 19. Lfg XI/93, Erich Schmidt Verlag, Berlin

[Wien-93b] Wiendahl, H.-P.; Kuhn, A.; Reinhardt, A. (Eds.): Handbuch Simulationsanwendung in Produktion und Logistik, Fortschritte in der Simulationstechnik, Vieweg & Sohn Verlagsgesellschaft, Braunschweig/Wiesbaden 1993.

[Wien-95a*] Wiendahl, H.-P.: Load-Oriented Manufacturing Control. Springer-Verlag, Berlin, 1995.

[Wien-95b] Wiendahl, H.-P.; Möller, J.; Nyhuis, P.: Engpaßorientierte Logistikanalyse auf der Basis von Betriebskennlinien. io Management Zeitschrift 64 (1995), No. 5, pp. 27–33.

[Wien-96a] Wiendahl, H.-P.; Kuhn, A.; Beckmann, H.; Fastabend, H.; Heiderich, T.; Helms, K.; Kloth, M.: Optimierung und Betrieb wandelbarer Produktionsnetze. In: Dangelmaier, W., Kirsten, U. (Eds.): Vision Logistik. PFT-Endbericht (FZKA-PFT 181), Karlsruhe, 1996.

[Wien-96b] Wiendahl, H.-P.; Dräger, H.; Franken, T.; Fastabend, H.: Überwachung und logistische Positionierung von Prozeßketten auf der Basis von Kosten-Durchlaufdiagrammen und Kennlinientechniken. In: Berkau, C. a.o. (Eds.): Kostenorientiertes Geschäftsprozeßmanagement. Vahlen-Verlag, Munich, 1996.

[Wien-97] Wiendahl, H.-P.: Fertigungsregelung: Logistische Beherrschung von Fertigungsabläufen auf Basis des Trichtermodells. Hanser-Verlag, Munich/Vienna, 1997.

[Wien-98] Wiendahl, H.-P.; Nyhuis, P.: Engpaßorientierte Logistikanalyse – Methoden zur kurzfristigen Leistungssteigerung in Produktionsprozessen. Munich, Transfer-Centrum GmbH, 1998.

[Wien-01*] Wiendahl, H.-P.; Schneider, M.: Development of a Logistic Operating Curve for an entire Manufacturing Department – Logistic Process Operating Curve (LPOC). In Inasaki, I. (Ed.): Initiatives of Precision Engineering at the Beginning of a Millenium, Springer Berlin Heidelberg, 2001.

[Wild-95] Wildemann, H.: Integriertes Qualitäts-Controlling logistischer Leistungen. In Wiendahl, H.-P. (Ed.): Erfolgsfaktor Logistikqualität. Springer-Verlag Berlin, Heidelberg, New York, 1995.

[Wild-98] Wildemann, H.: Marktführerschaft: Wege für ein profitables Unternehmenswachstum. Transfer-Centrum GmbH, Munich, 1998.

[Wind-01] Windt, K.: Engpaßorientierte Fremdvergabe in Produktionsnetzen, VDI-Verlag, Series 2, Düsseldorf, 2001.

[Wink-88] Winkelhake, U.: Permanente Maschinendatenerfassung automatischer Montageanlagen. Dissertation Universität Hannover, VDI Progress Reports, Series 2, No. 167, Düsseldorf, 1988.

[Wöhe-90] Wöhe, G.: Einführung in die Allgemeine Betriebswirtschaftslehre, 7. Edition, Handbuch der Wirtschafts- und Sozialwissenschaften, Verlag Vahlen, Munich 1990.

[Wyss-95] Wyssen, H.; Roder, B.; Scholtissek, P.: Regelmäßige Auftragsdurchlauf- und Arbeitssystemanalysen – ein Baustein der 'Leiterplattenstrategie 94–97'. Contribution to the seminar 'KVP in der Produktionslogistik' der Techno-Transfer GmbH, Hannover, November 2nd and 3rd 1995.

[Yu-01] Yu, K.-W.: Terminkennlinie – Eine Beschreibungsmethodik für die Terminabweichung im Produktionsbereich, Dissertation Universität Hannover, VDI-Verlag, Series 2, Düsseldorf, 2001.

[Zahn-94] Zahn, E.; Dillerup, R.: Fabrikstrategien und -strukturen im Wandel. In: Zülch, G. (Ed.): Vereinfachen und verkleinern – die neuen Strategien in der Produktion. Schäffer-Poeschel, Stuttgart, 1994.

[Zäpf-92a] Zäpfel, G.; Hödlmoser, P.: Läßt sich das KANBAN-Konzept bei einer Variantenfertigung wirtschaftlich einsetzen? ZfB 62. Jg. (1992), Issue 4, pp. 437-458.

[Zäpf-92b] Zäpfel, G.; Missbauer, H.; Kappel, W.: PPS-Systeme mit Belastungsorientierter Auftragsfreigabe. ZfB 62. Jg. (1992), Issue 8, pp. 897–919.

[Zülc-96a] Zülch, G.; Heel, J.; Krüger, J.: Simulation unterstützt die Parallelisierung der Produktionssystemplanung. VDI-Z 138 (1996), No.5, pp. 42–44.

[Zülc-96b] Zülch, G.; Heel, J.: Entwicklung von Suchstrategien. Beitrag zur 7. ASIM-Conference. Dortmund, May 11th–13th 1996.

Index

U

V

W

SPRINGER NATURE

GPSR Compliance

The European Union's (EU) General Product Safety Regulation (GPSR) is a set of rules that requires consumer products to be safe and our obligations to ensure this.

If you have any concerns about our products, you can contact us on ProductSafety@springernature.com

In case Publisher is established outside the EU, the EU authorized representative is:

Springer Nature Customer Service Center GmbH
Europaplatz 3
69115 Heidelberg, Germany

FSC
www.fsc.org
MIX
Papier | Fördert
gute Waldnutzung
FSC® C083411

Kolibri 360 GmbH
Ferdinand-Jühlke-Straße 7
99095 Erfurt, Deutschland
produktsicherheit@kolibri360.de